D0987049

Parasite Communities:
Patterns and Processes

Parasite Communities: Patterns and Processes

Edited by

Gerald W. Esch
*Wake Forest University,
North Carolina,
USA*

Albert O. Bush
*Brandon University,
Manitoba,
Canada*

John M. Aho
*University of Georgia,
South Carolina,
USA*

London New York
CHAPMAN AND HALL

First published in 1990 by
Chapman and Hall Ltd
11 New Fetter Lane, London EC4P 4EE

Published in the USA by
Chapman and Hall
29 West 35th Street, New York NY 10001

Typeset in 10pt Times by Scarborough Typesetting Services
Printed in Great Britain by St Edmundsbury Press, Bury St Edmunds, Suffolk

ISBN 0 412 33540 9

British Library Cataloguing in Publication Data

Parasite communities.
1. Animals. Parasites
I. Esch, Gerald W. II. Bush, Albert O.
III. Aho, John M.
591.52′49

ISBN 0–412–33540–9

Library of Congress Cataloging in Publication Data

Parasite communities: patterns and processes/
edited by Gerald W. Esch, Albert O. Bush,
John M. Aho.
p. cm.
Includes bibliographical references.

ISBN 0–412–33540–9
1. Helminths – Ecology. 2. Biotic communities.
I. Esch, Gerald W. II. Bush, Albert O.,
1948– . III. Aho, John M.
QL392.P37 1990
595.1′045249 – dc20 89–23867
CIP

Contents

Contributors

John M. Aho, Savannah River Ecology Laboratory, Aiken, SC, USA.

Albert O. Bush, Department of Zoology, Brandon University, Brandon, Manitoba, Canada.

Andy P. Dobson, Department of Biology, University of Rochester, Rochester, NY, USA.

Gerald W. Esch, Department of Biology, Wake Forest University, Winston-Salem, NC, USA.

Timothy M. Goater, Department of Biology, Wake Forest University, Winston-Salem, NC, USA.

John C. Holmes, Department of Zoology, University of Alberta, Edmonton, Alberta, Canada.

Clive R. Kennedy, Department of Biological Sciences, Hatherly Laboratories, University of Exeter, Exeter, Great Britain.

Armand Kuris, Department of Biological Sciences and Marine Science Institute, University of California, Santa Barbara, CA, USA.

David J. Marcogliese, Department of Fisheries and Oceans, Marine Fish Division, Bedford Marine Institute, Dartmouth, Nova Scotia, Canada.

Danny B. Pence, Department of Pathology, Texas Tech University Health Sciences Center, Lubbock, TX, USA.

Peter W. Price, Department of Biological Sciences, Northern Arizona University, Flagstaff, AZ, USA.

Allen W. Shostak, Maurice-Lamontagne Institute of Fisheries and Oceans, Division of Fisheries Research, Mont-Joli, Quebec, Canada.

Daniel Simberloff, Department of Biological Science, Florida State University, Tallahassee, FL, USA.

Wayne P. Sousa, Department of Zoology, University of California, Berkeley, CA, USA.

Preface

We first discussed the possibility of organizing a symposium on helminth communities in June, 1986. At that time, we were engaged in writing a joint paper on potential structuring mechanisms in helminth communities; we disagreed on a number of issues. We felt the reason for such debate was because the discipline was in a great state of flux, with many new concepts and approaches being introduced with increasing frequency. After considerable discussion about the need, scope and the inevitable limitations of such a symposium, we decided that the time was ripe to bring other ecologists, engaged in similar research, face-to-face. There were many individuals from whom to choose; we selected those who were actively publishing on helminth communities or those who had expertise in areas which we felt were particularly appropriate. We compiled a list of potential participants, contacted them and received unanimous support to organize such a symposium.

Our intent was to cover several broad areas, fully recognizing that breadth negates depth (at least with a publisher's limitation on the number of pages). We felt it important to consider patterns amongst different kinds of hosts because this is where we had disagreed among ourselves. We also recognized that different hosts are resources for the helminths which infect them; that models, so common in population ecology, might well be important for understanding community structure; and that it was time to draw meaningful parallels between helminth parasites and free-living communities. We intended to include a chapter on the importance of phylogenies as historical determinants, but the author of that chapter was forced to withdraw due to other obligations.

Parasitism represents a ubiquitous mode of life which frequently involves multispecies assemblages; it thus seems particularly appropriate for community level studies. To some, the complexity of many helminth systems make them seem intractable to understanding patterns and processes. To others, that complexity is an attractive feature. It is clear that many ecologists have accepted the challenge of studying helminth communities, taking advantage of the opportunities that studies on such organisms provide. This book is a first course, a primer if you will, on the community ecology of parasitic helminths. Many of the basic processes controlling patterns (e.g. life-history strategies, food-web topology, competitive interactions, local adaptations and dispersal between different habitats) are poorly understood. We had no a priori expectations that definitive answers would emerge. At best, our goal was to provide an overview of helminth communities. However, if they help to focus questions and stimulate interest, we feel we have accomplished our goal.

A symposium is only as good as its participants. We wish to sincerely thank all of the individuals who either participated in, or helped with the details of organizing, the symposium. Judging by the reception it received on the second of August 1988, at the annual meeting of the American Society of Parasitologists held at Wake Forest University in Winston-Salem, North Carolina, it was a resounding success. All manuscripts were reviewed by at least two of the editors; in addition, we acknowledge additional reviews provided by Tim Goater, John Holmes, Allen Shostak and Peter Price. We want to acknowledge, with our sincerest thanks, the secretarial assistance of Zella Crockett and Cindy Davis (Wake Forest University) for the volumes of correspondence which were required in arranging this symposium and to Jan Hinton (Savannah River Ecology Laboratory) for retyping an author's manuscript. We also sincerely thank both Tim Hardwick, formerly Senior Editor, and Bob Carling, Commissioning Editor, at Chapman and Hall who were most helpful in many ways. Support for the publication of this book was provided by the U.S. Department of Energy through contract DE-ACO99-76SR00-819 to the University of Georgia. The Wake Forest University Research and Publication Fund also provided support.

EDITORS' NOTE

It was our initial intention that each member of the symposium organizing committee would be a coeditor on this volume. However, other professional commitments prevented Dr Clive Kennedy from participating in an editorial capacity. We sincerely appreciate his initial participation and continuing encouragement in the preparation of this volume. From the outset, both in organizing the symposium and editing this volume, duties have been shared equally. The sequence of the editors is therefore random and does not reflect seniority.

Gerald W. Esch
Albert O. Bush
John M. Aho
Clive R. Kennedy

Symposium participants are: kneeling from left to right – John Aho, Wayne Sousa, Peter Price, Clive Kennedy. Back row, left to right – Al Bush, Jerry Esch, John Holmes, Armand Kuris, Danny Pence, Andy Dobson. Not pictured is Dan Simberloff (fondly referred to by one of his symposium colleagues as our 'empty niche').

1

Patterns and processes in helminth parasite communities: an overview

Gerald W. Esch, Allen W. Shostak, David J. Marcogliese and Timothy M. Goater

1.1 INTRODUCTION

The structure of helminth communities, their dynamic components and processes and their range of diversities, have long held fascination for parasitologists. Perhaps the earliest body of work in this area was generated by the great Russian academician, V. A. Dogiel and his colleagues in the 1930s (Dogiel, 1964). While it is generally agreed that Crofton (1971a, b) was largely responsible for introducing a quantitative approach to the study of helminth population dynamics, most agree that Holmes (1961, 1962) initiated a quantitative approach to the study of helminth community dynamics. Since these bench-mark publications, significant advances have been made in each area, both empirically and conceptually. Indeed, some of the terminology and ideas which will be highlighted throughout this book were not in the literature at the time the idea for this publication was conceived a little over two years ago.

A variety of schemes for classifying parasite communities and the processes that organize them, can be found in the literature. A primary objective of this introduction is to clarify the relationships among these schemes. In doing so, we will organize it around a hierarchical format which has evolved from studies at both the population and community levels. The community hierarchy follows the one initially developed for parasite populations (Esch *et al.*, 1975). Thus, as the infrapopulation was defined to include all members of a given species of parasite within a single host, a parasite infracommunity (Bush and Holmes, 1986b) includes all of the infrapopulations within an individual host. Of critical importance to this basic hierarchical conceptualization is the implication that infracommunities as well as infrapopulations may be replicated from one host to another which, in turn, has significant ramifications for analyses at both infracommunity and infrapopulation levels.

The next hierarchical level includes the metapopulation and the component community. The metapopulation represents all of the infrapopulations sampled from a given host species within an ecosystem (Riggs and Esch, 1987). Parallel to the metapopulation is the component parasite community which represents all of the infracommunities within a given host population (Holmes and Price, 1986).

The most complex population level of organization is the suprapopulation which represents all individuals of a given parasite species, regardless of life cycle stage, within an ecosystem (Esch *et al.*, 1975). Parallel to the suprapopulation is the compound community which consists of all the parasite communities within an ecosystem (Holmes and Price, 1986).

Many key concepts and principles are most appropriately discussed in relation to a single level in this scheme of organization, while others have application across more than one level of organization. These concepts and principles will be introduced and placed within the framework of this hierarchy in the remainder of this chapter.

1.2 INFRACOMMUNITY CONCEPTS

Competition

'Competition occurs whenever two or more organismic units use the same resources and when those resources are in short supply' (Pianka, 1983). Competition can take one of two forms: either exploitation or interference. In the former case, the interaction of two populations or individuals is mediated indirectly, through inhibitory effects such as utilization of a common resource. Exploitation competition, according to Halvorsen (1976), is uncommon among helminth parasites, occurring only when intestinal parasites compete for limited food resources. However, Dobson (1985) identified a large number of hosts in which he judged parasite co-occurrence to be frequent enough to set the stage for exploitation competition. Interference competition occurs when two individuals or species are vying for the same resource and there is some form of direct confrontation or interaction that reduces access to the resource for one or both individuals or species. An excellent example of interference competition was reported by Stock and Holmes (1987a) and involves a cestode of grebes. They reported that the cestode itself, or pathological changes induced by the cestode in the intestine, were responsible for affecting both species richness and linear distributions of smaller enteric helminths, especially species which absorb nutrients directly through their surfaces. A number of cases of interference competition between helminths, primarily in mammalian hosts, are described by Dobson (1985).

The outcome of competition can be considered in one of three ways. The first is competitive exclusion. Competitive exclusion infers that two species with very similar or identical requirements cannot coexist simultaneously in the same space (Gause, 1934). According to Holmes (1973), competitive exclusion among parasites is a common phenomenon. In his extensive review, he cited numerous field studies to support this contention and readers are referred to his paper. The second outcome, interactive site segregation, refers to the specialization or segregation of niches by two species in which the realized niche of one, or both, is reduced by the presence of the other species (Holmes, 1973). The third outcome, selective site segregation, must be considered within an evolutionary context. It is non-interactive and implies an absence of current competition; within an evolutionary time-frame, genetic changes in one or more of the formerly competing species have occurred, so that their resource requirements have diverged.

These definitions and concepts have had a significant impact on development of theory regarding the evolution of parasite communities. However, debate continues with respect to their general applicability. Much of this discussion stems from the debate regarding the structure of free-living communities (Diamond, 1975; Connell, 1980; Chapter 11). For example, selective site segregation has engendered 'the ghost of competition past' (Connell, 1980). As a consequence, selective site segregation is difficult to distinguish from other possible causal mechanisms of community organization. Rohde (1979) contends that interactive site segregation is not a significant operating force in the structuring of helminth infracommunities. He uses as evidence to support this assertion the vacant spaces on the gill surfaces of marine fish which he says reduces the importance of competition in site selection by monogenean trematode species. Instead, he favours a mating hypothesis which states that 'the intrinsic factor responsible for niche restriction in many parasites may be selection to increase intraspecific contact and thus the chances to mate'. He states further that selective site segregation is frequently not the outcome of 'selection for stabilizing segregation of competing species, but has the adaptive function of maintaining intraspecific contact in low density populations'.

Bush and Holmes (1986b) offer evidence for interactive site segregation as a force in structuring parasite infracommunities within lesser scaup. They observed exceptionally diverse infracommunities with 52 species of helminths and tight species packing. Similarly, potentially interactive infracommunities have been observed by Stock and Holmes (1988) in grebes. Lotz and Font (1985) determined that helminth infracommunities in bats were also diverse and tightly packed, but they concluded that contemporary interactions were of minor importance in structuring these complex infracommunities.

Schad (1966) suggested a novel sort of competitive interaction hypothesis for organizing helminth infracommunities. He proposed that non-reciprocal cross immunity evoked by one species may function to limit the infrapopulation of a competing species. Such a phenomenon would clearly have implications for the development of parasite infracommunities in hosts with appropriate immune systems and thus, is an argument in support of the idea of selective site segregation as an evolutionary force in structuring some parasite infracommunities.

The niche

Perhaps the most widely used definition of the niche concept is that of Hutchinson (1957). He defined the niche as an *n*-dimensional hypervolume, with multiple axes for all variables contributing to the success of a given species. For any parasitic species, these variables would include abiotic factors such as nutrients, moisture and temperature as well as factors affecting host specificity, the nature of the host's response in the form of immunity and conversely, the nature of the parasite's ability to counteract the host's immune response.

The fundamental niche is referred to by Solbrig and Solbrig (1979) as 'the hyperspace bounded by the values of the variables that organisms require to live'. Perhaps because of biotic interactions (competition, predation, etc.), many species occupy a subset of the fundamental niche and this is said to be their realized niche.

The niche concept is difficult to grasp with all of its ramifications. The concept becomes even more difficult to comprehend because of what some refer to as the empty or vacant niche. Price (1980, 1984, 1987) makes a cogent argument for the existence of vacant niches in his effort to describe the nature and evolution of parasite infracommunities. He alludes to the obvious fact that parasite communities have acquired new species over time, but that the pattern of acquisition has been quite variable. He then describes four possible models for evolution of parasite infracommunities. The Nonasymptotic Model (Southwood, 1980) assumes the presence of vacant niches, with the accumulation of new species in a continuous linear sequence over time and without saturation. The Theory of Island Biogeography (MacArthur and Wilson, 1967), or the Asymptotic Equilibrium Model, contends that communities are in equilibrium because of a balance between new colonization and extinction, and that extinctions are due to a combination of competition and stochastic events. This model assumes the existence of empty niches within an evolutionary context. A third approach is the Asymptotic Nonequilibrium Model (Connor and McCoy, 1979; Lawton and Strong, 1981). It proposes that a stable equilibrium is not achieved and that diversity could be easily enriched from other geographic localities. This model also assumes the existence of vacant niches. Finally there is the Cospeciation Model of Brooks (1980) and Mitter and Brooks (1983). This model suggests that the phylogenies of host and parasite communities have co-evolved for so long that competition is no longer a powerful organizing force and that, consequently, the infracommunities are no longer interactive.

Guild concept

Root (1967) coined the term guild to describe functionally similar species in a community, that is, species which share common resources. This concept has met with controversy in the free-living ecological literature (e.g. Hairston, 1984) due to difficulties in establishing exact ecological relationships among species. However, it holds great promise for parasitic organisms. For example, Schad (1963) described the radial distribution of pinworms in the European tortoise. Based on the mouthpart morphology of these parasites, he speculated that some species were indiscriminant feeders on lumenal contents while others fed primarily on fine particulate matter such as bacteria. Each group thus represented a separate and distinct feeding guild. Using a similar approach, Bush and Holmes (1986b) distinguished some species of cestodes and acanthocephalans in their lesser scaup study as 'small absorbers' (paramucosal) and others as 'large absorbers' (mid-lumenal). A third feeding guild within the scaup infracommunity included digenetic trematodes since these species not only absorb nutrients directly through their tegumental surface, but also actively ingest lumenal contents.

The significance of the guild concept rests with the idea that if competition is, or has been, important in structuring parasite infracommunities, then it is most likely to be demonstrable amongst members of the same guild. However, it should also be noted that when infracommunities are separated into different guilds, such as one for feeding mechanisms and one for microhabitat, some species which belong to one guild may diverge into another, such as in Pianka's (1983) guild organization of the lizard fauna in the Kalahari Desert. It becomes imperative that as many variables as possible of a species' niche be measured before reaching conclusions about potential interactions.

Isolationist/interactive communities

Holmes and Price (1986) synthesized current notions regarding competition, niche and guild as they apply to the level of helminth infracommunities and defined isolationist and interactive communities. The former is comprised of species with low colonizing ability. They usually have small infrapopulations where individualistic responses dominate, interspecific interaction is weak even among guild members, and vacant niches exist. Interactive communities are comprised of species with high colonizing abilities. They have large infrapopulations where interspecific interactions dominate individualistic responses, vacant niches do not exist, and species respond to the presence of other guild members.

Holmes and Price (1986) presented the isolationist/interactive community hypothesis as a dichotomy, but noted that a classification of communities into only two types was probably 'too crude to be of lasting utility'. Goater *et al.* (1987) interpreted isolationist and interactive communities as end points of a continuum. A number of studies (Lotz and Font, 1985; Shostak 1986; Jacobson, 1987; Stock and Holmes, 1987a, 1988) provide evidence that many communities exist which can be classified between the extremes.

1.3 COMPONENT COMMUNITY CONCEPTS

Core–satellite species

Caswell (1978) and Hanski (1982) introduced the notion of core and satellite species at the component community level. Core species are those which occur with relatively high frequencies and densities (Holmes and Price, 1986). Satellite species, on the other hand, occur with less frequency than their core counterparts and also are relatively less numerous. According to Holmes and Price (1986), the number and distribution of core species within a component community provide an indication of the potential for biotic interactions.

Bush and Holmes (1986a, b) described the highly complex infra- and component community structure of parasites within lesser scaup. Of the 52 species of helminths found in scaup, only eight were observed with a frequency high enough to be considered as core species. Six of the eight core species were described as specialists in scaup and two were generalists, being found in other waterfowl as well as in scaup. Another group of eight, described as secondary species, were intermediate in frequency and abundance and were positively associated with the core species, but not with each other. The remaining 36 species were considered as satellite species and were randomly distributed among the scaup infracommunities. The core species in scaup were described as forming two suites, one of which uses *Hyalella azteca* as an intermediate host, and the other *Gammarus lacustris*. They concluded that these core and secondary species of the absorbing guild were the primary determinants of an interactive community.

Stock and Holmes (1987b) indicated that the level of host specificity was an important determinant in structuring helminth infracommunities in four species of grebes. Thus, half of all the species recovered and eleven of 14 core species mature in grebes. Moreover, these species accounted for nearly 95% of the infrapopulations in each grebe infracommunity. Stock and Holmes demonstrated that satellite species in grebes were relatively unimportant in enriching the helminth diversity in all four species of grebe. They go on to say, 'this pattern suggests that species richness in these parasite communities is enhanced not by a high degree of specificity to a single host species, as Toft (1986) suggested, but by a lower degree of specificity which allows exchange among related host species'. Contributing to the parasite community structure is the nature of the foraging habits which widely overlap among the four grebe species, but also tend to be concentrated on certain food items, much like scaup (Bush and Holmes, 1986a). A similar pattern of infracommunity species association which could be related to prey (intermediate host) preferences was described in two species of insectivorous bats (Lotz and Font, 1985).

Goater and Bush (1988) described the infracommunities in long-billed curlews as interactive, yet quite simple in richness and abundance. Only two core species were recovered while three others were specialists in sympatric host species, and three species were generalists. They concluded that 'while high diversity may not be required for interactive communities, the presence of sympatric congeners might be crucial. The magnitude of their importance may be enhanced if they are host-specialists and they use the same intermediate host species'.

In contrast, the presence of congeneric species within the same infracommunity does not necessarily preclude the establishment of apparently isolationist communities. Jacobson (1987) examined the infra- and component communities of intestinal helminths in the yellow-bellied slider, *Trachemys scripta*. Using Brillouin's index, diversity was calculated to be $H = 0.46$, which places it well into the range for many bird species as reported by Kennedy *et al.* (1986). Mean densities of parasites in *T. scripta* were relatively low and more in line with those of fish rather than birds (Kennedy *et al.*, 1986). As Jacobson (1987) observed, many species co-occurred in the slider turtles, but there was no clear indication that the presence or absence of one species influenced another. The three congeners of the acanthocephalan *Neoechinorhynchus* were sympatric in their linear and circumferential distributions within the intestine and there was no evidence for competitive interactions affecting their spatial patterns.

Development of component communities

Holmes and Price (1986) reviewed several hypotheses which focus on the processes which ultimately determine species richness at the component community level. The Island Distance Hypothesis, which is based on the Theory of Island Biogeography (MacArthur and Wilson, 1967), suggests that parasite species richness is related to the difficulty of invasion. It predicts that isolationist communities will develop when parasite colonization is limited such as in the case of the depauperate infracommunities described for salamanders by Goater *et al.*, (1987), or for the many fish helminth infracommunities reviewed by Kennedy *et al.* (1986). Goater *et al.* (1987) contradicted the so-called 'Co-speciation' Model of Brooks (1980) which predicts 'parasite specialization caused by speciation in a host lineage resulting in co-speciation in their parasite lineage'. Goater *et al.* (1987) support their contention by noting that speciation of desmognathine salamanders did not result in co-speciation of parasites since none harbour their own 'unique' helminth species. They base their position on the premise that helminth communities, composed of phylogenetically unrelated species which often represent individual host captures, will be organized differently than those composed of phylogenetically related species, as with congeners for example. They predict that the former will be isolationist and the latter interactive. The phenomenon of 'host capture' implies that parasites are acquired from ecological associates and then subsequently become adapted to the new host, perhaps speciating in the process (Holmes and Price, 1980). Thus, in cases where several different species of helminths are transmitted by the same intermediate host upon which a given definitive host specializes as a prey item, the result will be a greater helminth species diversity and larger infrapopulations, e.g. the helminth infracommunities of scaup which are dominated by hymenolepid cestodes (Bush and Holmes, 1986b). In contrast, when transmission is limited and each helminth species represents an isolated 'capture' over evolutionary time, there is less potential for interactions being important in structuring the infracommunities in the past or the present. In such cases, isolationist communities are predicted. According to Price (1980), 'resources available are so diverse it is improbable that a new colonist carries a niche exploitation pattern very similar to a resident species'. Only in cases where transmission has not been limited in evolutionary and ecological time are interactive parasite communities predicted (Goater *et al.*, 1987).

Host age and the component community

While most studies which attempt to relate the structuring or development of component communities to host ageing have not used the replicate approach of Bush and Holmes (1986b), there is ample evidence to indicate that such a procedure would be most useful. These older studies are none the less instructive since they provide insight as to the manner in which component community structure may be shaped by the ageing process. Changes in community organization at the component level may be influenced by such variables as a shift in host diet or the volume of food consumed, ontogenetic changes in immunocompetency, and modification in the probability of contact with potential intermediate hosts (Esch, 1983).

One of the most comprehensive studies relating ageing and change in component community diversity was that of Humphrey *et al.* (1978). They examined and compared the helminth infracommunities in the brown pelican. Parasite diversity increased almost linearly for the first nine weeks following hatching. Diversity then fell sharply during and immediately following fledging, only to increase again in sexually mature birds. The changes in diversity were attributed to exposure to more taxonomically diverse prey (intermediate hosts) prior to fledging. The decline in diversity paralleled the period during which the young birds began feeding themselves. At this time, food consumption was either reduced in volume or in diversity, or both, and presumably resulted in a reduction in exposure to infective agents of the parasites. As the birds became more successful predators, diversity of prey in the diet again increased, resulting in greater exposure to infective agents and enhanced parasite diversity.

Migration and component community structure

The classical studies reported by Dogiel (1964) have provided the greatest impetus for examining the diversity changes of parasite component communities among hosts having wide migration ranges. Dogiel's efforts in this area resulted in the establishment of a classification scheme for parasites of migratory hosts in the Northern Hemisphere which is still in use (e.g. Hood and Welch, 1980). Parasites are divided into four categories, each based on a specific transmission process. 'Ubiquitous parasites' include ectoparasites which may be transmitted throughout the host's migratory range. Several enteric species are also included with this typically surface-dwelling group. 'Southern species' infect birds on their wintering sites and are not carried north successfully. 'Northern species' are acquired on the summer breeding grounds and are not carried south successfully during migration. 'Migration species' infect birds during the migration process and are usually not successfully transmitted at either the breeding or the wintering sites.

The component community diversity among migrating vertebrates, whether birds or other taxonomic groups, may change during migration for a variety of reasons. Among anadromous or catadromous fishes, the switch is primarily related to the impact of changing from freshwater to saltwater, or vice versa, and the ability of the parasite to withstand the osmotic shock of such a change. In such cases, the most vulnerable parasites are those which live on external surfaces or within the intestine. Parasites which are sequestered internally, either in organs, body cavity, or in the blood vascular system, are likely to be retained by anadromous or catadromous host species since, internally, environments remain relatively stable over the long term. Among migrating birds, two major factors play a role in affecting parasite diversity. The first is a shift in exposure to potential intermediate hosts. Second, profound changes in host diet during migration cause significant changes in intestinal physiology which, in turn, has a significant impact on parasite diversity.

In North America, the nature of change in component community diversity has been examined in a variety of migratory bird species. Hood and Welch (1980) compared the helminth diversity in populations of the red-winged blackbird on both their wintering and breeding grounds. Of the 17 species comprising the component community, two were ubiquitous, four were associated with breeding areas, six with wintering sites and two were migratory. They concluded that variation in diet and thus exposure to intermediate hosts, was the single most important factor in influencing both the diversity of infracommunities and the abundance of infrapopulations.

Seasonal changes in the component community

Changes in component community structure are, in many cases, inextricably linked to host ageing, diet and migration patterns, and all are related to the concept of temporal change in communities. There are, however, few studies to document such an assertion.

Chappell (1969) examined the parasite fauna in the three-spined stickleback from a pond in England. He attributed seasonal change in the parasite community to temperature-related effects on the timing, rates, and duration of cercariae production from molluscan hosts, as well as to diet shift. Owen and Pemberton (1962) observed significant changes in parasite community diversity among starlings in northern England and indicated that the major influence in these changes was the shift in diet from plant to animal material.

1.4 COMPOUND COMMUNITY CONCEPTS

The biocoenosis and parasite flow

According to Noble and Noble (1982), a biocoenosis is composed of an assemblage of organisms whose interests or needs are linked by requirements which are prescribed by a well-defined habitat and mutual interactions. It is occasionally referred to as a 'species network'. While the biocoenosis concept is not now widely used, it none the less, has value in terms of considering parasite flow within ecosystems and is thus of significance in discussion of certain dynamic qualities of parasite compound communities.

The earliest and still one of the most thorough studies to employ the concepts of biocoenosis and the compound community as it is now known was that of Wisniewski (1958) and co-workers in Lake Druzno, Poland. Their primary goal was to establish the direction of parasite flow within the lake. They sampled thousands of invertebrates and vertebrates for evidence of parasite transmission and flow. The major pattern of flow was from fish, amphibians and various invertebrate groups to birds. Of the 135 helminth species maturing in vertebrates within the lake, 99 were found in birds. Among 72 species of larval parasites identified, 22 were taken from fish and amphibians, underscoring the importance of these two vertebrate groups as transmitting agents in this particular system. A significant feature of this study was the proposed relationship between the pattern of parasite flow, the dominance of birds as definitive hosts, and the eutrophic character of the lake. It was concluded that the preponderance of the parasite fauna in birds would be characteristic of any eutrophic body of water.

A slightly different approach to the study of patterns involved in parasite flow was undertaken by Rodenberg and Pence (1978) in their investigation of six mammalian species. They observed parasite circulation to be greatest among species having some degree of niche overlap within the same habitat. They went on to conclude that an abundance of insects (potential intermediate hosts) in the diets of the ecologically similar rodent species was undoubtedly the key element contributing to the similarity of helminth infracommunities and thus to the flow of parasites at the compound community level.

Holmes and Podesta (1968) compared the helminth faunas of wolves and coyotes. Based on their effort and on other studies, they concluded that the helminth faunas of wolves in various regions throughout North America are basically similar and, consequently, that food habits of these hosts must also be similar. Coyote parasites, on the other hand, tended to vary extensively from area to area, both in terms of diversity and species composition. Generally, the parasite faunas in coyotes are distinctive from those in wolves. However, in two regions of their study, the diversity of parasitic helminths, especially cestodes, was as similar to those of wolves as it was to coyotes in other areas. This suggested a pattern of circulation of parasites between the two host species in these two areas, probably because of overlapping food resources.

Neraasen and Holmes (1975) provided an extensive and thorough analysis of helminth parasite flow through four species of geese. They developed the concept of an 'infective pool' to assist in explaining the pattern. Parasites release infective agents into a given environment and a 'pool' accumulates. Then, either directly or indirectly, a source of infection is provided to the same or other potential host species in the area. An individual 'host will contribute to, and draw from, the infective pool in accordance with its abundance and suitability as a host'. The circulation of parasites from the infective pool through the various host species depends to a great extent on the density of each host species and on the level of contact among potential hosts as well as with the infective pool. Within the goose populations, four groups of helminths, mostly cestodes, could be distinguished, each representing a different pattern of flow. The first group was comprised of cestodes which were not exchanged among the different species of geese, but which were transmitted primarily from other anseriforms to geese. These would constitute satellite species among the geese. The second group included cestodes brought to the breeding grounds by any of the geese species and then circulated to the others. In the third group were cestodes brought to the breeding grounds by one host species and then maintained only by the same host species. The last group included species which were brought to the breeding grounds by a single host species, but were subsequently maintained by another host group. They concluded that several factors were key to maintaining the circulation pattern among the geese populations. These included the density of host species, the migration route employed by the different host species, diet, and the degree of host specificity shown by the different species of cestodes.

All of these studies (Wisniewski, 1958; Rodenberg and Pence, 1978; Holmes and Podesta, 1968; Neraasen and Holmes, 1975) illustrate the nature and pattern of helminth circulation among ecologically overlapping host species. The most significant elements in establishing these patterns are similarity in diet preferences by the various host species and the degree of host specificity shown by the parasite species. In addition, 'differential rates of parasite recruitment and mortality, coupled with the nature of parasite transmission dynamics within communities of free-living organisms, will play an important role in ultimately determining parasite diversity in a given host species' (Esch, 1983).

Allogenic–autogenic concept

Recently, Esch *et al.* (1988) introduced the concept of allogenic versus autogenic species in an effort to better define parasite transmission dynamics. They used the concept to interpret helminth colonization patterns among parasites in British fish. An autogenic species was defined as one in which the entire life cycle of the parasite is completed within the confines of an aquatic ecosystem. An allogenic species was defined as one which uses fish or other aquatic vertebrates as an intermediate host and which matures in birds or mammals.

In Britain, three families of fish were examined within the framework of the allogenic-autogenic concept. In salmonids, helminth infracommunities were dominated by autogenic species which were also found to be responsible for most of the similarity within and between geographic localities. Cyprinids were dominated by allogenic helminth species which also accounted for most of the similarity within and between localities. Anguillids, on the other hand, exhibited features which were intermediate, with neither category consistently dominating nor providing for a clear pattern of similarity. The different colonization strategies of autogenic and allogenic species, the degree of host vagility as well as migratory habits, collectively provide a suitable explanation for the observed patchy distribution of many helminths in British fish.

Habitat stability

Without question, the most complicated form of host/parasite organization is the compound community. There are two problems involved in considering this level of organization. The first is related to the incredible complexity of parasite compound community structure. Of concern is not only the host community and its myriad of interacting abiotic and biotic variables, but that of the parasite's as well. Second, factors affecting the parasite compound community present a formidable situation since the parasite's immediate habitat is alive and capable of responding to the parasite in a variety of ways. Unfortunately, there are no studies which have attempted to deal with an entire compound community. There are, however, several which have dealt with some aspects of compound community structure and dynamics (Wisniewski, 1958; Chubb, 1963; Esch, 1971; Halvorsen, 1971; Wootten, 1973; Kennedy and Burrough, 1978; Holmes, 1979; Leong and Holmes, 1981). In almost all of these studies, the investigator's approach was to consider the compound community within the framework of ecological succession and attempt to focus on a single host group, primarily fish.

As previously alluded to, the most comprehensive study of compound communities was conducted in eutrophic Lake Druzno in Poland by Wisniewski (1958). The basic aim of this investigation was to examine the nature of transmission dynamics and parasite flow within the system. The most significant finding was that the parasite compound community in Lake Druzno was dominated by parasites which completed their life cycles in shore birds and waterfowl (allogenic species of parasites). This observation was attributed directly to the eutrophic nature of the lake which Wisniewski claimed would influence the character of the compound community in any similar body of water.

Chubb (1963) used the approach taken by Wisniewski in attempting to relate the nature of the compound community in fish to the trophic status of a lake in Wales. He asserted that the compound community pattern he observed was far different from the one in Lake Druzno and that it was related to the oligotrophic nature of the system in which he was working. However, he did go on to qualify his observations relating trophic status and compound community structure by stressing that the diversity and individuality of lakes will be reflected in the individuality of parasite faunas irrespective of trophic dynamics. Kennedy and Burrough (1978) reached similar conclusions after investigating the parasite faunas in brown trout and perch in a series of British lakes with differing trophic status. Thus, relationships and patterns could be established with respect to trophic dynamics, but the individual nature of each body of water could override any effects created by trophic status.

Esch (1971) reported on the compound community status among centrarchids and molluscs from an oligotrophic and a eutrophic lake. His conclusions regarding differences in the two systems led him to assert that the more closed nature of the oligotrophic lake would produce a compound community dominated by autogenic parasite species which would complete their life cycles for the most part in predatory fish and not in shore birds or waterfowl. The opposite would be true for the eutrophic lake where most allogenic parasite species would mature sexually in fish-eating birds and mammals. In this sense, his conclusions were similar to those of Wisniewski (1958) except that he proposed that the processes involved in creating patterns of change during ecological succession are most probably related to changes in food-web dynamics which go hand-in-hand with succession. This can be related to the work of Leong and Holmes (1981), who investigated the compound community within an oligotrophic system in Alberta, Canada. In doing so, they not only characterized and quantified the abundance of parasites, but the relative abundances of host species as well. They concluded that the nature of the parasite compound community was a function of the numerically dominant host species and not necessarily the trophic status of the lake.

The impact of habitat stability on compound communities is also well illustrated by a recent study on avocets (Edwards and Bush, personal communication). They collected avocets from both ephemeral and permanent bodies of water and determined the patterns of similarity and overlap of intestinal helminths. Among birds taken from ephemeral bodies of water, the infracommunities were composed mainly of specialist species.

Interactions between helminth species were limited. On the other hand, avocets collected from permanent bodies of water had infracommunities which were dominated by helminth species which were specialists in a variety of duck species, primarily lesser scaup. These helminths were superimposed on the normal parasite infracommunities, being positioned into linear gaps along the intestine, but also overlapping the normal avocet fauna. The lesser scaup specialists exhibited interactive patterns among themselves as well as with avocet specialists. They concluded that helminth communities from avocets in ephemeral bodies of water were mainly isolationist in character and that vacant niches were relatively common, while birds from permanent bodies of water possessed saturated and interactive communities.

SUMMARY

Clearly parasite community structure and organization are highly complex, undoubtedly the product of long and continuing interaction involving both host and parasite populations. Moreover, this particular area of parasite ecology is in a flux from the perspective of understanding those forces which operate to structure parasite communities, the mechanisms of evolution involving both hosts and parasites, and then in terms of how parasite communities are related to the organization and dynamics of free-living communities.

It can be concluded that infra-, component and compound parasite communities are influenced by the sum of a complex set of biotic and abiotic factors which directly impact on processes that will affect the transmission dynamics of helminth parasites. An understanding of these biotic and abiotic factors will, in the long term, provide a solid basis for properly assessing the nature of parasite community structure and function. The chapters which follow provide substantive new and integrative observations and thus will elevate our understanding and interest in the overall ecological and evolutionary relationships of parasite communities.

REFERENCES

Brooks, D. R. (1980) Allopatric speciation and non-interactive parasite community structure. *Syst. Zool.*, **30**, 192–203.

Bush, A. O. and Holmes, J. C. (1986a) Intestinal parasites of lesser scaup ducks: patterns of association. *Can. J. Zool.*, **64**, 132–41.

Bush, A. O. and Holmes, J. C. (1986b) Intestinal parasites of lesser scaup ducks: an interactive community. *Can. J. Zool.*, **64**, 142–52.

Caswell, H. (1978) Predator-mediated coexistence: a nonequilibrium model. *Am. Nat.*, **112**, 127–54.

Chappell, L. H. (1969) Competitive exclusion between two intestinal parasites of the three-spined stickleback' *Gasterosteus aculeatus* L. *J. Parasitol.*, **55**, 775–8.

Chubb, J. C. (1963) On the characterization of the parasite fauna of the fish of Lynn Tegid. *Proc. Zool. Soc. (Lond.)*, **141**, 609–21.

Connell, J. H. (1980) Diversity and the coevolution of competitors, or the ghost of competition past. *Oikos*, **35**, 131–8.

Connor, E. F. and McCoy, E. D. (1979) The statistics and biology of the species–area relationship. *Am Nat.*, **113**, 791–833.

Crofton, H. D. (1971a) A quantitative approach to parasitism. *Parasitology*, **62**, 179–93.

Crofton, H. D. (1971b) A model for host–parasite relationships. *Parasitology*, **63**, 343–64.

Diamond, J. M. (1975) Assembly of species communities. In *Ecology and Evolution of Communities*, (eds M. L. Cody and J. M. Diamond), Harvard University Press, Cambridge, pp. 342–4.

Dobson, A. P. (1985) The population dynamics of competition between parasites. *Parasitology*, **91**, 317–47.

Dogiel, V. A. (1964) *General Parasitology*, Oliver and Boyd, Edinburgh.

Esch, G. W. (1971) Impact of ecological succession on the parasite fauna in centrarchids from oligotrophic and eutrophic ecosystems. *Am. Midl. Nat.*, **86**, 160–8.

Esch, G. W. (1983) The population and community ecology of cestodes. In *Biology of the Eucestoda, vol. I* (eds C. Arme and P. W. Pappas), Academic Press, London and New York, pp. 81–137.

Esch, G. W., Gibbons, J. W. and Bourque, J. E. (1975) An analysis of the relationship between stress and parasitism. *Am. Midl. Nat.*, **93**, 339–53.

Esch, G. W., Kennedy, C. R., Bush, A. O. and Aho, J. M. (1988) Patterns in helminth communities in freshwater fish in Great Britain: alternative strategies for colonization. *Parasitology*, **96**, 519–32.

Gause, G. F. (1934) *The Struggle for Existence*, Hafner, New York, pp. 1–163, (reprinted 1964 by Williams and Wilkins, Baltimore, MD).

Goater, C. P. and Bush, A. O. (1988) Intestinal helminth communities in long-billed curlews: the importance of congeneric host-specialists. *Hol. Arctic Ecol.*, **11**, 140–45.

Goater, T. M., Esch, G. W. and Bush, A. O. (1987) Helminth parasites of sympatric salamanders: ecological concepts at infracommunity, component and compound community levels. *Am. Midl. Nat.*, **118**, 289–300.

Hairston, N. G. (1984) Inferences and experimental results in guild structure. In *Ecological Communities: Conceptual Issues and the Evidence* (eds D. R. Strong Jr, D. Simberloff, L. G. Abele and A. B. Thistle, Princeton University Press, Princeton, pp. 19–27.

Halvorsen, O. (1971) Studies on the helminth fauna of Norway. XVIII. On the composition of the parasite fauna of coarse fish in the River Glomma, Southeastern Norway. *Norw. J. Zool.*, **19**, 181–92.

Halvorsen, O. (1976) Negative interaction amongst parasites. In *Ecological Aspects of Parasitology*, (ed. C. R. Kennedy), North-Holland Publishing Co., pp. 99–114.

Hanski, I. (1982) Dynamics of regional distribution: the core and satellite species hypothesis. *Oikos*, **38**, 210–21.

Holmes, J. C. (1961) Effects of concurrent infections on *Hymenolepis diminuta* (Cestoda) and *Moniliformis dubius* (Acanthocephala). I. General effects and comparison with crowding. *J. Parasitol.*, **47**, 209–16.

Holmes, J. C. (1962) Effect of concurrent infections on *Hymenolepis diminuta* (Cestoda) and *Moniliformis dubius* (Acanthocephala), II. Effects on growth. *J. Parasitol.*, **48**, 87–96.

Holmes, J. C. (1973) Site selection by parasitic helminths: interspecific interactions, site segregation, and their importance to the development of helminth communities. *Can. J. Zool.*, **51**, 333–47.

Holmes, J. C. (1979) Parasite populations and host community structure. In *Host–parasite Interfaces* (ed. B. B. Nickol), Academic Press, New York, pp. 27–46.

Holmes, J. C. and Podesta, R. (1968) The helminths of wolves and coyotes from the forested regions of Alberta. *Can. J. Zool.*, **46**, 1193–204.

Holmes, J. C. and Price, P. W. (1980) Parasite communities: the roles of phylogeny and ecology. *Syst. Zool.*, **29**, 203–13.

Holmes, J. C. and Price, P. W. (1986) Communities of parasites. In *Community Ecology: Patterns and Processes* (eds, D. J. Anderson and J. Kikkawa), Blackwell Scientific Publications, Oxford, pp. 187–213.

Hood, D. E. and Welch, H. E. (1980) A seasonal study of the parasites of the red-winged blackbird, *Agelaius phoeniceus* in Manitoba and Arkansas. *Can. J. Zool.*, **58**, 528–37.

Humphrey, S. R., Courtney, C. H. and Forrester, D. H. (1978) Community ecology of the helminth parasites in brown pelicans. *Wilson Bull.*, **90**, 587–98.

Hutchinson, G. E. (1957) Concluding remarks. *Cold Spring Harbor Symposia in Quantitative Biology*, **22**, 415–27.

Jacobson, K. C. (1987) Infracommunity structure of enteric helminths in the yellow-bellied slider, *Trachemys scripta scripta*. Master's thesis, Wake Forest University, Winston-Salem, N.C., 46 p.

Kennedy, C. R. and Burrough, R. J. (1978) Parasites of trout and perch in Malham Tarn. *Field Studies*, **4**, 617–29.

Kennedy, C. R., Bush, A. O. and Aho, J. M. (1986) Patterns in helminth communities: why are fish and birds different? *Parasitology*, **93**, 205–15.

Lawton, J. H. and Strong, D. R. (1981) Community patterns and competition in folivorous insects. *Am. Nat.*, **118**, 317–38.

Leong, T. S. and Holmes, J. C. (1981) Communities of Metazoan parasites in open water fishes of Cold Lake, Alberta. *J. Fish Biol.*, **18**, 693–714.

Lotz, J. M. and Font, W. F. (1985) Structure of enteric helminth communities in two populations of *Eptesicus fuscus* (Chiroptera). *Can. J. Zool.*, **63**, 2969–78.

MacArthur, R. H. and Wilson, E. O. (1967) *The Theory of Island Biogeography*, Princeton University Press, Princeton.

Mitter, C. and Brooks, D. R. (1983) Phylogenetic aspects of coevolution. In *Coevolution* (eds, D. J. Futuyma and M. Slatkin), Sinauer, Sunderland, pp. 65–98.

Neraasen, T. G. and Holmes, J. C. (1975) The circulation of cestodes among three species of geese resting on the Anderson River Delta, Canada. *Acta. Parasitol. Pol.*, **23**, 277–89.

Noble, E. R. and Noble, G. A. (1982) *Parasitology: The Biology of Animal Parasites*, Lea and Febiger, Philadelphia.

Owen, R. W. and Pemberton, R. T. (1962) Helminth infection of the starling (*Sturnus vulgaris* L.) in northern England. *Proc. Zool. Soc. London*, **139**, 557–87.

Pianka, E. R. (1983) *Evolutionary Ecology*, Harper and Row Publishers, New York.

Price, P. W. (1980) *Evolutionary Biology of Parasites*, Princeton University Press, Princeton.

Price, P. W. (1984) Communities of specialists: vacant niches in ecological and evolutionary time. In *Communities: Conceptual Issues and the Evidence* (eds, D. R. Strong Jr, D. Simberloff, L. G. Abele and A. B. Thistle), Princeton University Press, Princeton, pp. 510–23.

Price, P. W. (1987) Evolution in parasite communities. *Int. J. Parasitol.*, **17**, 203–8.

Riggs, M. R. and Esch, G. W. (1987) The suprapopulation dynamics of *Bothriocephalus acheilognathi* in a North Carolina cooling reservoir: abundance, dispersion and prevalence. *J. Parasitol.*, **73**, 877–92.

Rodenberg, G. W. and Pence, D. B. (1978) Circulation of helminth species in a rodent population from the high plains of Texas. *Occasional Papers of the Museum Texas Tech University*, No. 56, 1–6.

Rohde, K. (1979) A critical evaluation of intrinsic and extrinsic factors responsible for niche restriction in parasites. *Am. Nat.*, **114**, 648–71.

Root, R. B. (1967) The niche exploitation pattern of the blue-green gnatcatcher. *Ecol. Monogr.*, **37**, 317–50.

Schad, G. A. (1963) Niche diversification in a parasitic species flock. *Nature*, **198**, 404–5.

Schad, G. A. (1966) Immunity, competition, and natural regulation of helminth populations. *Am. Nat.*, **100**, 359–64.

Shostak, A. W. (1986) Sources of variability in life-history characteristics of the annual phase of *Triaenophorus crassus* (Cestoda: Pseudophyllidea). PhD thesis, University of Manitoba, Winnipeg, Manitoba. 324 p.

Solbrig, O. T. and Solbrig, D. J. (1979) *Introduction to Population Biology and Evolution*, Addison-Wesley, Reading, MA.

Southwood, T. R. E. (1980) *Ecological Methods*, 2nd edn, Chapman and Hall, London.

Stock, T. M. and Holmes, J. C. (1987a) *Dioecocestus asper* (Cestoda: Dioecocestidae): an interference competitor in an enteric helminth community. *J. Parasitol.*, **73**, 1116–23.

Stock, T. M. and Holmes, J. C. (1987b) Host specificity and exchange of intestinal helminths among four species of grebes (Podicipedidae). *Can. J. Zool.*, **65**, 669–76.

Stock, T. M. and Holmes, J. C. (1988) Functional relationships and microhabitat distributions of enteric helminths of grebes (Podicipedidae). *J. Parasitol.*, **74**, 214–27.

Toft, C. A. (1986) Communities of species with parasitic life styles. In *Community*

Ecology (eds, T. Case and J. Diamond), Harper and Row, New York, pp. 445–63.

Wisniewski, W. L. (1958) Characterization of the parasitofauna of an eutrophic lake. *Acta. Parasitol. Pol.*, **6**, 1–64.

Wootten, R. (1973) The metazoan fauna of fish from Hanningfield Reservoir, Essex, in relation to features of the habitat and host populations. *J. Zool. (Lond.)*, **171**, 323-31.

2

Host populations as resources defining parasite community organization

Peter W. Price

2.1 INTRODUCTION

The food web is the basis for understanding community organization. All aspects of the web may play a role in the processes defining community structure: resources in the trophic level below (e.g. hosts), competition, commensalism and mutualism on the next trophic level (e.g. among parasites) and predation, parasitism and mutualism on the trophic level above. In communities of parasites all of these processes are intimately involved. The resource base is the host, a critical variable necessary for understanding parasite communities. Competition and commensalism are common and mutualists frequently mediate the host–parasite interaction. Predation is an integral part of transmission in many animal parasite systems. In addition to this tight linkage in food webs involving parasites, the component communities of one host frequently have cross-linkage into other component communities on other hosts. Therefore, host associations are critical to understanding parasite communities, but also parasite communities profoundly influence host interactions. This complexity of interactions has been largely ignored by both general ecologist and parasitologist alike. It is only the parasite community ecologists who have a broad perception of the intricacies of these food web interactions.

A generalized food web involving animal parasites must therefore involve at least several definitive host species often infected while eating parasitized prey, the parasite community, several intermediate hosts, and the resource bases on which these latter hosts feed (Fig. 2.1). Although such food webs are incredibly numerous and diverse, they have been ignored in

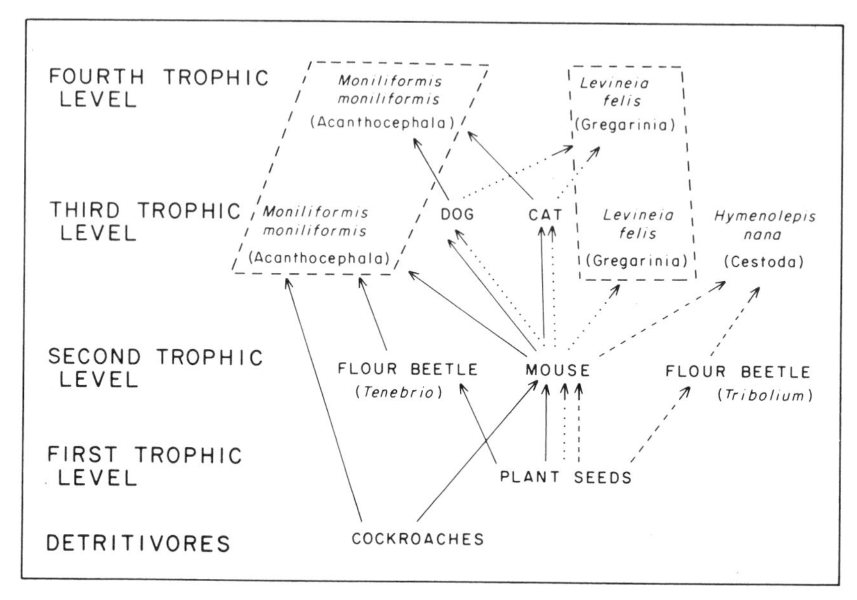

Fig. 2.1 A simplified food web involving only three parasite species of small rodents. Three trophic levels and detritivores are all linked by the movement of parasites between definitive and intermediate hosts. These life cycles are relatively simple because only two hosts are needed to complete a life cycle, and no transport hosts are required. Note that *Levineia* feeds at two trophic levels – mouse and its predators. *Moniliformis* feeds at two trophic levels (flour beetle, mouse, cat, dog) and on detritivores.

Arrows indicate movement of energy; solid lines indicate linkage in the *Moniliformis* trophic system, dotted lines, the *Levineia* system, and dashed lines, the *Hymenolepis* system.

studies of food web design (e.g. Cohen, 1978; Pimm, 1982). Such complex and extensive linkages also have dramatic effects from above with parasites exerting striking organizing power on host communities, another area in need of concerted exploration (e.g. Price *et al.*, 1986, 1988). For example, a parasite is likely to mediate host interactions, providing a net benefit for the more resistant host species and a negative effect for the more susceptible host (Fig. 2.2).

How then can we deal with all the complexity and intimacy of association? One approach is to study small compartments: the predator–prey transmission dynamics; competition among parasites in the definitive host; the process of colonization of hosts by parasites. All are essential components and have been studied with success. But ultimately a more complete understanding of the food web must be obtained, even though a daunting

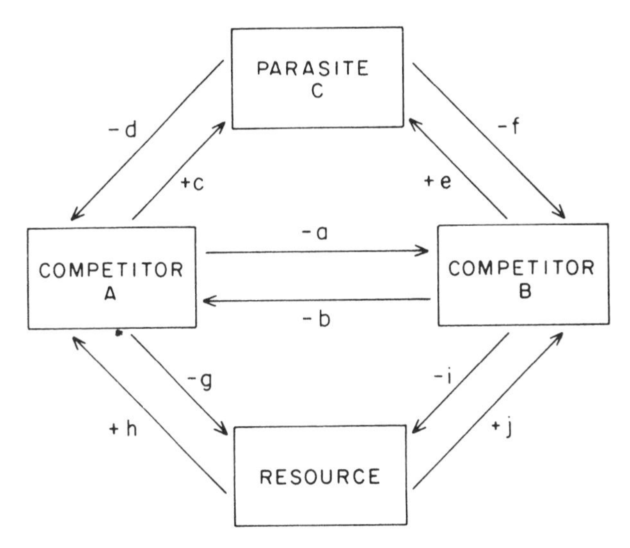

Fig. 2.2 A three-trophic-level system in which a parasite has the potential for mediating competition in the lower trophic level. Each interaction is given a sign indicating positive or negative impact in the direction of the arrow. For example, Competitor A benefits Parasite C (+c) and the parasite has a negative impact on the competitor (−d). However, the parasite may provide a net benefit to one competitor by having a more strongly negative impact on the other competitor ($-f \gg -d$, making $-b \ll -a$); from Price *et al.* (1986). Reproduced with permission from the *Ann. Rev. Ecol. Syst.*, **17**. © 1986 by Annual Reviews Inc.

complexity exists. One simplifying principle, which clarifies one avenue of entry to this complexity, is that trophic systems are driven from below. The resource base on which a community is built is likely to be the most critical organizing influence. Other interactions such as competition and predation are likely to be secondary in importance to community organization. For example, the nature of the resource base is likely to define the presence or absence of competition or its intensity (Price, 1984b) (Fig. 2.3).

For parasite communities the resource base is the host population, in many cases involving a definitive as well as one or more intermediate hosts. If trophic systems are driven from below, then we should concentrate some attention on host characteristics which influence parasite communities: host population density; host size; population growth rate; geographical distribution; and other hosts with which the parasites interact. Then we can examine the evidence for these host characteristics which influence parasite community structure. I will develop this theme in succeeding sections of this chapter.

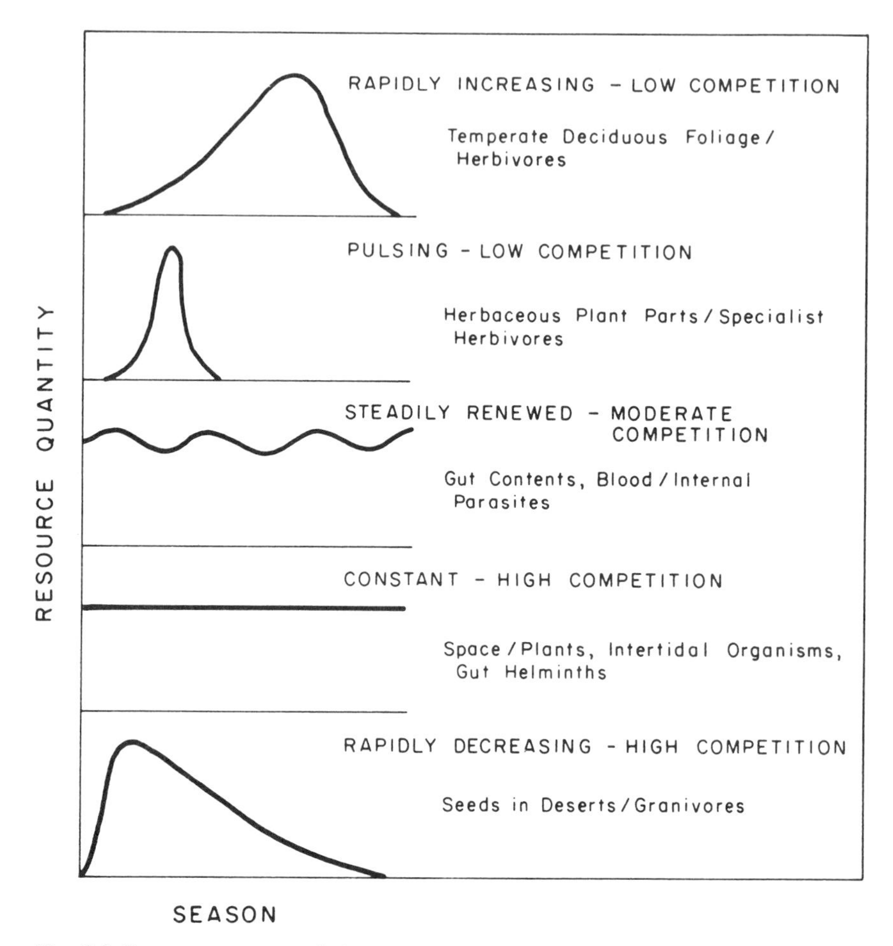

Fig. 2.3 Resource types and the consequences for the probability of interspecific competition on these types. For example, rapidly increasing resources during the active season of the exploiters will reduce the probability of competition being an important organizing force in a community. But intensity of competition will also be modified by uniform or patchy distribution of resources, and the rate of colonization of resources (after Price, 1984b).

2.2 THE RESOURCE BASE: PROPERTIES OF HOSTS

Beyond physiological properties of hosts which require specific attention for each parasite–host interaction, there are many host population properties which are more generalizable in terms of their impact on parasite communities. These population properties are intimately associated in predictable ways (Fig. 2.4) discussed below (see also Price *et al.*, 1988).

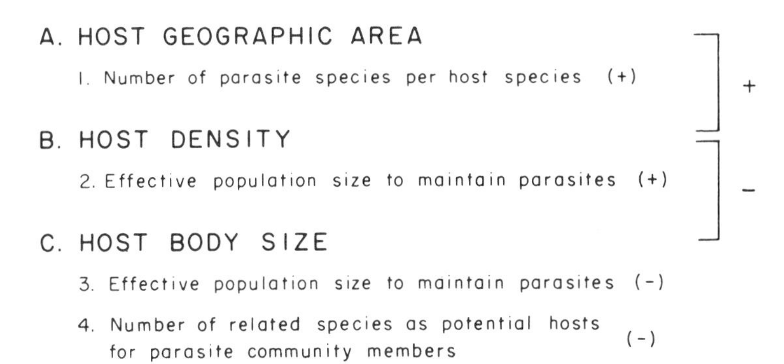

Fig. 2.4 General relationships between host characteristics A, B, and C, and how they interact (on right) and consequences for the parasite community, 1–4, and the slope of the relationship, + or −.

Host geographic area

A general positive relationship has been found frequently between host geographic range and the number of parasitic species recorded from that host (Price, 1980; Price *et al.*, 1988). This relationship holds for within one host species (Freeland, 1979) and for between different host species (Price and Clancy, 1983) (Fig. 2.5).

Some consequences for both host and parasites may be deduced. As host geographic range increases: (a) local community richness of parasites is likely to increase, (b) probability of competition among parasites increases, (c) probability of exchange of parasites within and between component communities increases, (d) probability of host genetic and immunological resistance to parasites increases because of exposure to more species, (e) probability that hosts acquire parasites more deleterious to the host's potential competitors simply as a function of the number of parasite species sustained, and (f) probability that hosts are exposed to novel parasite species is reduced because large areas are sampled by the host for parasites. The correlates of host geographic area are extensive but need more study in relation to the consequent effects on local community structure.

Host density

There is a positive correlation between host geographic area and local abundance (Hanski, 1982; Brown, 1984) (Fig. 2.6). Thus, in addition to the properties of communities relating to geographic area considered above, there are the effects of local abundance which probably include the

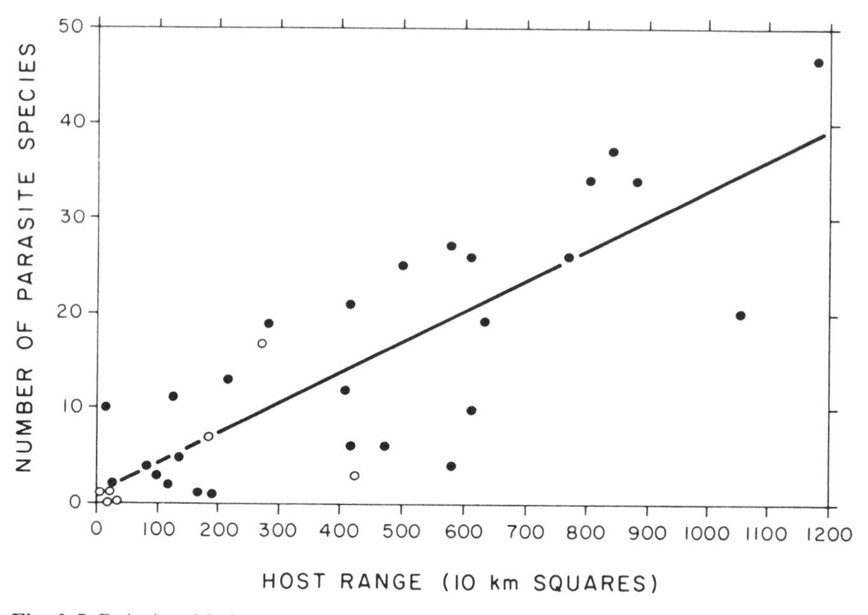

Fig. 2.5 Relationship between geographical range of British freshwater fish as hosts and the number of helminth parasite species per host. Solid circles – indigenous species; open circles – introduced species. The relationship accounts for 68% of the variance ($p < 0.05$). From Price and Clancy (1983). Reproduced with permission.

following. As local abundance increases: (a) the effective population size for maintaining parasite populations increases, resulting in more parasite species being maintained in communities at the local level, (b) lower local extinction rates of parasite species will result because host populations have a higher probability of remaining above threshold densities (Kermack and McKendrick, 1927), for example, persistence of a virulent disease such as measles in humans is limited to populations over 350 000 individuals (Black, 1966) (Fig. 2.7), (c) more pathogenic parasites, which require higher threshold densities, will be supported in communities, and (d) there is an increased probability of hosts acquiring parasites more deleterious to the host's potential competitors (as in Host Geographic Area (e) above). Again, the ramifications of host density are considerable.

Host body size

In general, as host density increases host body size declines, and as body mass declines population growth rate increases (Peters, 1983; Calder, 1984) (Fig. 2.8). These properties tend to reinforce those above: (a) a high

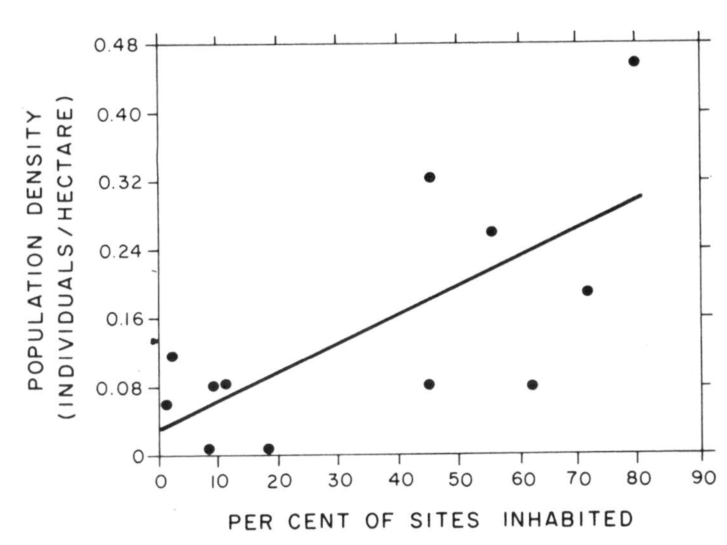

Fig. 2.6 The relationship between the breadth of distribution of desert rodent species represented by the per cent of sites inhabited, and the population density (individuals per hectare). The relationship accounts for 48% of the variance ($p < 0.01$). After Brown (1984).

reproductive rate maintains a population of new hosts available for colonization by parasites, (b) with rapid turnover of individual hosts there will be less time for equilibrium communities to develop, or for individual hosts to become saturated with parasite species, and parasite populations, (c) competition in such parasite communities may be reduced, and (d) the effective population size of hosts available for colonization (i.e. with low immunological resistance to parasites) is high so that highly virulent parasites can be maintained. This fosters the effects of increased probability of acquiring parasites more deleterious to the host's potential competitors, described above.

Another effect of host body size is that as body sizes of species decline, there will be more host species of that body size in the community (May, 1978; Peters, 1983). This increases the probability of there being species related to the host species which become available as alternative hosts. This extends the complexity of community dynamics: as body sizes of hosts decline the number of potential hosts increases. Also, as the number of potential hosts increases the complexity of parasite-mediated interactions between hosts will increase.

This body of relationships and predictions based on hosts as resources for parasite communities may serve as a general basis from which to attempt an arrangement of parasite community characteristics. A search for pattern in

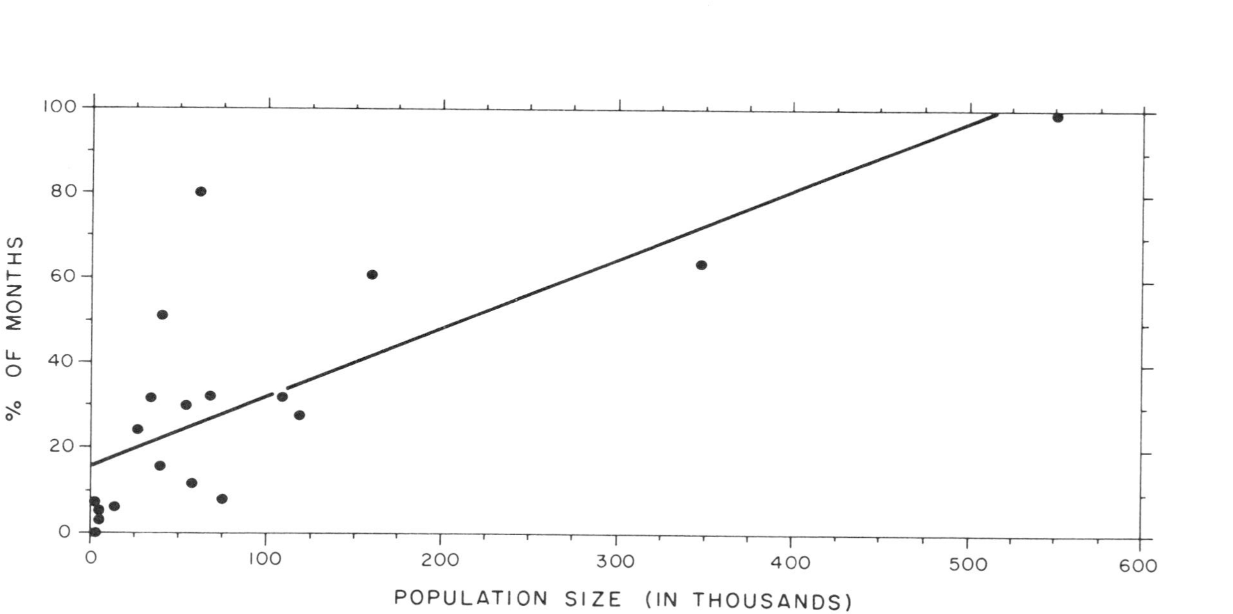

Fig. 2.7 The relationship between human population sizes on islands and the per cent of months in which measles persisted in the population. Note that only in the largest population, on Hawaii with 550 000 people, did measles persist throughout the year. In smaller populations, such as on Fiji with 346 000 people, the disease could not persist. The relationship accounts for 59% of the variance ($p < 0.01$). Based on data in Black (1966).

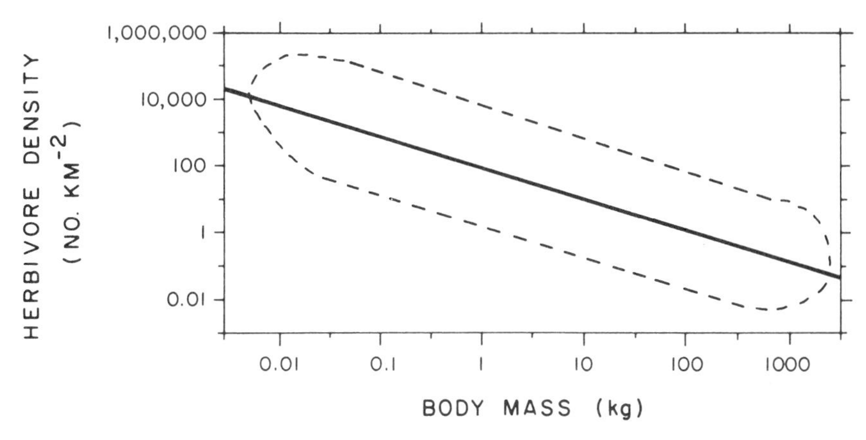

Fig. 2.8 The relationship between body mass of herbivorous mammals and their population density. Note the logarithmic scales on each axis. The linear regression line and the envelope enclosing the data points are provided. After Peters (1983).

community richness and organization may well use the host charactistics as a unifying theme. Some empirical studies lend support to this approach and will be discussed in section 2.3.

2.3 RESOURCE-RELATED PATTERNS

Several approaches to parasite community study reinforce the argument that entering the food web at the level of the resource base helps in synthesis and the recognition of pattern. That we can recognize any pattern in such complex and speciose communities is a major, rather recent development, one that should stimulate much more enthusiasm for the development of broad generalizations about the organization of parasite communities.

Community richness

The numbers of parasites in communities differ profoundly. For example, fish hosts and amphibian hosts have fewer species of intestinal helminth parasites than bird and mammal hosts (Lotz and Font, 1985; Kennedy *et al.*, 1986; Goater *et al.*, 1987) (Table 2.1). Kennedy *et al.* (1986) detected a set of characteristics which adequately account for these differences. All the major factors they identified that contributed to high community richness were host characters; they were determined by the resource base.

Table 2.1 Characteristics of some helminth intestinal parasite communities in fish, amphibians, birds and mammals

Host	*Range in number of parasite species per host individual*	*Proportion of host individuals with 1 or 0 parasite species*
Fish		
Atlantic eel, *Anguilla anguilla*	0–3	0.5–1.0
Barbel, *Barbus barbus*	1–2	0.9
Bream, *Abramis brama*	0–2	0.7
Brown trout, *Salmo trutta*	1–4	0.4
Chub, *Leuciscus cephalus*	1–3	0.7–0.9
Dace, *Leuciscus leuciscus*	0–5	0.5–1.0
Roach, *Rutilus rutilus*	1–3	0.7–0.9
Salmon, *Salmo salar*	0–4	0.7
Sea char, *Salvelinus alpinus*	1–2	0.8
Amphibians (salamanders)		
Black bellied, *Desmognathus quadramaculatus*	0–7	0.4
Mountain, *D. ochrophaeus*	0–4	0.7
Seal, *D. monticola*	0–6	0.4
Shovel-nosed, *Leurognathus marmorata*	0–3	0.9
Birds		
Bonaparte's gull, *Larus philadelphia*	0–9	0.1
Lesser scaup, *Aythya affinis*	8–28	0
Long-billed curlew, *Numenius americanus*	2–7	0
Marbled godwit, *Limosa fedoa*	2–5	0
Willet, *Catoptrophorus semipalmatus*	4–13	0
Mammals		
Big brown rat, *Eptesicus fuscus*	0–9	0.1
Rice rat, *Oryzomys palustris*	13–17	0

Sources: Lotz and Font (1985), Kennedy *et al.* (1986), Goater *et al.* (1987).

These major characters included the following: (a) complexity of host alimentary canal, that is, the more complex the habitat, the more diverse is the templet into which parasite niches fit (Southwood, 1977), (b) amount of food eaten (endotherms consume much more food than ectotherms), increasing the probability of infection of many parasite species, and large numbers of individuals per species, (c) host vagility, (host movement increases exposure to more and more parasite species and moreover, hosts sample the available habitats more effectively), and (d) host diet breadth (generalists are exposed to a greater variety of sources for parasites than specialists).

Kennedy *et al.* (1986) recognized that many other factors may modify these patterns, but the general patterns are evident even though such factors as competition in parasite communities are not included. Indeed, it is the host characters listed above that define the communities in which competition is likely to play an important role, especially in birds and mammals.

Goater *et al.* (1987) have provided further support for the generalizations above. 'The consequences of being ectothermic generalist insectivores, coupled with simple enteric systems and low vagility, has led to the development of depauperate helminth infracommunities in salamanders. These factors outweigh their having a broad diet . . .'.

Further studies of this kind, which enable a strong comparative approach among communities, are vital to research in this field. Standardization of reporting data on parasite communities would foster such comparisons enormously. The approach taken by Kennedy *et al.* (1986) was a major advance in the understanding of parasite communities. The unifying theme of their study was basic host biology. The resource base, to a large extent, drove the pattern. Organization of the community was directed from below.

Host as habitat

Southwood (1977) emphasized that the habitat forms the template into which the various attributes of species fit. For parasite communities the individual host forms a habitat available for colonization. Such habitats are replicated many times in the population of hosts, and very similar habitats exist in related species of potential hosts. Thus parasitologists deal with the most clearly structured habitats in any ecological system. The host as a habitat provides a remarkable opportunity for comparative community analysis, so much more difficult to explore for free-living organisms.

The host habitat defines the diversity of substrates which can be exploited by parasites. Using the concept of the ecological niche developed by Hutchinson – an n-dimensional hypervolume, every point in which enables persistence and reproduction of a species – allows the identification of places in that habitat actually available for parasite colonization. Every part of the habitat colonized in one host by one kind of parasite is presumably available in other related hosts. If it is not occupied in related hosts, then a vacant ecological niche exists (Fig. 2.9). The use of parasite communities and hosts as templates has provided the clearest analysis of species packing, niche availability, niche occupation and the existence of vacant niches (Rohde, 1978b, 1979; Lawton and Price, 1979; Price, 1980, 1984c; Lawton, 1984; Bush and Holmes, 1986; Stock and Holmes, 1988).

The identification of vacant niches is central to the study of community ecology. Questions on the extent to which communities are saturated, and

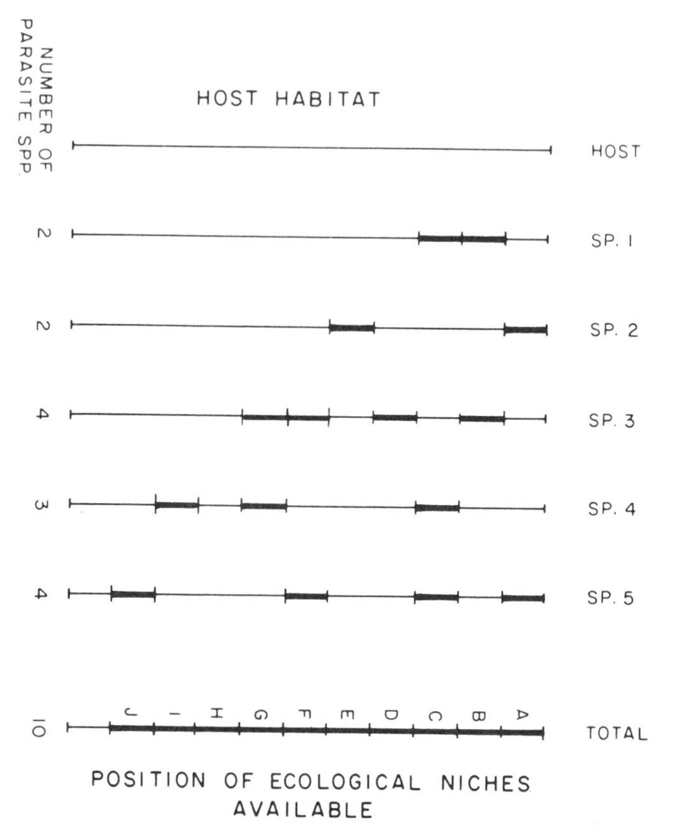

Fig. 2.9 The process of discovering vacant niches in parasite communities. As a simple case the alimentary canal of a host is regarded as the habitat available for colonization by a parasite species. The available ecological niches which have been colonized in evolutionary time are defined by the extant parasitic species across all relevant host taxa. Any one parasite community on one host species is saturated if all ecological niches are occupied, but many communities are likely to be depauperate.

whether communities are at equilibrium with respect to the number of species present have been difficult to address. But with such structured habitats as hosts, identification of vacant niches is relatively simple, although rigorous, and community saturation can be readily examined.

The concept of the vacant niche has been challenged (Herbold and Moyle, 1986; Colwell, 1989). Although it is clear that extinctions create empty niches, speciation events can only occur if niches are available for new species, and nobody can possibly claim that all communities are saturated with species. Herbold and Moyle (1986) offered no criticism of the approaches taken in this chapter which are reflected in the papers they cite

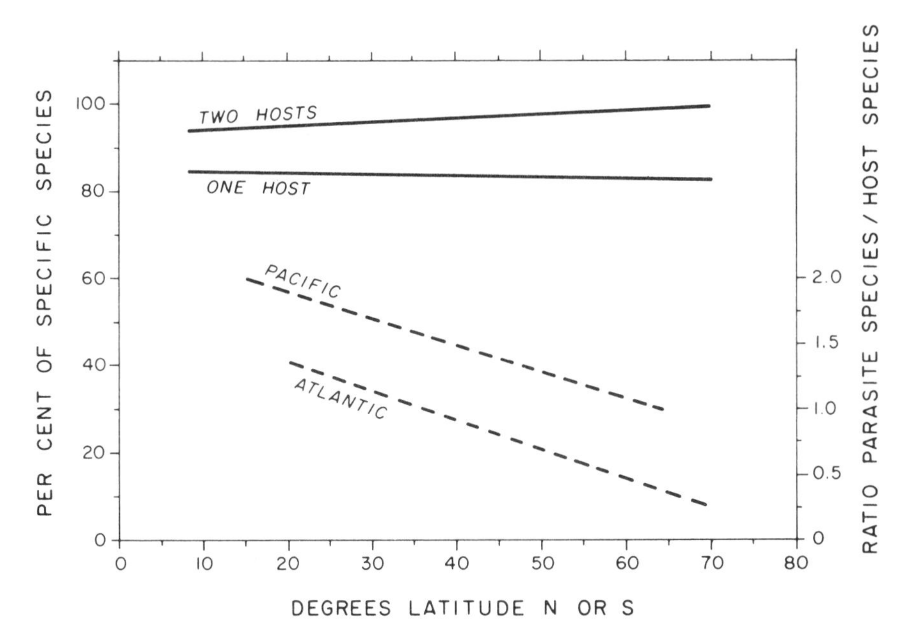

Fig. 2.10 The relationship between degrees of latitude and the per cent of specific monogenean parasite species, utilizing only one host or two hosts (left axis and solid lines), and the ratio of parasite species to host species in the Pacific and Atlantic oceans (right axis and dashed lines). Note that specificity remains very high even though parasite species at high latitudes commonly occur without other species to interact with. The ratio of parasite species to host species changes from a low of 0.3 in the Atlantic to a high of 2.0 in the Pacific, an increase in parasite diversity of 6.6 times on a per-host basis. Based on data in Rohde (1978a, 1986).

by Lawton (1984) and Price (1984c). Hutchinson's (1957) definition of the niche, including the phrase 'a state of the environment which would permit the species . . . to exist indefinitely', makes explicit the view that environments form niches, not species. Thus, empty niches can occur, as has been recognized throughout most ecological history by Grinnell (1914), Hutchinson (1957), Mayr (1963), Colbert (1980) and Strong *et al.* (1984).

The power of this comparative approach, using most habitats as templates for parasite niches, is illustrated well by Rohde's (1978a, b) analyses of monogeneans and other parasites on marine fish. Niche specificity is high in monogeneans, both in site specificity within the host habitat and specificity to a particular host species. This specificity does not change throughout the whole distributional range of marine fishes (Fig. 2.10). Rich communities exist in the tropics with up to nine monogenean species per host species and up to seven parasite species present on one host (Rohde, 1978a, b). However, in cold seas, most fish species have only one or no monogean parasites, so there is a strong latitudinal gradient in parasite species richness (Fig. 2.10) (Rohde, 1980, 1986). Yet parasite species are just as specific in depauperate cold-sea communities as they are in rich tropical-sea faunas. This result is contrary to the literature on competition driving narrow niche occupation in the tropics.

The evidence from Rohde's studies shows that vacant niches are common in tropical and temperate systems. Communities are unsaturated. Of 53 fish species around Lizard Island (15°S) 17% had no monogenean parasites, even though the habitat templates they provided for colonization were equally diverse compared to those with nine parasites per community. Around Heron Island (23°S) 31% of 74 host species had no monogenean parasites (Rohde, 1978b). Given that equivalent host habitats could support up to nine monogenean parasites in the community, this means that about 23 host species had nine vacant niches each. Even in the richest parasite assemblages hundreds of vacant niches exist (23 × 9 = 207 vacant niches in parasite communities on 74 hosts). In cold seas the vacancy rate is extreme. In the White Sea (65°N) 73% of 33 host species were free of parasite species, and the remainder supported one (21%), two (3%) or three (3%) parasite species (Rohde, 1978b). No more than three monogeneans existed in any parasite component community. Thus, vacant niches in the White Sea fish community vastly outnumbered occupied niches. Host species had habitats available for roughly 297 parasite species (33 × 9 niches each), but only 12 of these available niches were occupied – 4% occupancy. Further examples are given in Table 2.2.

I agree with Rohde's (1978b) interpretation of these results: 'Many (and perhaps all) habitats can accommodate more species than at present.' 'The

Table 2.2 Estimates of the ecological niche occupancy by monogenean parasite species on marine fish, and the extent to which communities are undersaturated in different locations

Locality	*Latitude*	*Number of fish species*	*Number of parasite species*	*Per cent of niches**	
				occupied	*vacant*
Barents Sea	70°N	47	19	4	96
White Sea	65°N	33	12	4	96
Behring Sea	63°N	25	24	11	89
Black Sea	45°N	34	6	2	98
Heron Island	23°S	74	98	15	85
Lizard Island	15°S	53	55	12	88

Based on Rohde (1978b)

* Calculations assume that there are nine ecological niches per host species available to parasites, and that parasite species are 100% host specific.

fact that many habitats are empty shows that many species have not had time to spread into less utilized habitats, and that MacArthur's "principle of equal opportunity", according to which habitats are filled with as many species as they can "hold" at any stage of evolution (MacArthur, 1972) does not apply to many of the little dispersing smaller species.' The argument is that no fish host species is immune to colonization by parasites, and no site is unavailable. If sites can be colonized on one host species they can be colonized on another. Thus, through long periods of evolutionary time all sites on all hosts have a certain probability of being colonized. These probabilities are modified by geographical range of hosts, population density, generalized or specialized habitat occupation, and the characteristics discussed under community richness. But none of these factors is likely to reduce the probability to zero, given enough evolutionary time.

Rohde's analysis and similar approaches provide the most cogent entry to the analysis of community saturation every achieved. The results demonstrate how far reality is from much theory, and how theory needs to be updated to become consistent with reality. These major contributions to general ecology stem from our perception that hosts as resources provide replicated habitats. Therefore we have very powerful comparative approaches to parasite community analysis rarely so easily available to the ecologist working with non-parasitic species.

Parasite mediation of host interactions

The three properties of hosts discussed above under section 2.2 provide predictions about the way in which parasite communities are assembled, and how these will influence interactions between host populations and host species. In relation to host geographic area one consequence is that, as area increases, the probability that the host acquires parasites more detrimental to the host's competitors increases. This is illustrated well by McNeill's (1976, 1980) scenario for the one-sided influence of disease when Europeans invaded Amerindian lands. McNeill explained that diseases were collected in urban centers in Europe from the hinterland. Such accumulation of parasites was facilitated by extensive and rapid travel across Asia to the Far East, so all the parasites of Europe, Asia and India were sampled. This began in the 13th and 14th centuries with travellers such as Marco Polo, and Ghengis Kahn's communication routes. The rapid travel and large urban populations provided the resource base for highly virulent diseases, which were also transported rapidly to North America. The impact on the small, isolated 'disease-inexperienced peoples' (McNeill, 1980) was devastating in every case. The effective area for collecting diseases in urbanized Europe was far greater than the small more sedentary populations in the New

Table 2.3 Some examples of parasitic mediation between potentially competing hosts in which winners and losers are predicted based on their geographic range and body mass as discussed in the text. Only the nine cases supporting the geographic range predictions are given of the 11 cases tested. The corresponding data for body mass for these cases are included. Predictions about body mass are not supported in examples, 2, 5, 8 and 9.

Parasite or disease	*Winner*	*Range (million ha) (mass g)*	*Loser*	*Range (million ha) (mass g)*
1. Meningeal Worm (*Parelaphostrongylus tenuis*)	Whitetail deer (*Odocoileus virginianus*)	1088 (136 000)	Moose (*Alces alces*)	843 (454 000)
2. Meningeal worm	Whitetail deer	1088 (136 000)	Mule deer (*Odocoileus hemionus*)	748 (136 000)
3. Meningeal worm	Whitetail deer	1088 (136 000)	Woodland caribou (*Rangifer caribou*)	535 (227 000)
4. Meningeal worm	Whitetail deer	1088 (136 000)	Elk (*Cervus canadensis*)	169 (409 000)
5. Meningeal worm	Whitetail deer	1088 (136 000)	Pronghorn antelope (*Antilocapra americana*)	313 (45 000)
6. Sylvatic plague (*Yersinia pestis*)	Deer mouse (*Peromyscus maniculatus*)	1518 (28)	Bushytail woodrat (*Neotoma cinerea*)	319 (425)
7. White pine blister rust (*Cronartium ribicola*)	Red currant (*Ribes triste*)	1227 (small)	White pine (*Pinus strobus*)	529 (large)
8. Parasites	Norway rat (New world) (*Rattus norvegicus*)	large (241)	Black rat (New world) (*Rattus rattus*)	small (213)
9. Parasites	Norway rat (Old world)	1718 (397)	Black rat (Old world)	787 (180)

Based on Price *et al.* (1988), which should be consulted for sources and methods.

World. This kind of interaction can also be seen as wild animal species interact.

The host properties of density and body size also relate to host species' interactions because effective population size for maintaining parasite species increases with density and with reproductive rate.

Therefore, the prediction can be made that when host species pairs interact, the species with the larger geographic area is likely to win because it has a larger store of parasite species with which to interact. Also the species with the smaller body size should win because geographic areas are usually larger and threshold densities for maintaining parasites are more commonly exceeded.

These predictions and tests were made by Price *et al.* (1988). The data available for tests were not extensive but there was a remarkable conformity to the predicted patterns. In the test of geographic range effect, nine of eleven cases supported the prediction (Table 2.3) and in the test of body size effect, twelve of fifteen cases were supportive. Therefore, we have the beginnings of a predictive biology on the ways in which parasites will mediate interactions in host communities. Again the patterns are observed by concentrating on hosts as resources for parasites. Freeland (1983) suggested a broader range of interactions which need more study. Moore (1987) provided an excellent overview of other ways in which parasites influence host populations.

2.4 CONCLUSIONS

Clearly, much more research is needed on other systems to examine the extent to which the predictions in section 2.2 and the patterns in section 2.3 are supported or underminded. But, such studies at the parasite community level can have an important impact on the understanding of host and parasite population dynamics. Working from the bottom of the trophic system, and concentrating on the resource base also clarifies the evaluation of such characters as interspecific and intraspecific competition. Kennedy *et al.* (1986) make clear predictions on where interspecific competition should be found in parasite communities, and to a large extent resolves the debate between the isolationists and the interactionists (Holmes and Price, 1986; Stock and Holmes, 1988). However, Rohde's (1979, 1980) arguments make it clear that alternative hypotheses to competition in species-rich communities should be examined.

In the general community of ecologists there is a growing recognition that parasite systems must be included in syntheses of ecology, and chapters of books are now commonly devoted to them (May, 1981; Futuyma and Slatkin, 1983; Price, 1984a; Diamond and Case, 1986; Kikkawa and

Anderson, 1986). But studies of parasites enable new approaches to analysis and the discovery of new and important mechanisms in natural systems. Thus parasitologists can take a leading role in advancing areas of ecology. The host, as habitat and resource base, provides such a strong comparative tool, about which so much is known relative to other kinds of habitats and resources, that parasite community studies will become recognized as essential models for certain kinds of critical analysis. The cases discussed in this chapter relating to community richness, host as habitat, and parasite mediation of host interactions fit this prediction. For ecologists studying free-living organisms the quantification of habitat and resources is frequently difficult and irksome. For parasitologists, there is well-developed theory and generalizations about host size, reproductive rate and so on (Peters, 1983; Calder, 1984), both of relevance to the development of a synthesis on parasite community patterns and processes.

ACKNOWLEDGEMENTS

This chapter was partially supported by a grant from the National Science Foundation BSR-8705302. I am grateful to Albert Bush and Gerald Esch for reviews of this chapter.

REFERENCES

Black, F. L. (1966) Measles endemicity in insular populations: critical community size and its evolutionary implication. *J. Theoret. Biol.*, **11**, 207–11.

Brown, J. H. (1984) On the relationship between abundance and distribution of species. *Am. Nat.*, **124**, 255–79.

Bush, A. O. and Holmes, J. C. (1986) Intestinal helminths of lesser scaup ducks: an interactive community. *Can. J. Zool.*, **64**, 142–52.

Calder, W. A. (1984) *Size, Function, and Life History*, Harvard University Press, Cambridge, Mass.

Cohen, J. E. (1978) *Food Webs and Niche Space*, Princeton University Press, Princeton, N.J.

Colbert, E. H. (1980) *Evolution of the Vertebrates*, Wiley, New York.

Colwell, R. K. (1989) Ecology and biotechnology: expectations and outliers. In (eds, J. Fiksel and V. T. Covello), *Risk Analysis Approaches for Environmental Releases of Genetically Engineered Organisms*, Springer-Verlag, Berlin, in press.

Diamond, J. and Case, T. J. (eds) (1986) *Community Ecology*, Harper and Row, New York.

Freeland, W. J. (1979) Primate social groups as biological islands. *Ecology*, **60**, 719–28.

Freeland, W. J. (1983) Parasites and the coexistence of animal host species. *Am. Nat.*, **121**, 223–36.

Futuyma, D. J. and Slatkin, M. (1983) *Coevolution*, Sinauer Associates, Sunderland, Mass.

Goater, T. M., Esch, G. W. and Bush, A. O. (1987) Helminth parasites of sympatric salamanders: ecological concepts at infracommunity, component and compound community levels. *Am. Midl. Nat.*, **118**, 289–300.

Grinnell, J. (1914) An account of the mammals and birds of the lower Colorado Valley. *University of California Pubs. Zool.*, **12**, 51–294.

Hanski, I. (1982) Dynamics of regional distribution: the core and satelite species hypothesis. *Oikos*, **38**, 210–21.

Herbold, B. and Moyle, P. B. (1986) Introduced species and vacant niches. *Am. Nat.*, **128**, 751–60.

Holmes, J. C. and Price, P. W. (1986) Communities of parasites. In *Community Ecology: pattern and process*, (eds, J. Kikkawa and D. J. Anderson), Blackwell Scientific Publications, Melbourne, pp. 187–213.

Hutchinson, G. E. (1957) Concluding remarks. *Cold Spring Harbor Symp. Quant. Biol.*, **22**, 415–27.

Kennedy, C. R., Bush, A. O. and Aho, J. M. (1986) Patterns in helminth communities: why are birds and fish different? *Parasitology*, **93**, 205–15.

Kermack, W. O. and McKendrick, A. G. (1927) A contribution to the mathematical theory of epidemics. *Proc. R. Soc. Lond. A. Math. Phys. Sci.*, **115**, 700–21.

Kikkawa, J. and Anderson, D. J. (eds) (1986), *Community Ecology: pattern and process*, Blackwell Scientific Publications, Melbourne.

Lawton, J. H. (1984) Non-competitive populations, non-convergent communities, and vacant niches: the herbivores of bracken. In *Ecological Communities: conceptual issues and the evidence* (eds, D. R. Strong, D. Simberloff, L. G. Abele and A. B. Thistle), Princeton University Press, Princeton, N.J., pp. 67–100.

Lawton, J. H. and Price, P. W. (1979) Species richness of parasites on hosts: agromyzid flies on the British Umbelliferae. *J. Anim. Ecol.*, **48**, 619–37.

Lotz, J. M. and Font, W. F. (1985) Structure of enteric helminth communities in two populations of *Eptesicus fuscus* (Chiroptera). *Can J. Zool.*, **63**, 2969–78.

MacArthur, R. H. (1972) *Geographical Ecology: patterns in the distribution of species*, Harper and Row, New York.

McNeill, W. H. (1976) *Plagues and Peoples*, Anchor Press, Garden City, N.J.

McNeill, W. H. (1980) *The Human Condition*, Princeton University Press, Princeton, N.J.

May, R. M. (1978) The dynamics and diversity of insect faunas. In *Diversity of Insect Faunas* (eds, L. A. Mound and N. Waloff), Blackwell Scientific Publications, Oxford, pp. 188–204.

May, R. M. (ed.) (1981) *Theoretical Ecology: principles and applications*, 2nd edn, Sinauer Associates, Sunderland, Mass.

Mayr, E. (1963) *Animal Species and Evolution*, Cambridge, Mass., Belknap Press of Harvard University Press.

Moore, J. (1987) Some roles of parasitic helminths in trophic interactions. A view from North America. *Rev. Chilena Hist. Natur.*, **60**, 159–79.

Peters, R. H. (1983) *The Ecological Implications of Body Size*, Cambridge University Press, Cambridge.

Pimm, S. L. (1982) *Food Webs*, Chapman and Hall, London.
Price, P. W. (1980) *Evolutionary Biology of Parasites*, Princeton University Press, Princeton, N.J.
Price, P. W. (1984a) *Insect Ecology*, 2nd edn, Wiley, New York.
Price, P. W. (1984b) Alternative paradigms in community ecology. In *A New Ecology: novel approaches to interactive systems*, (eds, P. W. Price, C. N. Slobodchikoff and W. S. Gaud), Wiley, New York, pp. 353–83.
Price, P. W. (1984c) Communities of specialists: vacant niches in ecological and evolutionary time. In *Ecological Communities: conceptual issues and the evidence*, (eds D. R. Strong, D. Simberloff, L. G. Abele and A. B. Thistle), Princeton University Press, Princeton, N.J., pp. 510–23.
Price, P. W. and Clancy, K. M. (1983) Patterns in number of helminth parasite species in freshwater fishes. *J. Parasitol.*, **69**, 449–54.
Price, P. W., Westoby, M., Rice, B. *et al.* Parasite mediation in ecological interactions. *Annu. Rev. Ecol. Syst.*, **17**, 487–505.
Price, P. W., Westoby, M. and Rice, B. (1988) Parasite-mediated competition: some predictions and tests. *Am. Nat.*, **131**, 544–55.
Rohde, K. (1978a) Latitudinal differences in host-specificity of marine Monogenea and Digenea. *Marine Biol.*, **47**, 125–34.
Rohde, K. (1978b) Latitudinal gradients in species diversity and their causes. II. Marine parasitological evidence for a time hypothesis. *Biol. Zbl.*, **97**, 405–18.
Rohde, K. (1979) A critical evaluation of intrinsic and extrinsic factors responsible for niche restriction in parasites. *Am. Nat.*, **114**, 648–71.
Rohde, K. (1980) Comparative studies on microhabitat utilization by ectoparasites of some marine fishes from the North Sea and Papua New Guinea. *Zool. Anz.*, **204**, 27–63.
Rohde, K. (1986) Differences in species diversity of Monogenea between the Pacific and Atlantic oceans. *Hydrobiologia*, **137**, 21–8.
Southwood, T. R.E. (1977) Habitat, the templet for ecological strategies? *J. Anim. Ecol.*, **46**, 337–65.
Stock, T. M. and Holmes, J. C. (1988) Functional relationships and microhabitat distributions of enteric helminths of grebes (Podicipedidae): the evidence for interactive communities, *J. Parasitol.*, **74**, 214–27.
Strong, D. R., Lawton, J. H. and Southwood, R. (1984) *Insects on Plants: community patterns and mechanisms*, Harvard University Press, Cambridge.

3

Spatial scale and the processes structuring a guild of larval trematode parasites

Wayne P. Sousa

3.1 INTRODUCTION

An individual host is a patch of habitat for a particular stage in the life cycle of a parasite (Price, 1980; Holmes and Price, 1986). It contains resources necessary for growth and development of this infecting stage, and for production of the next, usually dispersing, stage. Each individual host is an inherently bounded, discrete habitat, that is isolated from other similar habitat patches by an external environment that is inhospitable to the parasitic stage that infects the host.

As patches of parasite habitat, hosts are both self-reproducing and ephemeral. Excluding instances of vertical transmission and assuming life-long infection (as is the case for many parasitic infections of invertebrate hosts), empty patches are born via the recruitment of susceptible offspring to a host population, patches increase in size and change in a variety of other characteristics (e.g. morphology, biochemistry, etc.) during host ontogeny, and they disappear (along with their resident parasites) when the host dies. The rates at which these processes occur vary among host populations and in time.

In many ecological systems, the distribution and accessibility of resources vary with spatial and temporal scale, as do the processes that structure populations and communities (e.g. Andrewartha and Birch, 1954; Wiens, 1976; Price, 1980; Allen and Starr, 1982; Connell and Sousa, 1983; Dayton and Tegner, 1984; Sousa, 1984; Addicott *et al.*, 1987). Studies of processes operating at different scales within systems of divided habitat patches have provided substantial insight in this regard. When strong asymmetrical interactions on the small scale, i.e. within a patch, preclude the coexistence of competitors, or of predators and their prey, the existence of multiple patches coupled by dispersal often promotes their coexistence on the larger scale (Hutchinson, 1951; Skellam, 1951; Huffaker, 1958; Cohen, 1970; Levins and Culver, 1971; Horn and MacArthur, 1972; Levin, 1974, 1976; Slatkin, 1974; Armstrong, 1976; Hastings, 1977, 1980; Caswell, 1978; Shorrocks *et al.*, 1979; Sousa, 1979; Lloyd and White, 1980; Atkinson and Shorrocks, 1981; Hanski, 1981, 1983; Ives and May, 1985; Murdoch *et al.*, 1985). Differential rates of dispersal among species, independent aggregation of species among patches, and an increased number of patches in the system enhance the likelihood that diversity will be maintained on the large scale, i.e. across all patches in the system.

The spatial scales of resources provided by hosts are hierarchical (Esch *et al.*, 1975; Margolis *et al.*, 1982; Holmes and Price, 1986; see also Chapter 1), and patterns and outcomes of interaction among parasites may vary among these scales. Nested levels in this hierarchy include: (a) tissues within a host, (b) an individual of a particular host species, (c) populations of a particular host species, and (d) communities of host species. The population of a particular parasite species that infects an individual host is called an infrapopulation; the collection of populations of different parasite species within a single host is an infracommunity. The assemblage of parasite species that infect a population of a particular host species is called a component community.

Populations of invertebrates that serve as intermediate hosts are commonly infected by several species of parasitic helminths (Denny, 1969; Wright, 1971; Brown, 1978; Rohde, 1982; Lauckner, 1980, 1983). To understand the processes that structure such assemblages of larval parasites better, I investigated patterns of species diversity of the helminths that infect the salt marsh snail, *Cerithidea californica*, at two different spatial scales. This parasite assemblage is composed solely of larval digenetic trematodes. Because the members are taxonomically similar and exploit a common resource, the assemblage is more appropriately referred to as a guild (*sensu* Root, 1967) than a community. The smaller of the two scales examined in this study is that of the individual snail, which potentially supports an infraguild of larval trematodes. The larger scale is that of the local host population and its component guild of parasites.

This chapter primarily examines patterns and processes at the second, larger scale, but summarizes what is known concerning structure and dynamics at the individual host scale. The latter is the subject of Chapter 3. Here, I examine several characteristics of local host and parasite populations that may influence the diversity of component guilds of larval trematodes. These characteristics include host population density, size/age distribution, location, and rates of host and parasite recruitment. The study is restricted to infections by redia and sporocyst stages of trematodes that use *Cerithidea* as first intermediate host.

3.2 PATCH DYNAMICS AND COMPLEX PARASITE LIFE CYCLES

The digenetic trematodes that infect *Cerithidea* have life cycles typical of most parasitic helminths (see below). There is an obligate sequence of intermediate and definitive hosts; transmission is effected either by free-living motile larvae or by encysted larvae that are ingested by the host. The complex life cycles of Digenea effectively isolate the patches of habitat afforded by individuals of the same host species, since the stages that infect one such individual cannot be directly transmitted to another. Therefore, in the case of the intermediate snail host, the rate of establishment of new redial and sporocyst infections within a local snail population depends on: (a) the abundance of infective stages (i.e. miracidia) in the local environment, (b) the rate at which hosts come into contact with these stages or ingest them, and (c) the susceptibility of the individual snails that comprise the local host population. The availability of infective stages is mainly determined by processes external to the local host population, e.g. the abundance and habitat use of the definitive hosts, and physical characteristics of the local aquatic environment including water flow, chemistry, temperature, turbidity, depth, etc. These physical characteristics may influence movement and survival of the infective stages. Differential production or mortality during the dispersal phase will cause rates of establishment to differ among species of parasites.

Rates of contact between host and parasite may also be influenced by the behaviour of each. If infective stages are transmitted by ingestion, host feeding habits will affect the rate at which new infections are acquired. If infective stages actively seek hosts using environmental gradients (e.g. light intensity) that are correlated with host abundance, or chemical cues from the host, rates of encounter will be higher. Miracidia appear to employ both of these mechanisms, although behavioural responses to host exudates are only observed when the larva is in close proximity to its host (Wright, 1959, 1971; Ulmer, 1971; Cable, 1972; Chernin, 1974; Shiff, 1974; MacInnis, 1976; Brown, 1978; Smyth and Halton, 1983).

The susceptibility of hosts may vary with the species or genetic strain of parasite and with host characteristics such as age, sex, and genotype

(Richards, 1976; Meuleman *et al.*, 1987). Susceptibility may also be affected by host nutritional state, which may, in turn, be influenced by host population density.

3.3 THE SYSTEM: *CERITHIDEA CALIFORNICA* AND ITS TREMATODE PARASITES

Details of the life history of *Cerithidea californica* are summarized in Sousa (1983). Dense populations of this deposit-feeding gastropod inhabit pickleweed (*Salicornia virginica*) marshes and adjacent high intertidal (+ 1.2–2.1 m mean lower low water) mudflats and tidal creeks in protected bays and estuaries along the Pacific coast of North America. The species' range extends from Tomales Bay (Marin Co., California) to central Baja, California, Mexico (Macdonald, 1969a, b). The snail is iteroparous and its larvae undergo direct development within benthic egg strings. Egg laying begins in late March or April, hatching starts by June and continues into August.

Cerithidea is first intermediate host to at least 18 species of digenetic trematodes in California (Martin, 1955, 1972; Yoshino, 1975). In Bolinas Lagoon (Marin Co., California), the site of this study, 15 species of trematodes were found in the seven annual samples of snail populations on which this chapter is based (Table 3.1). The life cycles of all but one of these species appear to follow the typical digenean sequence (Shoop, 1988): egg, free-living miracidium, intramolluscan sporocyst or redia stage, free-living

Table 3.1 Larval trematodes that infect *Cerithidea californica* in Bolinas Lagoon. The identities of species marked with an asterisk have yet to be determined

Family	*Species*
Cyathocotylidae	*Mesostephanus appendiculatus*
	cyathocotylid #2*
Echinostomatidae	*Acanthoparyphium spinulosum*
	Echinoparyphyum sp.*
	Himasthla rhigedana
Heterophyidae	*Euhaplorchis californiensis*
	Phocitremoides ovale
Microphallidae	microphallid #1*
	microphallid #2*
Notocotylidae	*Catatropsis johnstoni*
Philopthalmidae	*Parorchis acanthus*
Renicolidae	*Renicola buchanani*
	renicolid #2*
	renicolid #3*
Schistosomatidae	*Austrobilharzia* sp.*

cercaria, encysted metacercaria stage in a poikilothermic second intermediate host (invertebrate or fish), and finally, development from an ingested metacercaria, of a parasitic adult worm in the definitive vertebrate host. The life cycle of *Austrobilharzia* sp. is the only exception to this pattern; the metacercaria stage is absent and the definitive host is infected directly by a cercaria. The definitive hosts for most of the trematodes are probably birds, but there is little information concerning the distribution of adult trematodes among potential avian or mammalian hosts that inhabit the study area.

3.4 METHODS

Study sites

The study was conducted in Bolinas Lagoon, located 24 km NW of San Francisco, California (37°55′N, 122°41′W). *Cerithidea* populations and their parasites were sampled at two sites within the lagoon. One site is adjacent to the mouth of Pine Gulch Creek (hereafter PGC site) which flows into the lagoon on its western edge, and the other is at the northeast corner of Kent Island (hereafter KI site). The sites are designated 'B' and 'C' respectively, on the map of the lagoon that appears in Stenzel *et al.* (1976), and are about 750 m apart. The freshwater flowing from the creek at PGC is an important resource for birds. Densities of wintering, migrant waterfowl and roosting gulls and terns are much higher at this site than at KI which lacks a source of freshwater. Some of these birds are probably definitive hosts of trematodes that infect *Cerithidea*.

The sites also differ in sediment characteristics. The surface sediment at PGC is a poorly sorted, very fine sandy mud and has a considerably higher organic content than the surface sediment at KI which is a well sorted, fine to medium sand (Ritter, 1969; Sousa, unpublished data). This variation in sediment quality is related to hydrological and biological differences between the sites. KI is closer to the mouth of the lagoon and to its main channel, so that tidal currents are relatively stronger, and deposition of fine particles is less, as compared to PGC. In addition, KI is inhabited by a dense population of ghost shrimp, *Callianassa californiensis*. While feeding and burrowing, *Callianassa* extensively rework the sediment, extracting or resuspending fine particulate matter, leaving a sandy, organically poor sediment (MacGinitie, 1934). In contrast, PGC has a depositional environment. Tidal currents at this site are slow and eddying. It receives a substantial input of allochthonous detrital matter and fine sediment from the creek, particularly following heavy winter rains. Bioturbation is minimal since *Callianassa* is not present at the site, possibly because the silty sediment at PGC is unsuitable for burrow construction, or because the ghost shrimp cannot tolerate the sharp reductions in salinity associated with high, winter, creek flow (Sousa and Gleason, 1989).

The species composition and abundance of foraging shorebirds also differs between the sites and appears to be related to the spatial variation in sediment and in the associated benthic invertebrates and fishes on which they prey (Stenzel *et al.*, 1976; Page *et al.*, 1979; Quammen, 1984). These birds are definitive hosts, and many of their prey are second intermediate hosts for some of the parasites that infect *Cerithidea* (Robinson, 1952; Russell, 1960a, b; Badley, 1979; Sousa, personal observation). Since shorebird species differ in diet, they probably harbour different infracommunities of parasites (Russell, 1960a, b). The differential use of lagoon habitats by bird species and variation in the abundances of second intermediate hosts between sites may combine to produce spatial differences in the abundance of infective stages of different trematodes.

At both study sites, snails are distributed as a series of subpopulations occupying shallow (5–15 cm deep) depressions or 'pans' in the surface of the mudflat which hold standing water at low tide. These pans range in size from slightly less than 1 m^2 to 20 m^2. During most months of the year they are flushed daily by high tides.

At PGC, the pans are located along the interface of the tidal mudflats and the higher elevation, *Salicornia*-dominated marsh. Snails rarely move between them (Sousa, unpublished data) presumably because the emergent mudflats (which often consist of hard, dried plates of sediment) and the dense stands of pickleweed that border the pans, represent physical barriers to snail movement. As a consequence of this lack of adult migration and the absence of a planktonic stage in the snails' life history, the demographies of different subpopulations of snails at PGC vary greatly. They differ in size-distribution, rate of recruitment, density, and rate of parasitic infection (Sousa, unpublished data).

At KI, pans occur across the upper tidal mudflat as well as along the mudflat–marsh boundary as at PGC. At the start of the study, densities of ghost shrimp were low in these pans; however, over the course of the study shrimp populations gradually invaded about half the pans, excluding the snails. Their recruitment to the pans followed two stormy winters (1982 and 1983) when a significant amount of sedimentation occurred at the site; however, the precise mechanism(s) responsible for these shifts in local distribution are unknown. Areas between the pans have remained pockmarked with the conical sediment mounds that mark the burrow openings of a dense population of ghost shrimp. Snails are present in these surrounding areas, but at much lower densities than in the pans which are foci of snail feeding and reproduction. The *Callianassa*-dominated areas between the pans remain moist during most low tides, and few of the KI pans have vegetation around their edges. These features, as well as the fact that the average distance between neighbouring KI pans is only half that of PGC pans (8 versus 16 m), probably account for the fact that rates of snail movement between subpopulation is at least three times greater at KI than PGC. As an apparent consequence of this greater exchange of migrants, demographic characteristics of different subpopulations at KI are very similar within any particular year, and they change quite synchronously over time (Sousa, unpublished data).

Sampling procedures

In August of 1981, a total of 34 subpopulations of *Cerithidea* were selected for long-term monitoring of demographic parameters and parasitic infection, 19 at PGC and 15 at KI. The chosen pans comprised almost all of those that contained snails along the 315 m and 70 m of marsh edge habitat studied at PGC and KI, respectively.

These subpopulations were sampled each August from 1981 through to 1987. The snails were sampled with a 225 cm^2 scoop core which collected all sediment and benthic invertebrates to a depth of 2 cm, sufficient to collect all snails in the area. The contents of each scoop were sieved through 1 mm mesh in the field and returned to the laboratory for analysis. The length (apex to aperture) of each snail was measured to the nearest 0.01 mm, then each was dissected to determine its sex and what species of trematode, if any, infected it.

Five or ten scoop samples were collected from each pan, depending on the density of snails; the greater number was taken in sparser populations. The scoops were made at regularly spaced intervals along the length of a metric tape transecting each pool, parallel to its long axis. These samples provided estimates of snail density (both young of the year and older individuals) and biomass. To ensure an adequate sample size of snails $\geq$ one year old, an additional sample of snails was collected from some pans. This supplemental sample was collected from one to three haphazardly chosen locations within a pool. Starting from the position of the first collected individual, all snails (excluding new recruits) were collected from the immediate area until the total sample numbered at least 100 individuals; a few collections fell short of this goal. Statistical tests verify that the size distributions of snails in these supplemental collections do not differ from those of $\geq$ one year old snails in scoop samples taken in the same pans. For the following analysis of parasite assemblages, the scoop and supplemental samples of $\geq$ one-year-old snails taken in a particular pan and year are pooled. Young-of-the-year snails collected in the scoop samples were never found to harbour trematode infections and are not considered in this analysis.

In most cases, snail subpopulations were dense enough that sampling was unlikely to have an impact on population dynamics. However, due to a variety of factors, but especially storm-related disturbance, the density of snails in certain pans sometimes fell to such low levels that I chose not to collect a sample for fear of affecting subsequent dynamics. Over the course of the seven years, a number of snail populations (and their component parasite assemblages) did become extinct due to physical and biological disturbances. For both these reasons, the number of subpopulations sampled at each site varied from year to year, and gradually diminished over the course of the study. By 1987, only 14 of the original pans sampled at PGC

Table 3.2 Annual rates of trematode infection in ≥ one-year-old *Cerithidea californica* at Pine Gulch Creek (PGC) and Kent Island (KI). Data from sampled subpopulations at each site are pooled

		Uninfected		*Infected by*				
				1 species		*> 1 species**		
Site	*Year*	%	(*N*)	%	(*N*)	%	(*N*)	total *N*
PGC	1981	84.34	(3502)	15.44	(641)	0.22	(9)	4152
	1982	87.96	(2571)	11.87	(347)	0.17	(5)	2923
	1983	72.17	(1079)	26.96	(403)	0.87	(13)	1495
	1984	81.85	(1768)	18.01	(389)	0.14	(3)	2160
	1985	85.80	(1837)	13.97	(299)	0.23	(5)	2141
	1986	74.50	(1461)	25.09	(492)	0.41	(8)	1961
	1987	66.33	(1186)	32.66	(584)	1.01	(18)	1788
KI	1981	74.07	(1631)	25.20	(555)	0.73	(16)	2202
	1982	89.96	(1505)	9.68	(162)	0.36	(6)	1673
	1983	77.16	(456)	21.83	(129)	1.01	(6)	591
	1984	83.19	(886)	16.81	(179)	0.00	(0)	1065
	1985	95.74	(1302)	4.19	(57)	0.07	(1)	1360
	1986	93.91	(1281)	6.09	(83)	0.00	(0)	1364
	1987	94.72	(932)	5.18	(51)	0.10	(1)	984

Average annual per cent uninfected and infected snails

	Uninfected		*Infected by*			
			1 species		*> 1 species**	
Site	*Mean*	*SD*	*Mean*	*SD*	*Mean*	*SD*
PGC	78.99	8.07	20.57	7.74	0.44	0.36
KI	86.96	8.86	12.71	8.55	0.32	0.40

* All are double infections except for one triple infection at KI in 1981.

and seven of those at KI remained. The numbers of ≥ one-year-old snails examined at each site during the seven censuses are listed in Table 3.2.

Measures of community structure

This chapter examines the structure of the larval parasite assemblages infecting subpopulations of *Cerithidea*. For each yearly, pooled collection from a subpopulation (hereafter referred to as a sample of snails), I have estimated several measures of parasite guild structure (Table 3.3). The raw data from which these measures are calculated were the number of infections of each parasite species found in the sample of snails. For mixed species infections, the occurrence of each species was counted as if it were a single infection of that species; i.e. each species' total for the particular sample was incremented by one for every mixed species infection in which it occurred. Since mixed infections are exceedingly rare (see below), this protocol had little influence on the results.

Since the rate of parasitic infection varied among snail subpopulations (see below), as did the number of snails collected from each, the total number of infections (NI) found in each sample of snails varied considerably. Species richness (S), the number of species in an assemblage, is strongly affected by sample size (Hurlbert, 1971; Heck *et al.*, 1975; Simberloff, 1979). In an effort to reduce the effect of small sample size *per se*, samples with fewer than 30 snails or five infections were excluded from the analysis. However, this modification of the data alone did not eliminate a significant correlation in several years at both sites between S and ln NI

Table 3.3 Measures of parasite guild structure within snail subpopulations. p_i is the proportion of infections by parasite species i

Measure	*Symbol*	*Method of computation*
Species richness	S	Total number of parasite species in sample
Expected number of species in a random sample of 15 infections	$E(S_{15})$	Calculated by rarefaction using multinomial formula of Heck *et al.* (1975)
Exponential of Shannon diversity index (Hill's (1973): N_1)	Exp (H′)	$\text{Exp}\,(-\Sigma p_i \ln p_i)$
Reciprocal of Simpson's diversity index (Hill's (1973): N_2)	$1/\Sigma p_i^2$	See symbol
Simple dominance (May, 1975)	Dom	p_i of most abundant species in sample

Table 3.4 Correlations between measures of species richness and the number of infections in a sample. See Table 3.3 for explanation of measures. Log transformation of number of infections improved linearity of the relationships. *n* is number of subpopulations sampled. A dash indicates that the sample size (*n*) was too small for statistical analysis

Site	*PGC*							*KI*						
Year	81	82	83	84	85	86	87	81	82	83	84	85	86	87
Correlation of:														
1. S and 1n (number of infections)														
r	0.65	0.87	0.42	0.78	0.38	0.47	−0.07	0.32	0.72	0.91	0.09	0.34	0.94	0.42
n	19	15	14	13	14	13	14	15	14	7	8	7	8	6
p[a]	**	***	+	***	+	+	ns	ns	**	**	ns	ns	***	ns
2. E (S_{15}) and 1n (number of infections)														
r	0.17	0.07	0.61	−0.07	−0.20	−0.53	−0.60	0.33	–	0.91	−0.10	–	–	–
n	15	11	13	8	10	10	13	15	2	4	8	0	1	0
p	ns	ns	*	ns	ns	+	*	ns	–	*	ns	–	–	–

[a] All probabilities in table are one-tailed: ns ≥ 0.10, + < 0.10, * < 0.05, ** < 0.01, *** < 0.001.

(Table 3.4). Rarefaction procedures were then used (Heck *et al.*, 1975) to estimate the expected number of trematode species in a sample of 15 infections (E (S_{15})) drawn randomly from the collection of infections found in each sample of snails. The multinomial-based formula of Heck *et al.* (1975) was applied since the infections within each collection of snails is likely to be a very small fraction of those within an entire subpopulation of snails, numbering in the thousands of individuals. In this situation, the successive collection of infected snails is unlikely to affect the relative abundance of infections by different parasites within the snail subpopulation. The sampling and estimation procedures also meet the criteria outlined by Sanders (1968) and Simberloff (1979): (a) the samples are taxonomically similar and come from the same habitat, (b) the method of sampling was consistent for all pans and years, and (c) (E (S_{15})) was computed by interpolation, not extrapolation. As a result of this rarefaction procedure, the sample sizes for analyses of relationships between (E (S_{15})) and host population characteristics were smaller than those for other measures of guild structure, since snail samples containing fewer than 15 infections were not included in the former data set. The correlation between (E (S_{15})) and ln NI is clearly weaker than between S and ln NI (Table 3.4). Partial correlations between (E (S_{15})) and ln Ni with mean length of snails in the sample held constant were not significant in any year, at either site.

The other measures of parasite guild structure (Table 3.3) were chosen for their ease of interpretation (Hill, 1973; Peet, 1974; May, 1975). The two diversity measures, Exp (H′) and $1/\Sigma p_i^2$, differ in their sensitivity to changes in the relative importance of species (Hurlbert, 1971; Peet, 1974). The first index is most sensitive to changes in rare species, while the second is more responsive to changes in common species.

3.5 THE ANALYSIS

Patterns at the level of the individual host

Several lines of evidence indicate that strong antagonistic interactions between larval parasite species occur within individual snails. Circumstantial evidence comes from the observation that mixed-species infections by larval trematodes are exceedingly rare in *Cerithidea* populations. Martin (1955) made 12 monthly collections of adult *Cerithidea* from a salt pond in Upper Newport Bay, California and reported the frequencies of different categories of infection for the pooled collection of monthly samples. A statistical analysis of these pooled data indicates that mixed-species infections were fewer than would be expected under the null hypothesis that parasites are randomly and independently distributed among snails

(Chapter 4). This result suggests negative interactions between species of parasites, but the pooled nature of the data set complicates interpretation of the statistical pattern of negative association. Heterogeneity in the relative abundance of parasite species among the monthly samples alone could produce such a pattern, without any direct interaction between species. Indeed, Martin (1955) found seasonal variation in the prevalance of infection, as did Yoshino (1975) for *Cerithidea* populations in Goleta Slough, California. Cort *et al.* (1937) were among the first to identify this problem with analyses of parasite associations based on heterogeneous data sets.

Mixed species were also very rare in the seven annual samples from Bolinas Lagoon (Table 3.2). Of 4462 infected snails examined during the seven censuses, only 91 (2%) were infected by more than one species. I have statistically compared the observed and expected numbers of mixed infections for each pan within a given year. This pan by pan analysis of samples collected in the same month of each year reduces the influence of spatial and temporal heterogeneity in parasite abundance discussed above. Due to the large number of parasite species involved and relatively low frequencies of infections per species, the expected numbers of mixed-species infections are often too small for the application of standard analyses for discrete data, e.g. contingency tables. Instead, I used Monte Carlo simulations (Sokal and Rohlf, 1981) to estimate the probability of the observed number of mixed infections under the null hypothesis of random, independent assortment of parasites among hosts. The results of this analysis agree with that of Martin's (1955) data discussed above; within a number of snail subpopulations in any given year, the frequency of snails infected by more than one species of larval trematode is less than would be expected by chance (Sousa, in prep).

Such negative associations are not proof of direct antagonistic interactions among species. One alternative explanation is that miracidia of one parasite species may actively avoid, or be unable to infect, snails that are already parasitized by a different species (for a discussion of possible mechanisms of indirect antagonism see Lim and Heyneman, 1972).

A second alternative explanation for the rarity of mixed infections is that parasite species may preferentially infect hosts of different size, although this pattern itself might be an evolutionary response to negative interactions in the past, i.e. niche partitioning (MacArthur and Levins, 1967; but see Connell, 1980). *Cerithidea* infected by different trematode species do differ in mean length (Sousa, 1983; Sousa and Gleason, 1989; see also Chapter 4), but the ranges of snail lengths in which the different species are found overlap considerably. While these distinctive patterns of distribution may reduce rates of interspecific interaction, two lines of evidence indicate that intramolluscan antagonistic interactions do occur, and that they are strongly hierarchical in outcome (Sousa, in prep.). While dissecting snails from the annual samples, I have observed a number of mixed infections in which rediae of one species are preying on rediae, sporocysts, or cercariae of another. These interactions are hierarchical: species with large rediae dominate, especially *Himasthla* and *Parorchis*. Redial species of intermediate size are, in turn, dominant over species with small rediae or sporocysts. I have also examined temporal patterns of parasite species replacement in marked snails carrying known infections that have been released in the field and recaptured at a later date. These sequences of replacement are also strongly hierarchical. Species with large rediae (*Himasthla* and *Parorchis*) most frequently invade and displace infections by other species; in contrast, infections by these two species are very rarely invaded and, if so, only by the other member of the pair.

In summary, the rarity of mixed-species infections, direct observations of hierarchical antagonism between co-occurring species, and the record of hierarchical species replacement over time, all suggest that negative interspecific interactions among larval trematodes occur at the scale of the individual snail, the infraguild (i.e. infracommunity) level. The next section addresses the question of whether these interactions are common enough to affect the structure of the component guild (i.e. component community) of parasites within a subpopulation of snails.

Patterns at the level of the host subpopulation

Hypotheses and predictions

In this section, I present two alternative hypotheses which may explain parasite guild structure at the level of the host population in this system. The first hypothesis is that the hierarchical, negative, interspecific interactions seen at the infraguild level strongly influence component guild structure. The second is that the spatial and temporal patterns of parasite recruitment and the duration of host exposure, rather than interspecific competition, are the primary determinants of guild structure at the component level.

Infraguild interactions will affect the structure of component parasite guilds if the host resource is sufficiently limiting that a few antagonistically-dominant species of parasites come to monopolize the infections. These conditions could result in two ways: (a) from high rates of recruitment and infection by these dominant species, or (b) from low rates of recruitment of new hosts to the population and the gradual accumulation of infections by the dominant parasites with time. If intramolluscan antagonism was the primary determinant of component-level structure, the following patterns would be predicted. First, assuming that the mean length of snails in a population is an index of mean age, species richness and diversity should exhibit a hyperbolic relationship when plotted against mean length of snail. Young populations composed predominantly of small snails will have only been exposed to infective parasite larvae for a short time. Therefore, only a small number of parasite species will have had the opportunity to infect them, and those parasite species with the highest recruitment rates will be most abundant. Species richness and diversity of parasite guilds infecting such snail populations will be low. At the opposite extreme, populations composed primarily of larger, older snails will have been exposed to infective miracidia for a relatively long time. For this reason, and the fact that infections appear to be life-long in *Cerithidea* (Sousa, 1983, in prep.), prevalence of infection should be high in these older populations and antagonistic interspecific interactions common. These interactions should result in most snails being infected by a few dominant trematode species, and diversity should decline. The highest species richness and diversity should be seen in populations of intermediately-sized snails which are old enough to have accumulated several parasite species, but are not so old that one or a few species have had sufficient time to dominate the majority of the infections. The pattern for species dominance should be the mirror image of that for species richness and diversity. This hypothesized hyperbolic relationship is analogous to that predicted by the Intermediate Disturbance Hypothesis for patterns of species diversity in assemblages of free-living organisms (Connell, 1978; Sousa, 1984).

The assumption that snail length is an index of snail age is certainly true in a relative sense (i.e. snails grow longer with time); however, the precise age of a snail cannot be predicted from its length. For example, snail growth can be stunted under conditions of high density. As a result, in populations of different density, snails of the same age may differ by several mm in length. In addition, parasitic infection can alter snail growth rate, either slowing or accelerating it depending on the species of trematode (Sousa, 1983, unpublished data). As a consequence, while a strong relationship between size and age exists, there is undoubtedly some variation in it.

A second prediction from the interspecific interaction hypothesis is that all else being equal, the species richness and diversity of parasites should increase with greater variation in host size within a snail subpopulation, since the distributions of different species among host size classes are heterogeneous (see references cited earlier). Partitioning of hosts by size may promote coexistence of parasite species by reducing the rates at which they antagonistically interact.

A third prediction is that, all else being equal, parasite species richness and diversity should be higher in host populations with a greater availability of uninfected, susceptible individuals since the frequency of mixed-species infections and antagonistic interactions should be lower under these conditions.

Under the second general hypothesis, a very different pattern of component guild structure is predicted. If rates of parasite recruitment and infection are low relative to the rate at which uninfected, susceptible new hosts recruit, the number of open patches of host resource may never become so limited that intramolluscan interactions would reduce the diversity of the component guild. Further, mortality of old, infected hosts may limit the accumulation of infections by dominant species. Under these conditions, an equilibrium assemblage, dominated by a few species would seldom if ever develop. Species richness and diversity would rise monotonically with mean snail age (length) as parasite species and infections accumulate with time. These indices of guild diversity might even display a faster than linear increase with mean length of snails in a subpopulation, if the range of mean host sizes is sufficiently large. This is because a snail's rate of growth slows with increasing length (Sousa, unpublished data); therefore, the mean age of a host population increases as a power function of mean length.

Conversely, species dominance should exhibit a monotonic decline with mean snail length. Neither variation in host size within a population, nor variation in the availability of uninfected, susceptible hosts would have a marked effect on guild structure, since the host resource is not limiting. Under these circumstances, guild structure would be determined primarily by temporal and spatial patterns of parasite recruitment. This situation is roughly equivalent to Wilson's (1969) non-interactive phase of community development. Price (1980) argues that this kind of nonequilibrium situation is typical of many, if not most, host-parasite systems.

Results

Scatterplots (Figs 3.1–3.3) and correlation analysis (Table 3.5) reveal that in five of the seven annual samples from PGC, rarefied species richness (E (S_{15})), and at least one of the two diversity indices (Exp(H′) and $1/\Sigma p_i^2$) exhibited a significant monotonic increase with the mean length of snails in a subpopulation. There was no significant correlation between these variables in 1982 or 1987. Simple dominance (Dom) declined significantly and monotonically with mean snail length in three of the seven years (Table 3.5; Fig. 3.4); similar, but non-significant, negative relationships were detected in the other four. The combined probability (Fisher, 1954) for all seven years was highly significant for each of the above correlations ($p < .001$ in each case).

Patterns at KI were similar to those at PGC, but not as strong. The smaller number of subpopulations sampled at this site reduces the statistical power of the correlation analysis. This problem is aggravated by the fact that within any particular year, KI subpopulations varied little in mean length (Figs

Table 3.5 Correlations between measures of parasite guild structure and average length of snails in a subpopulation. See Table 3.3 for explanation of measures. *n* is number of subpopulations sampled. Dom was normalized with an angular transformation. A dash indicates that the sample size (*n*) was too small for statistical analysis

Site	*PGC*							*KI*						
Year	81	82	83	84	85	86	87	81	82	83	84	85	86	87
Correlation of average length and														
1. E (S_{15})														
r	0.46	0.27	0.58	0.78	0.82	0.70	0.40	−0.20	–	0.11	0.90	–	–	–
n	15	11	13	8	10	10	13	15	2	4	8	0	1	0
p[a]	+	ns	*	*	**	*	ns	ns	–	ns	**	–	–	–
2. Exp (H′)														
r	0.55	0.40	0.40	0.61	0.73	0.66	0.34	−0.21	0.56	0.44	0.92	−0.44	0.66	0.65
n[b]	19	15	14	13	14	13	14	15	14	7	8	7	8	6
p	*	ns	ns	*	**	*	ns	ns	*	ns	***	ns	+	ns
3. $1/\Sigma p_i^2$														
r	0.48	0.31	0.30	0.69	0.74	0.67	0.43	−0.18	0.47	0.46	0.83	−0.43	0.57	0.75
p	*	ns	ns	**	**	*	ns	ns	+	ns	*	ns	ns	+
4. arcsin Dom														
r	−0.36	−0.26	−0.25	−0.69	−0.75	−0.63	−0.49	0.02	−0.24	−0.39	−0.54	0.25	−0.28	−0.79
p	ns	ns	ns	**	**	*	+	ns	ns	ns	ns	ns	ns	+

[a] All probabilities in table are two-tailed: ns ≥ 0.10, + < 0.10, * < 0.05, ** < 0.01, *** < 0.001.
[b] Sample sizes are the same for measures 2, 3, and 4 this table.

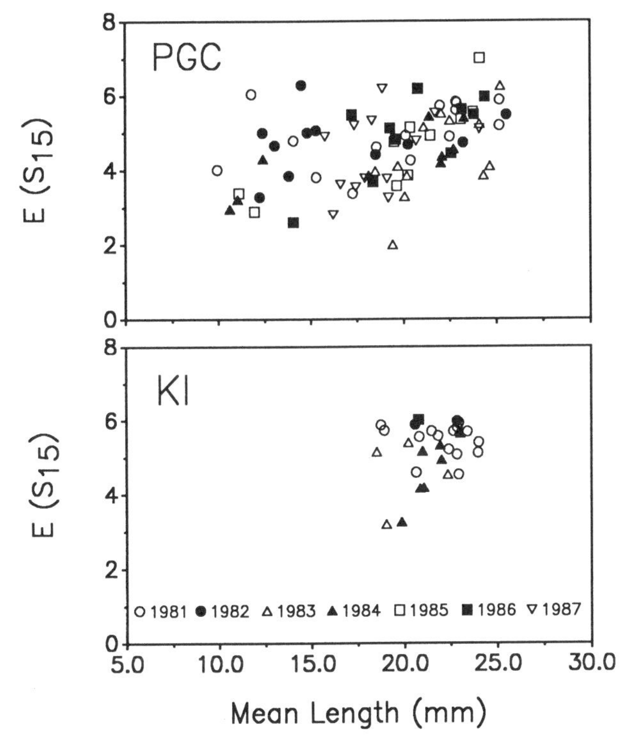

Fig. 3.1 Plot of rarefied number of trematode species versus mean length of snails in a subpopulation at each study site. Each symbol represents the pooled sample of snails collected from an individual pan in the indicated year (see text for further explanation).

3.1–3.4), as discussed earlier. Even so, significant positive correlations between rarefied species richness or the indices of diversity and mean snail length were detected in two of the seven years. Simple dominance appeared to be negatively correlated with mean snail length, but this relationship was never statistically significant.

Figure 3.5 is identical to Fig. 3.4, but indicates which species of parasite was the most prevalent in each pan. Note that the predominant species differed considerably between sites, even though the number of parasite species infecting a host subpopulation of a given mean length did not differ between sites (Fig. 3.1). Also note that *Echinoparyphium* was a conspicuous dominant in those PGC populations whose size distributions were dominated by small snails.

No significant relationship was found in any year, at either site, between the standard deviation of snail length and any of the measures of parasite guild structure. The same was true for a partial correlation analysis of these

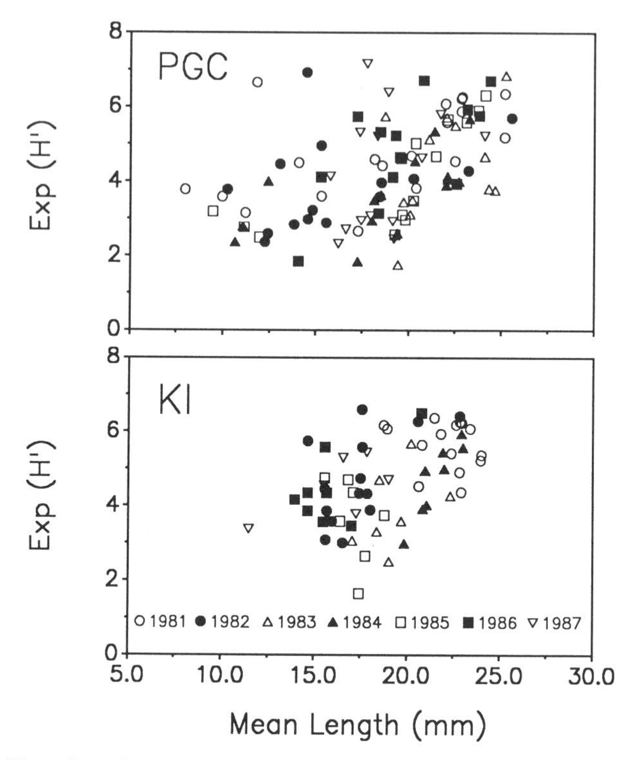

Fig. 3.2 Plot of antilogarithm of Shannon diversity index versus mean length of snails in a subpopulation at each study site. Symbols as in Fig. 3.1.

variables with mean snail length held constant. Scatterplots of the variables did not reveal any hidden, non-linear relationships.

Similarly, only three out of 38 partial correlations (mean length held constant) between rarefied species richness, or the diversity indices, and the mean density of uninfected snails were statistically significant. Two of these significant correlations were positive, favouring the Antagonistic Interaction Hypothesis; the third was in the opposite direction.

When no adjustment for differences in mean snail length is made, species richness and diversity were sometimes negatively related to the density of uninfected hosts. Significant negative relationships of these measures to the density of uninfected snails were found at PGC in three of the seven years (1984–86; r ranged from -0.57 to -0.76, $p < .05$ or $.01$). These were years in which the proportionately fewer infections in dense, younger, populations of small snails were dominated by one parasite species, *Echinoparyphium*. In each of these years, simple dominance was also significantly correlated with the density of uninfected snails, but in the opposite direction

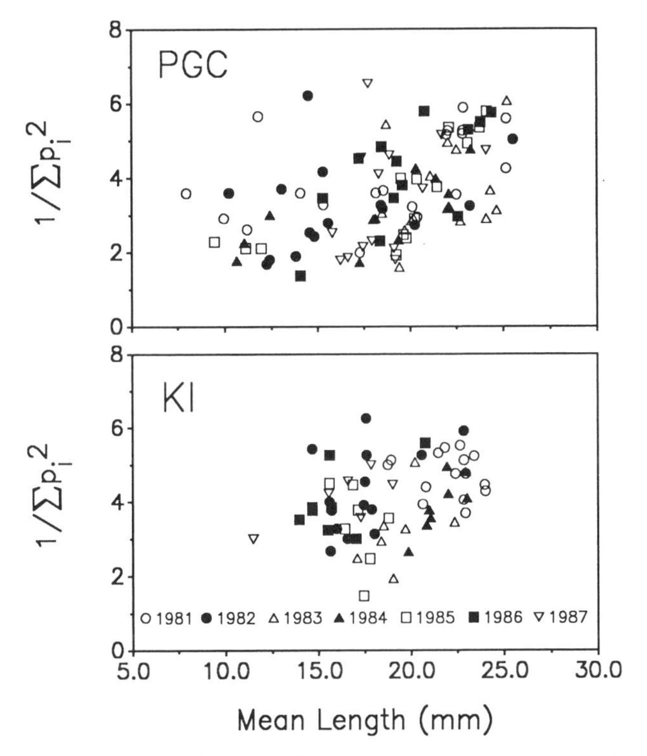

Fig. 3.3 Plot of reciprocal of Simpson diversity index versus mean length of snails in a subpopulation at each study site. Symbols as in Fig. 3.1.

to the diversity measures (r ranged from 0.65 to 0.73, $p < .05$ or $< .01$). None of these relationships were significant in any year at KI, where *Echinoparyphium* infections never dominated a component guild (Fig. 3.5).

3.6 DISCUSSION AND CONCLUSIONS

There is little evidence from this analysis that the antagonistic interactions between parasite species which occur within individual hosts have a strong impact on patterns of parasite species richness or diversity at the level of the host population. In particular, parasite diversity did not decline in older populations of hosts as would be predicted if uninfected hosts were a limited resource for parasites in such populations and infections came to be monopolized by a small number of antagonistically dominant species. In addition, neither the density of uninfected, susceptible hosts, nor the variation in host size within a snail subpopulation showed any relationship to

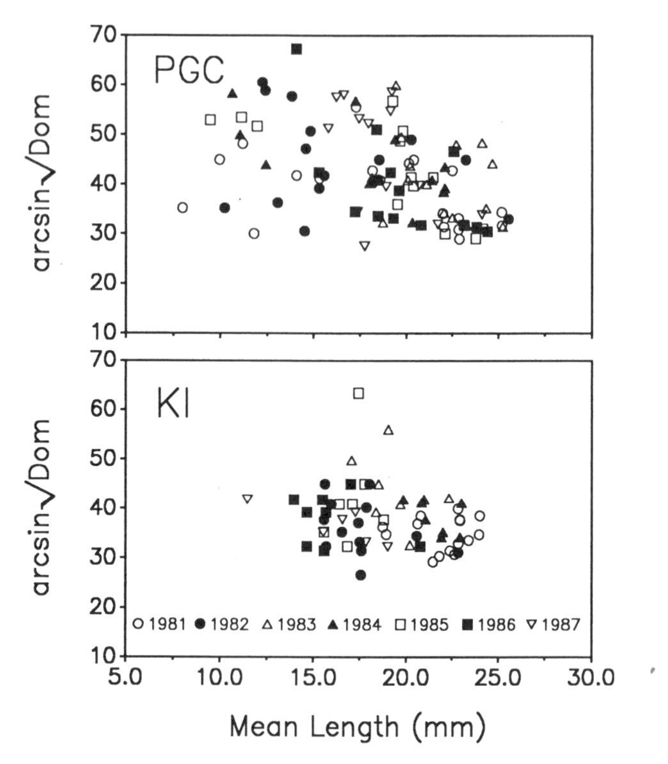

Fig. 3.4 Plot of arcsin-transformed simple dominance index versus mean length of snails in a subpopulation at each study site. Symbols as in Fig. 3.1.

parasite diversity. Indeed, the percentage of hosts infected by larval trematodes is generally greater in populations comprised of larger, older individuals (for full data set: PGC: $r = 0.46$, $n = 103$, $p < .001$; KI: $r = 0.86$, $n = 70$, $p < .001$). Apparently the number of uninfected, susceptible hosts never becomes sufficiently limiting, given the rates of snail and parasite recruitment in this system, to drive parasite diversity downward. Instead, species richness and diversity rise monotonically with mean snail length, or roughly, the mean age of the host population. This is not to say that competition has no influence on the structures of the component parasite guilds examined in this study. Indices of guild structure such as those computed for this analysis hide the population dynamics of individual species; some species may well be less abundant at the component level as a consequence of infra-level antagonistic interactions. It is clear, however, that any such reductions are more than compensated for by increases in both the number and equitability of other parasite species in older host populations. The competitive interactions that prevent the coexistence of

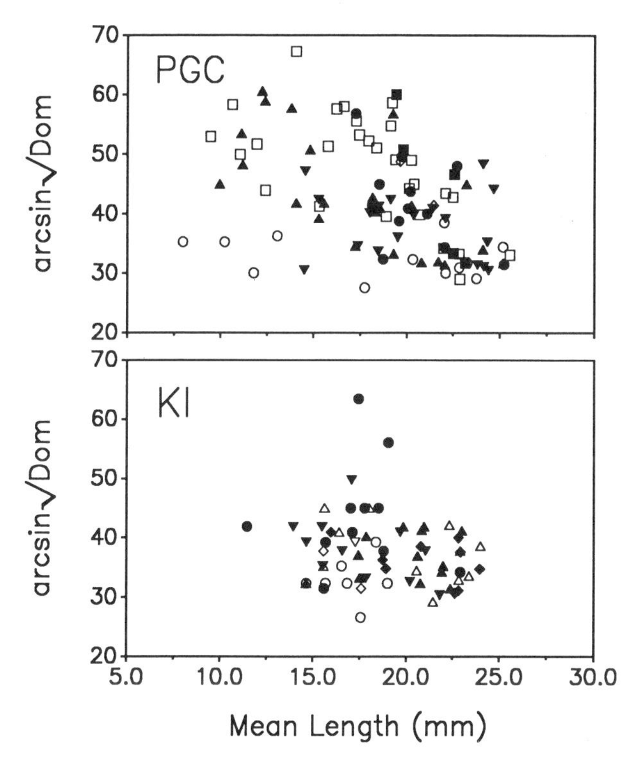

Fig. 3.5 Plot of arcsin-transformed simple dominance index versus mean length of snails in a subpopulation at each study site. Symbols indicate dominant species of parasite in each subpopulation: *Acanthoparyphium* (filled circle), *Euhaplorchis* (open triangle), *Parochis* (filled triangle), *Echinoparyphium* (open square), cyanthocotylid #2 (filled square), *Catatropis* (open inverse triangle), *Himasthla* (filled inverse triangle), renicolid #2 (open diamond), microphallid #1 (filled diamond), co-domination by two or more species (open circle).

parasite species within individual hosts have far less impact on the structures of the component parasite guilds infecting subpopulations of hosts. The population dynamics of the different parasite species in each of the host subpopulations will be examined separately (Sousa, in prep.).

Under a condition of relatively unlimited resources, which appears to characterize the component level of this system, spatial and temporal variation in the abundance and/or infectivity of miracidial stages will be the primary determinant of component guild structure. The observations that the dominant parasite species differ markedly between the sites (Fig. 3.5) and that the relative prevalence of different parasite species fluctuates greatly from year to year (Sousa, unpublished data) support this interpretation.

This study demonstrates that distinct processes are primarily responsible for the structure of assemblages of larval trematodes at different spatial scales; a similar conclusion has been reached for a number of assemblages of free-living organisms (see references cited earlier). If more than one species of parasite recruits to an individual snail, deterministic antagonistic interactions almost always result in the exclusion of one of the two species. Most larval trematodes exploit the gonad or digestive gland of their molluscan hosts (e.g. Wright, 1971; Brown, 1978; Kuris, 1974; Lauckner, 1980, 1983; Sousa, 1983), tissues which are not as easily partitioned as are the complex organ systems of the definitive vertebrate hosts (e.g. Crompton, 1973; Holmes, 1973; Kennedy, 1975; Bush and Holmes, 1983, 1986a, b). Consequently, parasite species rarely coexist on the small scale, and infraguilds are depauperate.

In contrast, the component guild of trematodes that infects populations of *Cerithidea* in Bolinas Lagoon appears to be structured largely by external processes, in particular those that determine spatial and temporal variation in the abundance of infective stages in the parasite life cycle. Interspecific antagonism on the small spatial scale does not have a detectable effect on patterns at this larger scale. Thus, it appears that very different processes determine the organization of larval parasite guilds at the two spatial scales examined in this study, the infra- and component guild levels of parasite resources.

Finally, it is apparent that the discrete, divided nature of resources provided by hosts promotes coexistence of parasite species on the large scale. This structure and the complex life cycle of the parasites preclude the direct spread between snails of asexual parasitic stages (rediae) of antagonistically dominant species. In effect, this provides a refuge for less aggressive parasites, similar to that demonstrated in Huffaker's (1958) classic study of predator–prey interactions in a patchy, experimental system. Independent aggregation of parasites among hosts, as a consequence of both asexual reproduction and differential exploitation of different-sized hosts, makes coexistence even more likely (Shorrocks *et al.*, 1979; Atkinson and Shorrocks, 1981; Dobson, 1985; Ives and May, 1985). In addition, the ephemeral nature of host patches, and the high rate at which new ones are created via host reproduction, make it all the more difficult for one or a few species to monopolize the host resource.

ACKNOWLEDGEMENTS

I am deeply indebted to many people for their assistance and support during this study. I owe a huge thanks to the seven annual 'snail crews' without whose help in the field and laboratory this study would not have been possible. G. Roderick and E. Adams wrote excellent, tailor-made computer programs for this analysis. B. Mitchell, A. Bush, G. Esch, and E. Grosholz provided many helpful comments on the manuscript. Parasite species indentifications were confirmed by H. Ching. Access to the study sites was kindly allowed by the Marin County Department of Parks and Recreation. This research was supported by National Science Foundation grants DEB-8004192, BSR-8300082, BSR-8505871, and BSR-8516522.

REFERENCES

Addicott, J. F., Aho, J. M., Antolin, M. F. *et al.* (1987) Ecological neighborhoods: scaling environmental patterns. *Oikos*, **49**, 340–46.

Allen, T. F. H. and Starr, T. B. (1982) *Hierarchy*, University of Chicago Press, Chicago.

Andrewartha, H. G. and Birch, L. C. (1954) *The Distribution and Abundance of Animals*, University of Chicago Press, Chicago.

Armstrong, R. A. (1976) Fugitive species: experiments with fungi and some theoretical considerations. *Ecology*, **57**, 953–63.

Atkinson, W. D. and Shorrocks, B. (1981) Competition on a divided and ephemeral resource: a simulation model. *J. Anim. Ecol.*, **50**, 461–71.

Badley, J. E. (1979) *The Endohelminth Fauna of Willets, Catoptrophorus semipalmatus (Gmelin 1789) (Charadriformes: Scolopacidae) from West Bay, Texas.* MS Thesis, Texas A and M University.

Brown, D. S. (1978) Pulmonate molluscs as intermediate hosts for trematodes. In *Pulmonates, vol. 2A. Systematics, Evolution, and Ecology* (eds, V. Fretter and J. Peake), Academic Press, New York, pp. 287–333.

Bush, A. O. and Holmes, J. C. (1983) Niche separation and the Broken-stick Model: use with multiple assemblages. *Am. Nat.*, **122**, 849–55.

Bush, A. O. and Holmes, J. C. (1986a) Intestinal helminths of lesser scaup ducks: patterns of association. *Can. J. Zool.*, **64**, 132–41.

Bush, A. O. and Holmes, J. C. (1986b) Intestinal helminths of lesser scaup ducks: an interactive community. *Can. J. Zool.*, **64**, 142–52.

Cable, R. M. (1972) Behavior of digenetic trematodes. In *Behavioural Aspects of Parasite Transmission*, (eds, E. U. Canning and C. A. Wright), Suppl. 1 to the Zool. J. Linnean Soc., Academic Press, London, pp. 1–18.

Caswell, H. (1978) Predator-mediated coexistence: a non-equilibrium model. *Am. Nat.*, **112**, 127–54.

Chernin, E. (1974) Some host-finding attributes of *Schistosoma mansoni* miracidia. *Am. J. Trop. Med. Hyg.*, **23**, 320–27.

Cohen, J. E. (1970) A Markov Contingency Table Model for replicated Lotka-Volterra systems near equilibrium. *Am. Nat.*, **104**, 547–59.

Connell, J. H. (1978) Diversity in tropical rainforests and coral reefs. *Science*, **199**, 1302–10.

Connell, J. H. (1980) Diversity and the coevolution of competitors, or the ghost of competition past. *Oikos*, **35**, 131–8.

Connell, J. H. and Sousa, W. P. (1983) On the evidence needed to judge ecological stability or persistence. *Am. Nat.*, **121**, 789–824.
Cort, W. W., McMullen, D. B. and Brackett, S. (1937) Ecological studies on the cercariae in *Stagnicola emarginata angulata* (Sowerby) in the Douglas Lake region, Michigan. *J. Parasitol.*, **23**, 504–32.
Crompton, D. W. T. (1973) The sites occupied by some parasitic helminths in the alimentary tract of vertebrates. *Biol. Rev.*, **48**, 27–83.
Dayton, P. K. and Tegner, M. A. (1984) The importance of scale in community ecology: a kelp forest example with terrestrial analogs. In *A New Ecology: Novel Approaches to Interactive Systems*, (eds, P. W. Price, C. N. Slobodchikoff and W. S. Gaud), John Wiley and Sons, New York, pp. 457–81.
Denny, M. (1969) Life-cycles of helminth parasites using *Gammarus lacustris* as an intermediate host in a Canadian lake. *Parasitology*, **59**, 795–827.
Dobson, A. P. (1985) The population dynamics of competition among parasites. *Parasitology*, **91**, 317–47.
Esch, G. W. Gibbons, J. W. and Bourque, J. E. (1975) An analysis of the relationship between stress and parasitism. *Am. Midl. Nat.*, **93**, 339–53.
Fisher, R. A. (1954) *Statistical Methods for Research Workers*, 12th edn, Oliver and Boyd, Edinburgh
Hanski, I. (1981) Coexistence of competitors in patchy environment with and without predation. *Oikos*, **37**, 306–12.
Hanski, I. (1983) Coexistence of competitors in patchy environment. *Ecology*, **64**, 493–500.
Hastings, A. (1977) Spatial heterogeneity and the stability of predator–prey systems. *Theor. Popul. Biol.*, **12**, 37–48.
Hastings, A. (1980) Disturbance, coexistence, history, and competition for space. *Theor. Popul. Biol.*, **18**, 363–73.
Heck, K. L. Jr, Van Belle, R. and Simberloff, D. (1975) Explicit calculation of the rarefaction diversity measurement and the determination of sufficient sample size. *Ecology*, **56**, 1459–61.
Hill, M. O. (1973) Diversity and evenness: a unifying notation and its consequences. *Ecology*, **54**, 427–32.
Holmes, J. C. (1973) Site segregation by parasitic helminths: interspecific interactions, site segregation, and their importance to the development of helminth communities. *Can. J. Zool.*, **51**, 333–47.
Holmes, J. C. and Price, P. W. (1986) Communities of parasites. In *Community Ecology: Patterns and Process*, (eds, J. Kikkawa and D. J. Anderson), Blackwell Scientific Publications, Melbourne, pp. 187–213.
Horn, H. S. and MacArthur, R. H. (1972) Competition among fugitive species in a harlequin environment. *Ecology*, **53**, 749–52.
Huffaker, C. B. (1958) Experimental studies on predation: dispersion factors and predator–prey oscillations. *Hilgardia*, **27**, 343–83.
Hurlbert, S. H. (1971) The non-concept of species diversity: a critique and alternative parameters. *Ecology*, **52**, 577–86.
Hutchinson, G. E. (1951) Copepodology for the ornithologist. *Ecology*, **32**, 571–7.
Ives, A. R. and May, R. M. (1985) Competition within and between species in a patchy environment: relations between microscopic and macroscopic models. *J. Theor. Biol.*, **115**, 65–92.

Kennedy, C. R. (1975) *Ecological Animal Parasitology*, John Wiley and Sons, New York.
Kuris, A. M. (1974) Trophic interactions: similarity of parasitic castrators to parasitoids. *Q. Rev. Biol.*, **49**, 129–48.
Lauckner, G. (1980) Diseases of mollusca: gastropoda. In *Diseases of Marine Animals, vol. I: General aspects, protozoa to gastropoda*, (ed. O. Kinne), John Wiley and Sons, New York, pp. 311–424.
Lauckner, G. (1983) Diseases of mollusca: bivalvia. In *Diseases of Marine Animals, vol. II: Introductions, bivalvia to scaphopoda* (ed. O. Kinne), Biologische Anstalt Helgoland, Hamburg, pp. 477–961.
Levin, S. A. (1974) Dispersion and population interactions. *Am. Nat.*, **108**, 207–28.
Levin, S. A. (1976) Population dynamic models in heterogeneous environments. *Ann. Rev. Ecol. Syst.*, **7**, 287–310.
Levins, R. and Culver, D. (1971) Regional coexistence of species and competition between rare species. *Proc. Nat. Acad. Sci. USA.*, **68**, 1246–8.
Lim, H. K. and Heyneman, D. (1972) Intramolluscan inter-trematode antagonism: a review of factors influencing the host–parasite system and its possible role in biological control. *Adv. Parasitol.*, **10**, 191–268.
Lloyd, M. and White, J. (1980) On reconciling patchy microspatial distributions with competition models. *Am. Nat.*, **115**, 29–44.
MacArthur, R. H. and Levins, R. (1967) The limiting similarity, convergence, and divergence of coexisting species. *Am. Nat.*, **101**, 377–85.
Macdonald, K. B. (1969a) Molluscan faunas of Pacific coast salt marshes and tidal creeks. *Veliger*, **11**, 399–405.
Macdonald, K. B. (1969b) Quantitative studies of salt marsh mollusc faunas from the North American Pacific coast. *Ecol. Monogr.*, **39**, 33–60.
MacGinitie, G. E. (1934) The natural history of *Callianassa californiensis*. *Am. Midl. Nat.*, **15**, 166–77.
MacInnis, A. J. (1976) How parasites find hosts, some thoughts on the inception of host–parasite integration. In *Ecological Aspects of Parasitology*, (ed. C. R. Kennedy), North-Holland Publ. Co., Amsterdam, pp. 3–20.
Margolis, L., Esch, G. W., Holmes, J. C. *et al.* (1982) The use of ecological terms in parasitology. *J. Parasitol.*, **68**, 131–3.
Martin, W. E. (1955) Seasonal infections of the snail *Cerithidea californica* Haldeman, with larval trematodes. In *Essays in the Natural Sciences in Honor of Captain Allan Hancock*, Allan Hancock Foundation, University of Southern California Press, Los Angeles, pp. 203–10.
Martin, W. E. (1972) An annotated key to the cercariae that develop in the snail *Cerithidea californica*. Bull. South. Cal. Acad. Sci., **71**, 39–43.
May, R. M. (1975) Patterns of species abundance and diversity. In *Ecology and Evolution of Communities*, (eds, M. L. Cody and J. M. Diamond), Belknap Press, Cambridge, Mass., pp. 81–120.
Meuleman, E. A., Bayne, C. J. and Van der Knaap, W. P. (1987) Immunological aspects of snail–trematode interactions. In *Developmental and Comparative Immunology* (eds, E. L. Cooper, C. Langlet and J. Bierne, Progress in Clinical and Biological Research, vol. 233, Alan R. Liss, Inc, New York, pp. 113–27.
Murdoch, W. W., Chesson, J. and Chesson, P. L. (1985) Biological control in theory and practice. *Am. Nat.*, **125**, 344–66.

Page, G. W., Stenzel, L. E. and Wolfe, C. M. (1979) Aspects of the occurrence of shorebirds on a central California estuary. *Stud. Avian Biol.*, **2**, 15–32.

Peet, R. K. (1974) The measurement of species diversity. *Ann. Rev. Ecol. Syst.*, **5**, 285–307.

Price, P. W. (1980) *Evolutionary Biology of Parasites*, Princeton University Press, Princeton.

Quammen, M. L. (1984) Predation by shorebirds, fish, and crabs on invertebrates in intertidal mudflats: an experimental test. *Ecology*, **65**, 529–37.

Richards, C. S. (1976) Genetics of the host–parasite relationship between *Biomphalaria glabrata* and *Schistosoma mansoni*. In *Genetic Aspects of Host–parasite relationships*, (eds, A. E. R. Taylor and R. Muller), Symposia of the British Society for Parasitology, vol. 14, Blackwell Scientific Publications, pp. 45–54.

Ritter, J. R. (1969) *Preliminary Studies of the Sedimentology and Hydrology in Bolinas Lagoon, Marin County, California, May 1967–June 1968*, U.S. Department of the Interior, Geological Survey, Water Resources Division, Menlo Park, California.

Robinson, E. J. (1952) A preliminary report on the life cycle of *Cloacitrema michiganensis* McIntosh 1938. *J. Parasitol.*, **38**, 368.

Rohde, K. (1982) *Ecology of Marine Parasites*, University of Queensland press, St. Lucia.

Root, R. (1967) The niche exploitation pattern of the blue-grey gnatcatcher. *Ecol. Monogr.*, **37**, 317–50.

Russell, H. T. (1960a) Trematodes from shorebirds collected at Morro Bay, California. *J. Parasitol.*, **46** (5 : 2), 15 [Abstr.].

Russell, H. T. (1960b) *Trematodes From Shorebirds Collected at Morro Bay, California*, PhD. thesis, University of California, Los Angeles, C.A.

Sanders, H. L. (1968) Marine benthic diversity: a comparative study. *Am. Nat.*, **102**, 243–82.

Shiff, C. J. (1974) Seasonal factors influencing the location of *Bulinus (Physopsis) globosus* by miracidia of *Schistosoma haematobium* in nature. *J. Parasitol.*, **60**, 578–83.

Shoop, W. L. (1988) Trematode transmission patterns. *J. Parasitol.*, **74**, 46–59.

Shorrocks, B., Atkinson, W. and Charlesworth, P. (1979) Competition on a divided and ephemeral resource. *J. Anim. Ecol.*, **48**, 899–908.

Simberloff, D. (1979) Rarefaction as a distribution-free method of expressing and estimating diversity. In *Ecological Diversity in Theory and Practice*, (eds, J. F. Grassle, G. P. Patil, W. Smith and C. Taillie), International Co-operative Publishing House, Fairland, Maryland, pp. 159–76.

Skellam, J. G. (1951) Random dispersal in theoretical populations. *Biometrika*, **38**, 196–318.

Slatkin, M. (1974) Competition and regional coexistence. *Ecology*, **55**, 128–34.

Smyth, J. D. and Halton, D. W. (1983) *The Physiology of Trematodes*, 2nd edn, Cambridge University Press, Cambridge.

Sokal, R. R. and Rohlf, F. J. (1981) *Biometry*, 2nd edn, W. H. Freeman and Co., San Franscisco.

Sousa, W. P. (1979) Disturbance in marine intertidal boulder fields: the nonequilibrium maintenance of species diversity. *Ecology*, **60**, 1225–39.

Sousa, W. P. (1983) Host life history and the effect of parasitic castration on growth: a field study of *Cerithidea californica* Haldeman (Gastropoda: Prosobranchia) and its trematode parasites. *J. Exp. Mar. Biol. Ecol.*, **73**, 273–96.

Sousa, W. P. (1984) The role of disturbance in natural communities. *Ann. Rev. Ecol. Syst.*, **15**, 353–91.

Sousa, W. P. and Gleason, M. (1989) Does parasitic infection compromise host survival under extreme environmental conditions: the case for *Cerithidea californica* (Gastropoda: Prosobranchia)? *Oecologia*, in press.

Stenzel, L. E., Huber, H. R. and Page, G. W. (1976) Feeding behavior and diet of the long-billed curlew and willet. *Wilson Bull.*, **88**, 314–32.

Ulmer, M. J. (1971) Site-finding behavior in helminths in intermediate and definitive hosts. In *Ecology and Physiology of Parasites*, (ed. A. M. Fallis), University of Toronto Press, Toronto, pp. 123–60.

Wiens, J. A. (1976) Population responses to patchy environments. *Ann. Rev. Ecol. Syst.*, **7**, 81–120.

Wilson, E. O. (1969) The species equilibrium. *Brookhaven Symp. Biol.*, **22**, 38–47.

Wright, C. A. (1959) Host location by trematode miracidia. *Ann. Trop. Med. Parasit.*, **53**, 288–92.

Wright, C. A. (1971) *Flukes and Snails*, Hafner Press, New York.

Yoshino, T. P. (1975) A seasonal and histologic study of larval digenea infecting *Cerithidea californica* (Gastropoda: Prosobranchia) from Goleta Slough, Santa Barbara County, California. *Veliger*, **18**, 156–61.

4

Guild structure of larval trematodes in molluscan hosts: prevalence, dominance and significance of competition

Armand Kuris

4.1 INTRODUCTION

Descriptive analyses of parasite community structure are now available for a wide variety of vertebrate hosts (see Chapters 1, 5, 6, 7, 8 and 9; see also Choe and Kim, 1987). Parasite communities of invertebrate hosts have received less attention. Yet, important features distinguish invertebrates as hosts for parasites. These attributes may impart some unique characteristics to the structure or organization of parasite communities of invertebrate hosts in contrast to vertebrate hosts. Firstly, invertebrates are usually smaller than vertebrates. Secondly, many invertebrates, particularly arthropods and molluscs, serve as intermediate hosts for parasites that complete their development in a vertebrate host. These multiple host life cycles considerably complicate the nature of parasite community organization. Consequently events in the vertebrate hosts, remote in space and time, may play a direct role in the structure of parasite communities of invertebrate intermediate hosts. The reverse is also possible; dynamics in the invertebrate host may affect parasite community structure in the vertebrate hosts. Further, invertebrate populations are generally more amenable to experimentation and field manipulation than are vertebrates. Thus, the organizational features of parasite communities in invertebrate hosts may be studied using the most powerful methodologies available to community ecologists (Castilla and Paine, 1987).

Arthropods and molluscs are the invertebrate groups in which species-rich assemblages of parasites have been most often reported (e.g. Price, 1973; Askew and Shaw, 1986; Martin, 1955; Holliman, 1961; Loker *et al.*, 1981). Most of the parasite species in these assemblages are parasitoids and parasitic castrators (Kuris, 1973, 1974; Combes, 1982).

These trophic interactions are defined on the basis of the interaction between the individual consumer (or its asexual progeny) and its host (Kuris, 1974). An individual parasitoid always kills its host. Pathogenicity is not intensity dependent and parasitoids are commonly large in body size relative to host size, often reaching from 10% to 50% the combined weight of host and parasitoid. Well known examples include ichneumonid hymenopterans and mermithoid nematodes. An individual parasitic castrator always blocks reproduction of its host. As with parasitoids, pathogenicity is not intensity dependent and body size of parasitic castrators is large compared with host size. Examples include rhizocephalan barnacle parasites of crustaceans and larval trematodes in their first intermediate molluscan hosts. With respect to host–parasite interactions, parasitoids and parasitic castrators are similar from a evolutionary standpoint. Both eliminate the reproductive future of the host.

Other parasites, termed typical parasites (Kuris, 1974), usually cause little or no pathology in individual infections. Pathology is intensity dependent, and at high intensities damage may be considerable. Body size is very small; typical parasites being less than 1% the weight of the host. Examples include adult trematodes and cestodes in their vertebrate hosts.

Because both parasitic castrators and parasitoids use much of the energy resources of their hosts and are so large relative to host size, the energy available to an infracommunity of parasitic castrators or parasitoids is quite likely to be a limiting resource. Thus, habitat displacement within the host is unlikely to ameliorate potentially competitive demands for resources within an individual host. Hence, all the parasitoid and parasitic castrator members of an infracommunity can be treated as a guild (*sensu* Root, 1973), using and potentially limited by the same resource, the host.

To examine community structure and function, I selected the guild of larval trematodes parasitizing the horn snail, *Cerithidea californica*. This abundant salt-marsh snail ranges geographically from Tomales Bay, California to Laguna San Ignacio, Baja, California (Abbott and Haderlie, 1980). It is often parasitized by a species-rich assemblage of larval trematodes (Martin, 1955; Yoshino, 1975; Sousa, 1983).

4.2 GUILD STRUCTURE OF LARVAL TREMATODES

Patterns of association of larval trematodes

Martin (1955) summarized an intensive survey of larval trematodes in the horn snail, *Cerithidea californica*, in Upper Newport Bay, California. His records of prevalence of both single and multiple infections provide data suitable for an analysis of parasite co-occurrence. Even some rare combinations could be analysed, as 12 995 snails were examined. His detailed taxonomic and life-cycle studies of trematodes at this locality (see Martin,

Table 4.1 The species of trematodes in *Cerithidea californica* observed at Upper Newport Bay (Martin, 1955; systematics updated according to Martin and Adams (1961), Adams and Martin (1963), Martin (1972)). Species are grouped by type of larval life cycle (i.e. those with redial stages and those lacking rediae, having sporocyst stages only). The four letter code will designate these species in subsequent figures and tables. Prevalence is from Martin (1955). Note that the overall prevalence is not a sum; multiple infections are included in the prevalence data for each species

Type of life cycle and family	*Code*	*Species*	*Prevalence (%)*
REDIAL SPECIES			
Heterophyidae	EUHA	*Euhaplorchis californiensis*	17.4
	STIC	*Stictodora hancocki*	4.0
	PHOC	*Phocitromoides ovale*	0.5
	PYGI	*Pygidiopsoides spindalis*	0.1
Echinostomatidae	HIMA	*Himasthla rhigedana*	4.0
	ECHI	*Echinoparyphium* sp.	2.5
	ACAN	*Acanthoparyphium spinulosum*	0.0*
Philophthalmidae	CLOA	*Cloacitrema michiganensis*	0.6
	PARO	*Parorchis acanthus*	1.8
Notocotylidae	CATA	*Catatropis johnstoni*	1.6
SPOROCYSTS ONLY			
Strigeidae	SMST	small strigeid	8.1
	MESO	*Mesostephanus appendiculatus*	0.0*
Schistosomatidae	AUST	*Austrobilharzia* sp.	1.2
Microphallidae†	SMXI	small xiphidiocercaria	10.5
	LGXI	large xiphidiocercaria	5.5
Renicolidae	RENC	*Renicola cerithidicola*	6.0
	RENB	*R. buchanani*	0.2
All Species			68.2

* Prevalence less than 0.05%.
† One of the microphallids is probably *Probolocoryphe uca* (Sarkisian, 1957; Martin, 1972).

1972) make this an exceptionally well-known larval trematode assemblage. The species of trematodes observed at Upper Newport Bay and the overall prevalence of each species are shown in Table 4.1.

If double infections of a snail result from independent random co-occurrence of two species of trematodes, then the expected frequency, f_c, of such double infections is

$$f_c \times \frac{A \times B}{N}$$

where A and B are the observed frequencies of species A and species B respectively and N is the total number of snails examined. Departures from random association may be detected by comparing f_c with the observed frequency of double infections, f_o.

The pattern of development of larval trematodes has been associated with their potential competitive ability (Lim and Heyneman, 1972; Kuris, 1973; Lie, 1973; Combes, 1982). For most species of trematodes, development in the first intermediate host (usually a snail) begins when the infecting miracidium transforms to a sporocyst. The sporocyst reproduces asexually, producing rediae. These latter proliferate asexually, producing numerous rediae and ultimately cercariae. Rediae are morphologically distinct from sporocysts (Schmidt and Roberts, 1985). The former have a mouth and gut and ingest snail tissues. The latter lack a mouth and gut and absorb nutrients through the body surface. The remaining species of trematodes produce only asexual sporocyst generations. Accordingly, the infections of trematodes in *Cerithidea californica* were analysed in relation to their life-cycle pattern (larval development with a rediae stage versus development with only sporocyst stages).

For those species whose larval development included rediae, all interspecific interactions were less frequent than expected on a random basis

Table 4.2 Number of observed (before slash) and expected (after slash) double infections among species of larval trematodes with redia stages in the snail, *Cerithidia californica*, from Upper Newport Bay, California (observed data from Martin, 1955)

	STIC	*PHOC*	*HIMA*	*ECHI*	*PARO*	*CLOA*	*CATA*
EUHA	0*/91	0*/11	3*/92	1*/58	0*/40	0*/14	3*/35
STIC			0*/21	0*/13	0*/9		0*/8
HIMA				0*/13	0*/9		0*/8
ECHI					0†/6		0†/5

G-test, * negative interaction, $P < 0.01$;
† negative interaction, $0.01 < P < 0.05$;
blank spaces represent combinations for which expected values were so low that an observed value of zero would not be significant. Species code as in Table 4.1

(G-test, Table 4.2). Double infections were exceedingly rare among species with redia stages.

When the frequency of double infections was examined for species with rediae interacting with species having only sporocyst stages, a more complicated pattern emerged (Table 4.3). If the redia species was an echinostomatid, double infections were almost always rare with significantly fewer observed than expected. The exception was for concomitant infections with the schistosomatid, *Austrobilharzia* sp., which occurred as frequently as expected. In one case (with *Acanthoparyphium spinulosum*), double infections were significantly more frequent than expected. Generally, for interactions involving the small strigeid and the microphallids, f_o was significantly less than f_c. The renicolids and the notocotylid (*Catatropis johnstoni*) had a variable pattern of association; some combinations were less frequently observed than expected while, for other combinations, f_o was not significantly different from f_c. *Renicola cerithidicola* was significantly

Table 4.3 Number of observed (before slash) and expected (after slash) double infections of larval trematodes with redia stages versus species with only sporocyst stages in the snail, *Cerithidea californica*, from Upper Newport Bay, California (observed data from Martin, 1955)

Redia Species	Sporocysts-only					
	Strigeid	*Schisto-somatid*	*Microphallids*		*Renicolids*	
	SMST	*AUST*	*SMXI*	*LGXI*	*RENC*	*RENB*
Heterophyids						
EUHA	96*/184	33^{NS}/28	0*/237	0*/125	170‡/136	6^{NS}/5
STIC	12†/42	3^{NS}/6	0*/55	0*/29	8*/31	
PHOC	0†/5		0*/7		2^{NS}/2	
Echinostomatids						
HIMA	0*/43	4^{NS}/7	0*/55	0*/29	2*/32	
ECHI	0*/27	8^{NS}/4	0*/35	0*/18	1*/20	
ACAN		1‡/0				
Philophthalmids						
PARO	0*/19		0*/24	0*/13	0*/14	
CLOA	0†/7		0*/8			
Notocotylid						
CATA	3*/17	3^{NS}/3	34‡/21	15^{NS}/11	7^{NS}/12	

G-test, * negative interaction, $P < 0.01$;
† negative interaction, $0.01 < P < 0.05$;
NS not significantly different from random expectation;
‡ positive interaction, $P < 0.01$;
blank spaces represent combinations for which expected values were so low that an observed value of zero would not be significant.

more frequently associated with *Euhaplorchis californiensis* than was expected on a random basis, as was *C. johnstoni* with the small xiphidiocercaria.

Consideration of pairs of species having only sporocyst larval stages (Table 4.4) showed that most combinations werre found as frequently as expected, based on the random model. Double infections involving the microphallids, strigeids or *Austrobilharzia* sp. were generally less frequently observed than expected. However, some double infections were usually recorded. The renicolids were always randomly associated with other trematodes having only sporocyst larval stages.

To summarize, among species with rediae, double infections were always found less frequently than expected on a random basis. Interactions between species with rediae and species with only sporocyst stages were variable; many negative and random associations were detected, as were three positive associations. Interactions among species producing only sporocyst stages were usually found as often as predicted by the random model. Several species exhibited consistent patterns of occurrence in double infections against all or most other species (e.g. the echinostomatids, *Austrobilharzia* sp., and the renicolids).

Geographic distribution

The snail, *Cerithidea californica*, is geographically widespread and the guild of larval trematodes parasitizing *C. californica* has been quantitatively analysed at four locations in southern and central California (Fig. 4.1). Of the 17 species collected at Upper Newport Bay (Martin, 1972), ten occur at Carpinteria Salt Marsh (Kuris, present study; 958 snails examined); nine at Goleta Slough (Yoshino, 1975; 2910 snails examined), and 13 at Bolinas Lagoon (Sousa, 1983, and Chapter 3; more than 25 000 snails examined).

Table 4.4 Number of observed (before slash) and expected (after slash) double infections among species of larval trematodes with only sporocyst stages in the snail, *Cerithidea californica*, from Upper Newport Bay, California (observed data from Martin, 1955)

	AUST	*SMXI*	*LGXI*	*RENC*	*RENB*
SMST	3*/13	24*/111	$48^{NS}/58$	$60^{NS}/64$	$2^{NS}/2$
AUST		3*/17	$4^{NS}/9$	$13^{NS}/10$	
SMXI			0*/75	$79^{NS}/82$	$2^{NS}/3$
LGXI				$46^{NS}/43$	$2^{NS}/2$

All symbols as in Table 4.3.

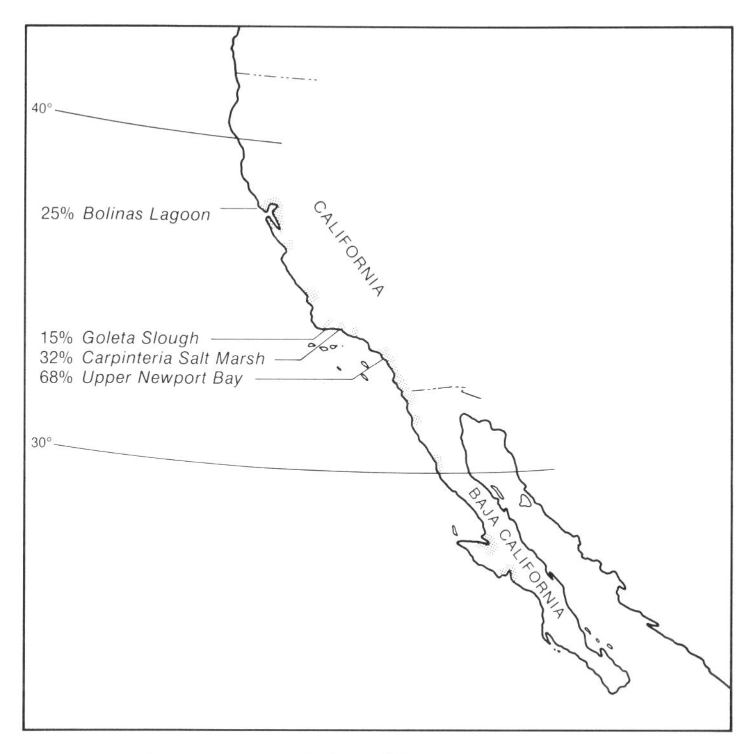

Fig. 4.1 Geographic range of *Cerithidea californica* with locations where the guild of larval trematodes parasitizing this snail has been quantitatively analysed. Percentages are for overall prevalence of the trematodes. Stippled areas represent locations of salt marshes where *C. californica* are present.

Prevalence of these trematodes has generally been high (Fig. 4.1, range 15–68%). At least one common species at Upper Newport Bay (one of the microphallids) may not occur at the other locations, as its intermediate host (*Uca crenulata*) has a restricted geographic range (Sarkisian, 1957). A few species not recorded at Upper Newport Bay were recovered from other locations (two at Bolinas, two at Carpinteria, one at Goleta).

Temporal and spatial heterogeneity

Seasonality

When samples from different seasons are combined, significant negative associations may be detected for non-interactive species if considerable temporal variation in prevalence is evident. Such negative associations would be spurious, as a biological interaction did not occur. Perhaps less obviously, variation of infectivity over time could also increase the likelihood of antagonistic interactions. If infections of antagonistic species are synchronous, then combining samples over time will diminish the predicted frequency of such interactions in a random model. Most of the larval trematodes in *Cerithidea californica* show a similar pattern of temporal fluctuation (Martin, 1955; Yoshino, 1975). This temporal concordance is probably associated with parasitization of shorebirds as adults. Infection dynamics of these trematodes may be similarly influenced by shorebird migratory behaviour.

Hence, temporal heterogeneity could affect the quantification of competitive interactions in guilds of larval trematodes. A method to analyse such a source of heterogeneity is presented below.

Variation in space

On a small scale, there is substantial variation in the prevalence of trematodes in *Cerithidea californica* (Chapter 3). The importance of this source of variation to the interactive dynamics of the guild can be assessed in a manner directly analogous to that presented for the impact of temporal variation on this system (see below).

Quantitative assessment of scale effects

To quantify the contribution of temporal (or spatial) heterogeneity to the pattern of association observed at a larger scale, the expected frequency, for subsamples, $\hat{f}$ (e.g. monthly) may be compared to f_c for the total sample (e.g. yearly) by summarizing the monthly samples ($\Sigma\hat{f}$) and comparing this with f_c of the total sample. If Σf is greater than the total sample f_c, then an analysis based on accumulated samples (f_c) will underestimate the importance of interactions in the dynamics of the system. If $\Sigma\hat{f}$ is less than f_c, then an analysis based on f_c will overestimate the importance of interactions in the dynamics of the system. If f_c and $\Sigma\hat{f}$ are not significantly different, then small scale heterogeneity may not be an important determinant of community structure. Of course, examination of $\hat{f}$ values is necessary for the recognition of transmission foci and seasonal factors.

A study of three species of larval trematodes in the marine snail, *Littorina littorea*, by Robson and Williams (1970) provided an excellent opportunity for an analysis of the impact of heterogeneity. Three species of trematodes, *Renicola roscovita, Cryptocotyle lingua* and *Hismasthla leptosoma*, were sufficiently common for quantification of $\hat{f}$ and f_c (Table 4.5). For interactions between *R. roscovita* and either of the other two species, the overall calculations were in very close agreement with the summed monthly samples

Table 4.5 Monthly analysis of observed (f_0) and expected ($\hat{f}$) frequencies of double infections of three common larval trematodes, *Renicola roscovita*, *Cryptocotyle lingua* and *Himasthla leptosoma* in the snail, *Littorina littorea*

	Date	N	*R. roscovita and C. lingua*			*R. roscovita and H. leptosoma*			*H. leptosoma and C. lingua*		
			f_0	$\hat{f}$	P	f_0	$\hat{f}$	P	f_0	$\hat{f}$	P
1966	June	104	0	1	--	0	0	--	0	0	--
	July	133	0	4	†	0	1	--	0	2	--
	Aug	216	1	10	†	3	3	NS	0	4	†
	Aug	175	1	6	†	5	3	NS	0	2	--
	Sept	394	2	27	†	5	8	NS	0	7	†
	Nov	313	3	16	†	8	5	NS	0	2	--
	Dec	289	7	9	NS	6	4	NS	0	1	--
1967	Jan	423	4	20	†	8	8	NS	0	1	--
	Feb	304	1	7	†	5	2	NS	0	5	†
	Mar	408	0	6	†	3	2	NS	0	0	--
	Apr	341	3	10	†	5	2	‡	0	0	--
	May	475	3	10	*	9	5	‡	0	1	--
	June	436	2	16	†	14	8	§	0	2	--
	July	568	5	17	†	9	5	‡	0	1	--
	Aug	361	2	8	†	5	3	NS	0	1	--
	Sept	353	2	12	†	7	4	‡	0	2	--
	Oct	353	4	9	*	4	1	§	0	0	--
	Nov	232	7	7	NS	2	1	NS	0	0	--
$\Sigma \hat{f}$		5878	47	195		98	65		0	31	
Overall sample (f_e)		5878	47	197	†	98	65	§	0	27	†

N is number of snails examined. Significant interactions (G-test) are indicated by

* negative interaction, $0.01 < P < 0.05$; † negative interaction, $P < 0.01$;
‡ positive interaction, $0.01 < P < 0.05$; § positive interaction, $P < 0.01$;
NS = no significant interaction, -- = $\hat{f}$ so low that significance could not be determined. To analyse for temporal effects f_e, based on the overal sample, is compared with Σf for the monthly samples.

Data from Robson and Williams, 1970

(differing by 0–1%), indicating that temporal heterogeneity did not affect the frequency of interspecific interactions. For *H. leptosoma* versus *C. lingua*, the monthly samples predicted 15% more negative interactions than did the total sample. The analysis based on the total sample underestimated the frequency of this potentially competitive interaction.

Causes of negative interactions

If sources of heterogeneity do not account for the absence of multiple infections when they are predicted on a random basis, then negative interspecific interactions must result from competitive exclusion at the infracommunity level. The nature of interference interactions between larval trematodes has been revealed using a reciprocal double infection experimental design developed by Lie (1966, 1973). In an extensive series of experiments on freshwater snail–trematode systems, Lie and co-workers (Lie, 1966; 1973; Basch and Lie, 1966; Lie *et al.*, 1968; Lim and Heyneman, 1972; Heyneman *et al.*, 1972) have shown that rediae of many echinostome species fed on and killed rediae and sporocysts of subordinate species. This direct type of interference competition resulted in the elimination of prior infections of subordinate species from snail hosts. Subsequent infections of subordinate species could not become established.

Interference competition has also been reported between species with only sporocyst stages and results in the elimination of a subordinate species by a dominant species (Basch *et al.*, 1969). The actual mechanism of this sort of dominance is unknown, but is considered indirect, because sporocysts, lacking mouths, cannot ingest other competitors. However, the net effect is similar to direct redial antagonism.

Direct and indirect competitive interactions may be either strong or weak. If strong, the dominant species rapidly eliminates the subordinate species. If weak, the subordinate species may persist for some time (Lie, 1973). In the guild of trematodes parasitizing *Cerithidea californica*, the interactions of the heterophyids with the microphallids appear to be strong (Table 4.3); although a total of 453 double infections were expected on a random basis, none were reported. Similarly, echinostomatids and philophthalmids generally engaged in strong negative interactions with other species. In contrast, the interactions between the heterophyids and the strigeid were relatively weak, with 108 double infections being observed (231 being expected).

In several systems, priority of occupancy determines dominance. Lie (1969) has shown that if an infection of the echinostome, *Hypoderaeum dingeri*, is present, then *Echinostoma audyi* cannot become established. In the reciprocal sequence of infection, *E. audyi* prevents the establishment of *H. dingeri*. Other examples of dominance based on priority are given by Lie *et al.* (1966) and Anteson (1970).

Neutral interactions

Using a random infection model for prediction of the number of expected double infections may underestimate the frequency of infracommunity interactions. In certain cases, prior infection with one species may predispose that host to infection by a second species. The latter may coexist with, or be dominant over, the former. Such an effect has been shown for *Echinostoma liei* which may preferentially infect snails previously parasitized by *Schistosoma mansoni* (Heyneman *et al.*, 1972). Suppression of host cellular defense mechanisms by *S. mansoni* accounts for the synergistic increased rate of establishment of echinostomes in previously infected snails (Lie *et al.*, 1976). In this specific instance, dominance of *E. liei* is strong and *S. mansoni* is eliminated in a few weeks (Heyneman *et al.*, 1972). Other examples of enhanced infection rates for dominant species superimposed on subordinate species have been reported (Lie *et al.*, 1973a; Boss, 1977; Jourdane, 1980).

In the analysis of dominance among the guild of trematodes parasitizing *Cerithidea californica*, a conservative approach was adopted. All combinations of trematode for which f_o was not significantly different from f_e were considered neutral associations even though the fitness of one of the species may have been much reduced.

Positive interactions

Obligate secondary invaders are known. Walker (1979) has shown that *Austrobilharzia terrigallensis* in Australia can only parasitize snails harbouring a previous infection of another trematode species. Such species are likely to be found in double infections and would therefore be most prevalent where parasitism by other trematodes is common. *Austrobilharzia* sp. at Upper Newport Bay were often found in double infections and may also be an obligatory secondary invader.

In snails that were secondarily parasitized by *A. terrigallensis*, production of cercariae and size of the rediae of the prior infection were always diminished, even though these other trematodes persisted for the duration of the infection by the schistosome (Walker, 1979). Hence, these obligate interactions are actually negative; *A. terrigallensis* being dominant over the other species.

Whether increased parasitism of previously infected snails is obligate or not, such an effect may yield a more frequent pattern of double infections than expected on a random basis, particularly if dominance is weak. Examination of the relative abundance and condition of larval stages and the quantity of cercariae shed indicated that one species is often partially suppressed. Cercarial production by these species is minimal; dead or damaged larval stages are often seen, and site displacement within the snail is observed (Robson and Williams, 1970; DeCoursey and Vernberg, 1974).

Weak dominance interactions may account for the three combinations in which double infections were more frequent than expected in the random model (Table 4.3). However, a conservative assumption was used for the analysis of interactions in the guild of trematodes parasitizing *Cerithidea californica*. Positive associations were treated as neutral interactions.

Evidence for dominance

None of the trematodes from *Cerithidea californica* are convenient laboratory systems. *C. californica* itself has not been cultured. So, determining competitive dominance using the double infection experimental approach of Lie and his co-workers is precluded here. However, using analogies with Lie's work and relying upon some reasonable inferences, I propose some relatively direct lines of evidence that enable postulation of dominance relationships among trematode species. Further, there are also some biological attributes that indirectly correlate with dominance.

Direct evidence

Snails collected from Carpinteria Salt Marsh were used to provide direct evidence for determination of dominance relationships. They were placed in filtered marsh water and observed shedding cercariae in the laboratory. Cercariae were identified (Martin, 1972). Shells were marked with Martek ink, colour-coded by species of trematode. Snails were then released, collected again on later dates, and re-examined for identification of cercariae. Dominance was inferred if a snail parasitized by a species of trematode became infected with another species. The latter species was inferred to be dominant. Similarly, if a snail having a double infection lost one of the species, the surviving species was inferred to be dominant. Note that no marked *C. californica* parasitized by any of the species of larval trematodes in this study was ever observed to lose the infection (Sousa, 1983; confirmed in the present study). Loss of infection followed by a later new infection cannot account for observations of species replacement.

Additional direct evidence for dominance was obtained by internal examination of snails shedding two species of cercariae. Twenty-one such snails were cracked open. If ingestion of larval stages was observed, the consumer species was considered dominant over the ingested species.

Table 4.6 Summary of criteria to postulate dominance relationships among a guild of larval trematodes

Direct evidence used to infer dominance relationships
1. Dominant trematode species observed eliminating subordinate species in double infections.
2. Dominant trematode species observed infecting a snail having a previous trematode infection.
3. Rediae of dominant trematode species observed ingesting larval stages of subordinate species.

Indirect evidence used to hypothesize dominance relationships when direct evidence is lacking
4. Rediae species are dominant over species with only sporocyst stages.
5. Closely related species (same genus or family) are likely to have similar interactive characteristics
6. Trematode species found in relatively small snails may be subordinate to trematode species found in larger snails.
7. Species with relatively large rediae may be dominant to species with small rediae.
8. Species principally undergoing larval reproduction in the ovotestis may be dominant over species undergoing development in the digestive gland, which may in turn be dominant over species developing in the mantle.
9. If a species in a double infection is displaced from its normal site within the snail it is probably subordinate.

The relatively direct evidence used to infer dominance is listed in Table 4.6. Table 4.7 summarizes the evidence obtained from the mark–recapture study and the examination of snails with double infections. In general, the echinostomes and philophthalmids were dominant over other species. In addition to these observations, Yoshino (1975) noted consumption of *Euhaplorchis californiensis* by *Parorchis acanthus*.

Two inconsistent direct observations were made. Both involved *Echinoparyphium* sp. and either another echinostome or a philophthalmid (Table 4.7). Two explanations could account for these discrepancies. The first possibility is that a trematode replacement was recorded that did not actually occur. This type of error could arise because a methodological problem was encountered in the cercaria identification mark–recapture study. All trematodes did not shed cercariae every day. In particular, snails infected with *Echinoparyphium* sp. shed cercariae reluctantly. So, *Echinoparyphium* sp. may have been present on the first collection date, but may not have released cercariae. The other possibility is that this was a genuine reversal of dominance. A few examples of mixed or incomplete dominance have been reported (Lie, 1973).

Table 4.7 Comparison of direct and indirect evidence for dominance relationships

Species pairs (dominance)	*Direct evidence*		*Indirect evidence*				
	(1 or 2)	(3)	(4, 5)	(6)	(7)	(8)	(9)
HIMA > ECHI	3	2*	=	YES	YES	=	NA
HIMA > EUHA	8	4	YES	YES	YES	=	YES
HIMA > SMST	1	0	YES	YES	NA	YES	NA
HIMA > RENC	2	0	YES	NA	NA	YES	NA
PARO > ECHI	3*	1	=	YES	YES	=	NA
PARO > EUHA	2	3	YES	YES	YES	=	YES
ECHI > EUHA	2	4	YES	NO	YES	=	NA
ECHI > RENC	1	0	YES	NA	NA	YES	NA
ACAN > EUHA	1	0	YES	=	YES	=	NA
EUHA > SMXI	2	0	YES	YES	NA	YES	NA

In the first column, species pairs with significant negative associations (Tables 4.2, 4.3, 4.4) are listed, a > b means a is dominant over b. Numbers in parenthesis refer to Table 4.6. Numbers in direct evidence columns are the number of direct observations consistent with the indicated dominance relationship;

* an observation counter to this dominance relationship was observed.
In the indirect evidence columns YES means agreement with direct evidence, NO means disagreement, = means no prediction can be made according to this criterion, NA means not applicable, data not available or not observed.

Indirect evidence

Based on reports in many different host systems involving different trematode taxa (Lim and Heyneman, 1972; Lie, 1973; Combes, 1982), species of trematodes having rediae were assumed to be dominant over species having only sporocyst intramolluscan stages. Table 4.1 lists developmental patterns for the guild of trematodes in *Cerithidea californica*.

If evidence on dominance status was available for a given species, similar relationships were assumed for closely related species (same genus or family) in the absence of evidence to the contrary. Echinostome trematodes are the best studied taxon with respect to intramolluscan competitive dominance (reviewed in Lim and Heyneman, 1972; Kuris, 1973; Lie, 1973). Although variation exists within the family, they have frequently been shown to be dominant over other trematodes. Likewise, *Schistosoma* is generally an inferior competitor (Kuris, 1973; Combes, 1982). Table 4.1 groups the guild of trematodes in *C. californica* by family.

Trematodes found in relatively small snails were assumed to be subordinate to species found in larger snails. If intramolluscan antagonism was operating in a trematode guild, subordinate species might be replaced by

Table 4.8 Size of infected *Cerithidia californica* at Carpinteria Salt Marsh

Trematode	*N*	*Mean shell length (mm ± s.d.)* *
HIMA	22	29.0 ± 4.0
PARO	14	28.1 ± 2.4
PHOC	8	27.0 ± 3.3
ACAN	8	25.4 ± 2.4
EUHA	177	25.9 ± 2.9
RENB	11	25.3 ± 2.8
ECHI	65	24.9 ± 2.5
STRI	7	23.9 ± 3.4
SMXI	7	22.7 ± 2.1
TOTAL	319	25.9 ± 3.1

* Vertical bar connects homogeneous groups (Sidak's test for inequality, following ANOVA ($F = 6.98$, $P < 0.01$) and Bartlett's test for equal variance (not significant).

dominant species as the snail grew and the time for exposure to infection by a dominant species increased. This effect would be most marked in situations where overall prevalence of trematodes increased with snail size. The sizes of snails parasitized by different trematodes showed significant variation at Carpinteria Salt Marsh (Table 4.8) and at Bolinas Lagoon (Sousa, 1983). Similar size distributions were observed at both these widely separated localities. *Himasthla* and *Parorchis* most often parasitized large snails, *Euhaplorchis* parasitized intermediate-sized snails, and *Echinoparyphium* and the xiphidiocercaria parasitized small snails. Overall prevalence increased dramatically with increasing snail size at both localities (Table 4.9, G-test, $P < 0.01$; Sousa, 1983), suggesting that hosts may become a limiting resource for trematodes that parasitize larger snails.

In general, large organisms eat smaller organisms. Hence, species with larger mature rediae were assumed to be dominant over species with smaller rediae. This is supported by limited empirical evidence. Lie and co-workers examined several pair-wise interactions among trematodes parasitizing two hosts, *Biomphalaria glabrata* and *Lymnaea rubiginosa*. The dominant trematode in the former snail is *Paryphostomum segregatum* (Lie and Basch, 1967). It has very large rediae compared with other echinostome species that parasitize *B. glabrata*. Another strong competitor in that guild, *Echinostoma liei*, has the next largest rediae (Jeyarasasingam *et al.*, 1972). In *L. rubiginosa*, four echinostomes have rediae of similar size and for three of these species, dominance is based on priority of arrival (Lie and

Table 4.9 Relationship between prevalence of all trematode infections and size of *Cerithidia californica* at Carpinteria Salt Marsh

Snail size class (in mm)	*N*	*Prevalence* (%)
16–23	442	18
24–29	471	40
30–36	45	89
all	958	32

Umathevy, 1965a, 1965b; Lie *et al.*, 1965). However, if a trematode species is dominant, even young, small rediae of such a species prey upon mature, larger rediae of a subordinate species. Also, rediae of a dominant species in a double infection may be larger than rediae of that species in a single-species infection (Lie, 1973; Lie *et al.*, 1973b). Donges (1972) has observed that mother rediae of *Isthmiophora melis* are more effective predators than are daughter rediae. Redia size varies considerably for the species of trematodes parasitizing *C. californica* (Table 4.10).

Assuming that the richest and presumably most preferred site within a snail is the gonad, I postulated that dominant species would occupy such a preferred site. Similarly, the digestive gland would be preferred over the mantle wall. All of the trematodes of *C. californica*, for which development

Table 4.10 Size of mature rediae for trematode species in *Cerithidia californica*

Trematodes	*Volume* (Mean length × width^2, mm^3)	*Source*
Echinostomatid/ Philophthalmids		
PARO	0.500	Stunkard and Cable, 1932
HIMA	0.207	Adams and Martin, 1963
ECHI	0.115	Present study
ACAN	0.020	Martin and Adams, 1961
Heterophyids	0.018	Martin, 1950b
STIC		
PHOC	0.008	Martin, 1950c
EUHA	0.004	Martin, 1950a
Notocotylid	0.017	Martin, 1956
CATA		

sites are known, develop primarily in gonadal tissue except the microphallids which develop in the digestive gland (Sarkisian, 1957) and *Renicola* spp. and *Catatropis johnstoni* which develop in the mantle wall (Yoshino, 1975).

If a species in a double infection was displaced from its normal site within the snail, it was probably subordinate. This line of reasoning would be relatively weak if site displacement occurred without a substantial reduction in cercarial production. However, observations on habitat displacement were usually accompanied by reports of sharp reductions in abundance of intramolluscan stages and cercarial output of the displaced species (DeCoursey and Vernberg, 1974). In double infections of *Euhaplorchis californiensis* and the small strigeid, the latter was displaced from the gonad to the digestive gland (Yoshino, 1975). Similarly, *E. californiensis* was displaced in a double infection with *Parorchis acanthus*.

Table 4.6 summarizes the direct and indirect evidence used to support the dominance hypothesis. The direct evidence of intramolluscan competitive dominance observed at Carpinteria is presented in Table 4.10 which also shows that the indirect evidence is almost always consistent with the direct observations (24 consistent observations, 1 inconsistent observation). This concordance supports use of the indirect evidence to erect a dominance hierarchy. Table 4.11 lists the species pairs which occur less frequently than expected, but for which no direct evidence of dominance was obtained. Accordingly, where no direct evidence was available, dominance was postulated on the basis of the indirect criteria.

The dominance hierarchy

Using the direct evidence where available, and the indirect criteria where direct evidence was not available, a dominance hierarchy for the guild of trematodes was constructed (Fig. 4.2). The ecological significance of dominance patterns among a guild of potential interference competitors depends, in part, on whether the interspecific interactions are linear hierarchies or competitive networks in which A dominates B, B dominates C, but C dominates A (Buss and Jackson, 1979). In general, the dominance patterns among the guild of larval trematodes of *Cerithidea californica* are linear. The two most dominant species were a philophthalmid (*Parorchis acanthus*) and an echinostome (*Himasthla rhigedana*). Both have very large rediae and parasitize large snails. Smaller echinostomes and philophthalmids were next most dominant, followed by heterophyids. While negative interactions were common within these groups, information was not available to recognize dominance relationships among the small echinostomes/philophthalmids or among the heterophyids. The most subordinate trematodes were the strigeids, microphallids, and the notocotylid.

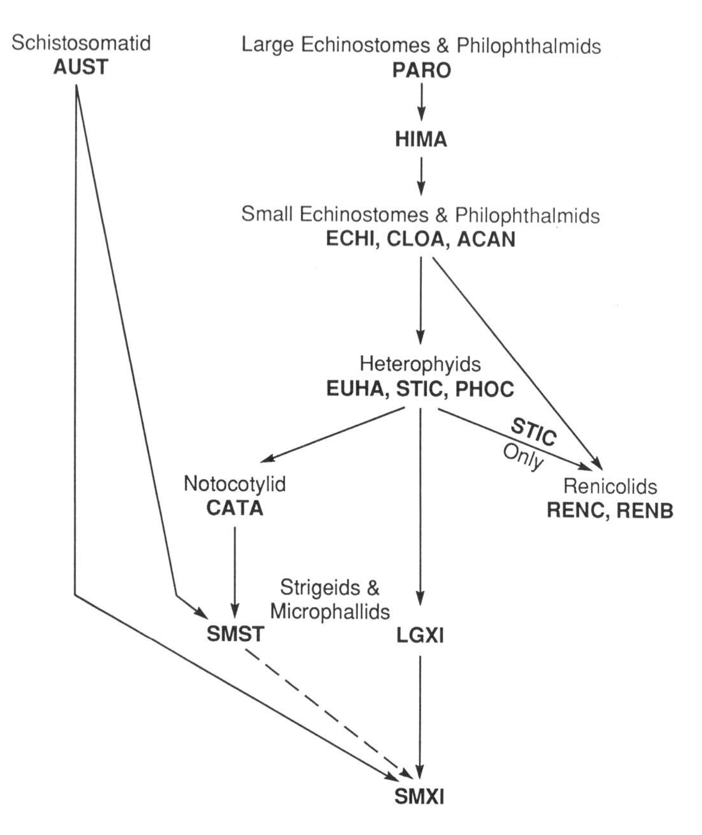

Fig. 4.2 Hypothesized dominance relationship for the guild of larval trematodes parasitizing *Cerithidea californica*. See Table 4.1 for four-letter code to trematode species. Solid arrows represent dominance relationships, all are transitive (all species subordinate to species B are also subordinate to species that dominate species B). The interrupted arrow indicates that the dominance relationship is not transitive (SMST is dominant to SMXI but CATA is not).

Austrobilharzia sp. is interesting. It is effectively a co-dominant with *H. rhigedana* and *P. acanthus* (Fig. 4.2), not being eliminated by either of these species with large rediae. Yet, it is only dominant over the most subordinate species. It is possible that some subordinate species may be protected from the most dominant species, when the subordinate species co-occur with *Austrobilharzia* sp. Of the 23 triple infections recorded by Martin (1955), nine involved *Austrobilharzia*. In two of these cases, a protective influence *Austrobilharzia* sp. is suggested. The only co-occurrence of *Echinoparyphium* sp. with *Euhaplorchis californiensis* (Table 4.2) was in a triple infection with *Austrobilharzia* sp. Similarly, one of only three records of *E. californiensis* co-occurring with *Catatropis johnstoni* was in a triple infection with the schistosome. Direct evidence indicates *Echinoparyphium* sp. to be strongly dominant over *Euhaplorchis californiensis*, and indirect evidence suggests that the latter is dominant over *C. johnstoni*. Thus, *Austrobilharzia* may be blocking or delaying these dominance effects in triple infections. Alternatively, in double infections, *Austrobilharzia* sp. may be slowly suppressing development of these species as has been shown for *A. terrigalensis* (Walker, 1979).

This analysis does not permit a determination of dominance among the small echinostomes and philophthalmids, nor among the heterophyids. Yet, they are strongly interactive (Table 4.2). Dominance based on prior occupancy may be an important interactive mechanism in such cases.

The renicolids do not interact with the subordinate species (Fig. 4.2). They are, however, dominated by most of the species with rediae. Renicolids develop in the mantle tissues. This remote site may provide a partial refuge from interactions with potentially dominant species. Their pattern of passivity and subordinance contrasts with that of *Austrobilharzia* sp., which is passive and dominant.

An interesting non-linear aspect of this dominance hierarchy involves *C. johnstoni*. This trematode is dominant to the small strigeid but not to the small microphallid, although the small strigeid is itself dominant over the small microphallid (Fig. 4.2). Hence, the small microphallid may be protected from the small strigeid when in double infections with *C. johnstoni*. The elements of non-linearity and passivity make this dominance hierarchy one of the most complex proposed.

Estimation of competitive impact

With a postulated dominance hierarchy, it is now possible to estimate the potential impact competitively dominant species have on subordinate species. If A is the number of snails infected with a dominant trematode (species A), B is the number of snails infected with a subordinate trematode (species B), and N is the total number of snails, then $N - A$ is the number of snails that B could potentially parasitize and

$$\frac{B}{N - A} \tag{1}$$

is the proportion of available snails actually infected by species B.

$$\frac{A\left(\frac{B}{N - A}\right)}{B} = \frac{A}{N - A} \tag{2}$$

estimates the proportion of the population of species B that has been lost due to the competitive impact of species A. Note that the proportional

Table 4.11 Dominance relationships postulated on indirect evidence

Indirect evidence					
(4, 5)	(6)	(7)	(8)	(9)	Postulated dominance
=	=	YES	=	NA	PARO > HIMA
YES	NA	YES	YES	NA	PARO > CATA
YES	YES	NA	YES	NA	PARO > SMXI
YES	YES	NA	YES	YES	PARO > SMST
YES	NA	NA	YES	NA	PARO > RENC
YES	YES	NA	YES	NA	HIMA > SMXI
YES	NA	YES	YES	NA	HIMA > CATA
YES	NA	YES	YES	NA	ECHI > CATA
YES	=	NA	YES	YES	ECHI > SMXI
YES	=	NA	YES	NA	ECHI > SMST
YES	YES	NA	=	YES	EUHA > SMST
YES	NA	NO	YES	NA	EUHA > CATA
YES	NA	NA	NA	NA	AUST > SMST
YES	NA	NA	YES	NA	AUST > SMXI
YES	NA	YES	NO	NA	CATA > SMST
=	=	NA	YES	YES	SMXI > SMST

Numbers in parenthesis refer to Table 4.6, symbols as in Table 4.7.

impact of the dominant species on species B is independent of B. Further, the proportional impact can be expressed solely in terms of the prevalence of the dominant, as

$$\frac{A}{N-A} = \frac{\text{prevalence of species A}}{1-\text{prevalence of species A.}} \tag{3}$$

An estimate of L, the actual number of infections of species B that have been lost to dominance effects may be obtained by

$$L = B\left(\frac{\text{prevalence of species A}}{1-\text{prevalance of species A}}\right) - \begin{array}{l}\text{number of observed}\\ \text{double infections of}\\ \text{species B and species A.}\end{array} \tag{4}$$

To estimate the number of infections of species B that have been lost to all dominant species, the total prevalence of all dominant species (corrected for observed double infections among the dominant species) may be substituted for prevalence of A in eq. (4).

Table 4.12 gives the numerical estimates of the effect of dominant species on subordinate species for the guild of larval trematodes parasitizing *Cerithidea californica* at Upper Newport Bay. For the poorest competitors, such as the small microphallid, this effect may be quite considerable.

Although this trematode was quite common at Upper Newport Bay, it appears that it would have been almost twice as abundant in the absence of competition.

4.3 DISCUSSION

Infracommunity interactions

The analysis of interspecific interactions among the guild of larval trematodes parasitizing *Cerithidea californica* suggests some general characteristics that may predict whether a particular parasite community will be interactive, with competition being an important process. At the infracommunity level, competition can be an important structuring force if three conditions are met. Firstly, the overall prevalence of the parasitic assemblage must be high. The quantitive assessment of competition will depend on the relative abundance of dominant and subordinate species. Secondly, a dominance hierarchy or network must exist among the potential competitors. Thirdly, within an individual host, interference competition mechanisms will be more important than exploitative mechanisms. The dominance hierarchy and interference competition conditions will generally be satisfied if potential competitors are limited by resources within a host such that fitness is sharply reduced in multiple infections. Indeed, subordinate species will often die. The evidence for these generalizations will first be reviewed for the *Cerithidea* system and for larval trematode assemblages in other snails. Then, the importance of these conditions will be discussed in the context of the classification of parasitic interactions as parasitoids, parasitic castrators or typical parasites.

For guilds of larval trematodes in snail hosts, high overall prevalence provides a necessary condition for an interactive assemblage. If dominant species have a relatively high prevalence, they can exert competitive pressure sufficient to alter the abundance of subordinate species. Consistent with this, among the four locations where the trematodes of *Cerithidea californica* have been studied, interactions are probably numerically most important at Upper Newport Bay where prevalence is highest. The most abundant species at Upper Newport Bay is *Euhaplorchis californiensis*, a species of intermediate dominance. At Bolinas Lagoon, the dominant echinostomes and philophthalmids are most abundant (Sousa, 1983). Competitive effects should be an important organizational force for the trematode infracommunities at Bolinas. Goleta Slough has the lowest abundance of trematodes (Fig. 4.1), with only *E. californiensis* exceeding 5% in prevalence. Hence, it is not surprising that Yoshino (1975) reported no significant interactions. He also noted that overall prevalence did not increase with snail size, suggesting that hosts did not become a limiting resource.

The dominance relationships exhibited by the guild of trematodes in *Cerithidea californica* may also be similar to trematode associations reported from other snail hosts (e.g. Ewers, 1960; Werding, 1969; Robson and Williams, 1970; Vaes, 1979; Vernberg *et al.*, 1969; DeCoursey and Vernberg, 1974; Combescot-Lang, 1976; Rhode, 1981). Congeners showed similar patterns of co-occurrence. For instance, species of *Himasthla* were present in several studies and were, in general, negatively associated with heterophyids and microphallids, while *Renicola* spp. occurred in double infection at least as frequently as predicted by a random model (Vernberg *et al.*, 1969; Werding, 1969; Robson and Williams, 1970; Combescot-Lang, 1976).

Studies in Malaysia, by Lie and colleagues (Lie *et al.*, 1965, 1966, 1973a, 1973b, 1973c; Basch and Lie, 1966) on a guild of trematodes parasitizing the freshwater snail, *Lymnaea rubiginosa*, provide the most complete experimental information on the nature of interspecific interactions among trematodes in a molluscan host. In a quantitative examination of 1092 snails, no double infections were observed among four common echinostomes (Lie *et al.*, 1966). Figure 4.3 shows the dominance hierarchy I developed based on their experimental observations. As with the guild of trematodes in *Cerithidea californica*, echinostomes are generally dominant. However, the

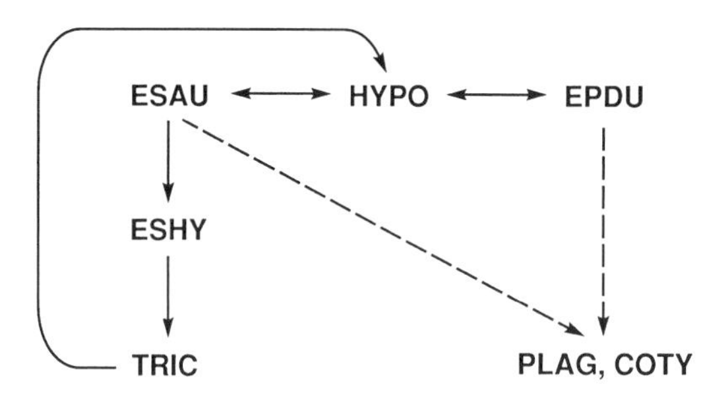

Fig. 4.3 Postulated dominance hierarchy for a guild of trematodes whose interspecific intra-molluscan interactions have been experimentally demonstrated by Lie and colleagues. Double headed arrows indicate that prior occupant is dominant, other notations as in Fig. 4.2.

Species code: ESAU = *Echinostoma audyi*, ESHY = *E. hystricosum*, EPDU = *Echinoparyphium dunni*, HYPO = *Hypoderaeum dingeri*, TRIC = *Trichobilharzia brevis*, PLAG = unidentified plagiorchiid, COTY = *Cotylurus lutzi*. Interactions between HYPO, ESHY and TRIC versus PLAG and COTY have not been reported; hence the known dominance relationships between ESAU and EPDU versus PLAG and COTY are shown as intransitive (interrupted arrows).

structure of the guild of trematodes in *L. rubiginosa* has some interesting functional differences from that of *C. californica*. Three echinostomes, *Echinostoma audyi, Hypoderaeum dingeri* and *Echinoparyphium dunni*, exert dominance based on priority of occupancy. Prior infections prevent the establishment of subsequent parasites. Thus, seasonality of exposure to miracidia will determine which species are effectively dominant among these three echinostomes. Differential ability to infect young snails would establish priority and promote a competitive advantage among these trematodes. Also of interest, a circular competitive network (Buss and Jackson, 1979) has been demonstrated. Degenerating rediae of *H. dingeri*, killed by sporocysts of the schistosomatid, *Trichobilharzia brevis*, are able to protect *T. brevis* from the predatory effects of rediae of *Echinostoma audyi* (Lie, 1973).

As suggested above, a parasite community organization characterized by strong interference competition, and a dominance hierarchy or network, may be anticipated when resources are limited such that fitness of potential competitors would be sharply reduced in multiple (either intraspecific or interspecific) infections. These conditions are often met where site displacement of subordinate species is not possible or cannot alleviate a substantial reduction in fitness.

These conditions are generally satisfied for symbiotic trophic interactions that characterize the relationships between rhizocephalan barnacles and epicaridean isopods and their crustacean hosts, larval trematodes and their first intermediate molluscan hosts, and hymenopterans, dipterans, nematomorphs, and mermithoid and steinernematid nematodes and their insect hosts. All these parasites are parasitoids or parasitic castrators as defined by Kuris (1974). All use a high proportion (often all) of the energy resources of their hosts to complete development (Kuris, 1973, 1974). Thus, they effectively treat the entire host, not just a site within the host, as a resource unit. These parasitoids and parasitic castrators have been shown to engage, at least occasionally, in intense interspecific competition, causing the death or complete suppression of inferior competitors (Fisher, 1962; Kuris, 1974). The overall importance of competition among such parasites thus depends primarily on prevalence. Of the parasitoid and parasitic castrators listed above, only the hymenopterans, dipterans and larval trematodes are often reported as species-rich assemblages with high prevalences sometimes being attained. Competition has often been considered to be important for parasitoid insects (e.g. Ehler, 1978; Force, 1980; Ehler and Hall, 1982) although opposing views are held (Dean and Ricklefs, 1979).

In contrast, guilds of typical parasites, *sensu* Kuris (1974), are unlikely to be structured by dominance hierarchies, interference competition and death of subordinate parasites. For typical parasites, high intensity, as well as high prevalence, is required for competition to contribute to community organization. If competition occurs it is likely to be manifested by site displacement and crowding effects (reduction in size, delayed maturity, reduced fecundity) of subordinate species. Sites within the host are the effective resource units.

Component community organizations

Sousa (Chapter 3) reported a similar, highly interactive infracommunity of larval trematodes in populations of *Cerithidea californica* from Bolinas Lagoon. He demonstrated considerable heterogeneity in trematode species composition from spatially separated snail populations. Subordinate competitors were more likely to parasitize small snails (see also Table 4.8). He sought evidence for competition at the component level by comparing a battery of species diversity indices over snail size for numerous snail populations. If trematode species diversity were to increase in intermediate size snails but decrease in the largest snails, then strong evidence for competition would be demonstrated. His data do not support such a pattern. In nearly all populations, trematode species diversity continued to increase (or plateaued) with increasing snail size. This analysis is fully consistent with the evidence for interaction and the quantification of competitive impact presented herein. Carpinteria Salt Marsh species diversity also generally increases with increasing snail size (Kuris, unpublished information). What the analyses of diversity indices show are that, in this highly interactive system, dominant species are generally not able to sequester most of the hosts even in the largest snail size classes. The limiting case of resource monopolization is not reached, even when most of the large hosts are parasitized (Table 4.9). Note that the hypothesized change in community structure (diversity) indices would be sufficient evidence for an interactive component community. However, these diversity indices, lacking information on species composition, are insensitive indicators of competitive interactions (see also Bouton *et al.*, 1980). They could not detect interactions sufficient to eliminate 30–40% of the populations of some subordinate species (Table 4.12). With Sousa (Chapter 3), I note that considerably more information is needed on demographic aspects (snail mortality) and infection dynamics from the perspective of hosts as patches, before interactive effects can be fully quantified.

It might be argued that the importance of competition in guilds of parasitic castrators or parasitoids is scale-dependent. It may be important locally, but on a global scale, the system is non-interactive, being driven by stochastic factors. Consistent with this, at an intermediate scale, there is considerable variation between sites within a salt marsh (Sousa, 1983 and Chapter 3;

Table 4.12 Estimation of effect of interspecific competition on subordinate trematode species parasitizing *Cerithidia californica* (Upper Newport Bay)

Subordinate sp.	*Number of dominant spp.*	*Total prevalance of dominant spp.**	*Number of infections of the subordinate sp.*		
			Present (observed)	*Lost to competition† [L, estimated using eq. (4)]*	*% lost*
SMXI	11	0.446	1360	1068	44
SMST	10	0.333	1059	416	28
LGXI	8	0.309	716	320	31
CATA	8	0.309	204	88	30
RENC	6	0.130	780	106	12
EUHA	5–7‡	0.090–0.135	2261	224–354	9–14
ECHI	2–4‡	0.058–0.065	331	20–23	6

* Corrected for observed double infections among dominant spp.
† Corrected for observed double infections of subordinate sp. with dominant spp.
‡ Dominance could not be evaluated within heterophyid and small echinostome/philophthalmid categories.

Kuris, unpublished observation). Further, individuals of *Cerithidea californica* usually remain at one location for several years (Sousa, 1983). However, the impact of this variation on the overall dynamics of the system is strongly dependent on the vagility of second intermediate and definitive hosts. If, at any life-cycle stage of the trematodes, the transmission of parasites to their next host (second intermediate host, definitive host or *C. californica*) is conducted at a larger scale (the marsh or a large portion of the marsh), then the component community of trematodes in *C. californica* will be integrated over these larger scales. Hence, while *C. californica* behaves as though the habitat was fine-grained, the other hosts reduce or eliminate these patchy effects by treating the habitat as coarse-grained. Second intermediate hosts (often crabs or fishes (references in Martin, 1972)) may be particularly important in this respect, distributing the trematodes much more widely and integrating many patches of *C. californica* into interactively larger units.

Variation in space is also relatively unimportant at large scales. On a geographic scale, membership in the guild of trematodes parasitizing *Cerithidea californica* is generally similar, and overall prevalence is usually high. Hence, competitive effects on the subordinate species appear to be widespread. The chief difference between locations is that a different suite of competitively dominant species may exert competitive pressure on subordinate components of the guild. In contrast, for other guilds of larval trematodes there may be considerable variation in prevalence between geographic locations (Robson and Williams, 1970; Loker *et al.*, 1981).

Samples over time from the same locality also do not account for appreciable heterogeneity in the extent of interspecific interactions as most of the trematodes of *C. californica* and *Littorina littorea* seem to have generally similar patterns of temporal fluctuations (Martin, 1955; Robson and Williams, 1970; Yoshino, 1975). This may be attributed to the similar seasonal behaviour of the vertebrate hosts, which in the cases analysed here, are migratory water birds. Other systems may well differ in this respect.

Some implications for compound communities

For trematodes, having complex life cycles, the nature of the trophic interaction between host and parasite changes. Larval stages in the first intermediate (molluscan) host are parasitic castrators, while larval stages in second intermediate hosts, and adults in definitive hosts, are typical parasites (Kuris, 1974). If competitive interactions in the molluscan host are important, this may have a cascading effect on the structure of parasitic assemblages involving trematodes in second intermediate and definitive hosts.

In particular, if subordinate parasites in the molluscan host are relatively pathogenic in subsequent hosts, their intensity-dependent pathogenicity may be significantly diminished by interactions with dominant trematodes in the molluscan host. Most noteworthy are compound communities involving human schistosomes. Schistosomes are often subordinate to other trematodes in intra-molluscan interactions (Lim and Heyneman, 1972). Suprapopulations (Margolis *et al.*, 1982) of these pathogenic parasites may be under partial control by competitively dominant echinostomes, cathaemasiid and paramphistome trematodes in the snail hosts (Kuris, 1973; Lie, 1973; Combes, 1982). Schistosome prevalence has been negatively associated with high prevalence of these other trematodes (Bayer, 1954; El-Gindy, 1965; Chu *et al.*, 1972; Loker *et al.*, 1981; Pointier *et al.*, 1985).

Severe habitat disturbance, often associated with increasing human populations, may reduce or eliminate vertebrate hosts for trematodes. The larval stages of such trematodes may have provided partial natural (biological) control of schistosome suprapopulations through competitive effects in snail intermediate hosts. These interactions may have important implications for the control of trematodes of medical and veterinary importance. Thus, appropriate habitat manipulations, such as wildlife restoration (Bayer, 1954; Gordon and Boray, 1970) and the use of other larval trematodes as biological control agents (Lie, 1973; Kuris, 1973; Nassi *et al.*, 1979; Combes, 1982) may be useful control measures.

The potential importance of interactions between larval trematodes on community structure of these parasites in the vertebrate hosts emphasizes the need for study of host–parasite systems at the suprapopulation and compound community levels. For species-rich assemblages, we now have detailed analyses of larval trematodes in snails (e.g. Robson and Williams, 1970; Sousa, 1983 and Chapter 3; Kuris, present study) and of several vertebrate systems (e.g. Bush and Holmes, 1986; Kennedy *et al.*, 1986; Goater *et al.*, 1987). However, there are, as yet, few studies that integrate these two (or three) spatially, temporally and physiologically separated, but necessarily linked, parts of a compound community. Such challenging studies may disclose whether biotic interactions among larval stages structure compound communities with respect to parasite abundance. Analyses of hosts for only one parasitic life-cycle stage (be it larval or adult) may not reveal significant structuring interactions, or at best, only provide information on fine-tuning (e.g. site specificity, crowding effects).

This emphasis on scale, and on interactions between the host population components of compound parasite communities, parallels the recent upsurge in interest on scale and coupling of spatially separated sites in marine ecosystems (Roughgarden *et al.*, 1988). Parasites offer a distinct advantage in such studies as the habitat units (hosts) are discrete and readily sampled at all scales by similar methodologies.

ACKNOWLEDGEMENTS

I thank D. Heyneman, Lie Kian Joe, H.-K. Lim and U. Jeyarasasingam for introducing me to the fascinating world of larval trematode interactions; J. D. Shields, K. Lafferty, U. Kitron, C. Ashbaugh, J. Aho, W. Sousa, G. Esch and R. R. Warner for their useful criticisms and comments, and W. Sousa for sharing information during the course of his studies on parasites of *Cerithidea*.

REFERENCES

Abbott, D. P. and Haderlie, E. C. (1980) Prosobranchia: marine snails. In *Intertidal Inverterbrates of California* (eds, R. H. Morris, D. P. Abbott and E. C. Haderlie), Stanford University press, Stanford, California, pp. 230–307.

Adams, J. E. and Martin, W. E. (1963) Life cycle of *Himasthla rhigedana* Dietz, 1909 (Trematoda: Echinostomatidae). *Trans. Am. Micros. Soc.*, **82**, 1–6.

Anteson, R. K. (1970) On the resistance of the snail, *Lymnaea catascopium pallida* (Adams) to concurrent infection with sporocysts of the strigeid trematode, *Cotylurus flabelliformis* (Faust) and *Diplostomum flexicaudum* (Cort and Brooks). *Ann. Trop. Med. Parasitol.*, **64**, 101–7.

Askew, R. R. and Shaw, M. R. (1986) Parasitoid communities: their size, structure and development. In *Insect Parasitoids*, (eds, J. Waage and D. Greathead), Academic Press, London, pp. 225–64.

Basch, P. F. and Lie, K. J. (1966) Infection of single snails with two different trematodes. I. Simultaneous exposure and early development of a schistosome and an echinostome. *Z. Parasitenk.*, **27**, 252–9.

Basch, P. F., Lie, K. J. and Heyneman, D. (1969) Antagonistic interaction between strigeid and schistosome sporocysts within a snail host. *J. Parasitol.*, **55**, 753–8.

Bayer, F. A. A. (1954) Larval trematodes found in some freshwater snails: a suggested method of bilharzia control. *Trans. Roy. Soc. Trop. Med. Hyg.*, **48**, 414–18.

Boss, J. M. (1977) Synergism between *Schistosoma mansoni* and *Echinostoma paraensei* in the snail *Biomphalaria glabrata*. PhD thesis, University of California, Berkeley, 151 p.

Bouton, C. E., McPheron, B. A. and Weis, A. E. (1980) Parasitoids and competition. *Am. Nat.*, **116**, 876–81.

Bush, A. O. and Holmes, J. C. (1986) Intestinal helminths of lesser scaup ducks: an interactive community. *Can. J. Zool.*, **64**, 142–52.

Buss, L. W. and Jackson, J. B. C. (1979) Competitive networks: nontransitive competitive relationships in cryptic coral environments. *Am. Nat.*, **113**, 223–34.

Castilla, J. C. and Paine, R. T. (1987) Predation and community organization on Eastern Pacific, temperature zone, intertidal shores. *Rev. Chilena Hist. Nat.*, **60**, 131–51.

Choe, J. C. and Kim, K. C. (1987) Community structure of arthropod ectoparasites on Alaskan seabirds. *Can. J. Zool.*, **65**, 2998–3005.

Chu, K. Y., Dawood, I. K. and Nabi, H. A. (1972) Seasonal abundance of trematode cercariae in *Bulinus truncatus* in a small focus of schistosomiasis in the Nile Delta. *Bull. W.H.O.*, **47**, 420–2.

Combes, C. (1982) Trematodes: antagonism between species and sterilizing effects on snails in biological control. *Parasitology*, **84**, 151–75.

Combescot-Lang, C. (1976) Etude des trématodes parasites de *Littorina saxatilis* (Olivi) et de leurs effets sur cet hôte. *Ann. Parasitol.*, **51**, 27–36.

Dean, J. M. and Ricklefs, R. E. (1979) Do parasites of Lepidoptera larvae compete for hosts? No! *Am. Nat.*, **113**, 302–6.

DeCoursey, J. and Vernberg, W. B. (1974) Double infections of larval trematodes: competitive interactions. In *Symbiosis in the Sea* (ed. W. B. Vernberg), University of South Carolina Press, Columbia, S.C., pp. 93–109.

Donges, J. (1972) Double infection experiments with echinostomatids (Trematoda) in *Lymnaea stagnalis* by implantation of rediae and exposure to miracidia. *Intern. J. Parasitol.*, **2**, 409–23.

Ehler, L. E. (1978) Competition between two natural enemies of Mediterranean black scale on olive. *Environ. Entomol.*, **7**, 521–3.

Ehler, L. E. and Hall, R. W. (1982) Evidence for competitive exclusion of introduced natural enemies in biological control. *Environ. Entomol.*, **11**, 1–4.

El-Gindy, M. S. (1965) Monthly prevalence rates of natural infection with *Schistosoma haematobium* cercariae in *Bulinus truncatus* in Central Iraq. *Bull. Endem. Dis.*, **7**, 11–31.

Fisher, R. C. (1962) The effect of multiparasitism on populations of two parasites and their host. *Ecology*, **43**, 314–16.

Force, F. D. (1980) Do parasitoids of Lepidoptera larvae compete for hosts? Probably! *Am. Nat.*, **116**, 873–5.

Goater, T. M., Esch, G. W. and Bush, A. O. (1987) Helminth parasites of sympatric salamanders: ecological concepts of infracommunity, component and compound community levels. *Am. Midl. Nat.*, **118**, 289–300.

Gordon, H. McL. and Boray, J. C. (1970) Controlling liver fluke: a case for wildlife conservation? *Vet. Rec.*, Mar., **7**, 288–9.

Heyneman, D., Lim, H. K. and Jeyarasasingam, U. (1972) Antagonism of *Echinostoma liei* (Trematoda: Echinostomatidae) against the trematodes *Paryphostomum segregatum* and *Schistosoma mansoni*. *Parasitology*, **65**, 223–33.

Holliman, R. B. (1961) Larval trematodes from the Apalachee Bay area, Florida, with a checklist of known marine cercariae arranged in a key to their superfamilies. *Tulane Stud. Zool.*, **9**, 2–74.

Jeyarasasingam, U., Heyneman, D., Lim, H. K. and Mansour, N. (1972) Life cycle of a new echinostome from Egypt, *Echinostoma liei* sp. n. (Trematoda: Echinostomatidae). *Parasitology*, **65**, 203–22.

Jourdane, J. (1980) Interference by *Schistosoma mansoni* with the natural resistance to *Echinostoma togoensis* in *Biomphalaria glabrata*. *Proc. 3rd Euro. Multicolloquium Parasit.* Cambridge, p. 39.

Kennedy, C. R. Laffoley, D. d'A, Bishop, J. *et al.* (1986) Communites of parasites of freshwater fish of Jersey, Channel Islands. *J. Fish. Biol.*, **29**, 215–26.

Kuris, A. M. (1973) Biological control: implication of the analogy between the tropic interactions of insect pest–parasitoid and snail–trematode systems. *Exp. Parasitol.*, **33**, 365–79.

Kuris, A. M. (1974) Trophic interactions: similarity of parasitic castrators to parasitoids. *Q. Rev. Biol.*, **49**, 129–48.

Lie, K. J. (1966) Antagonistic interaction between *Schistosoma mansoni* sporocysts and echinostome rediae in the snail *Australorbis glabratus. Nature (Lond.)*, **211**, 1213–15.

Lie, K. J. (1969) A possible biological control of schistosomiasis and other trematodes in snails. In *Proc. Seminar on Trematode Diseases* 4th Southeast Asian Regional Meeting on Parasitology and Tropical Medicine, Bangkok, pp. 131–41.

Lie, K. J. (1973) Larval trematode antagonism: principles and possible application as a control method. *Exp. Parasitol.*, **33**, 343–9.

Lie, K. J. and Basch, P. F. (1967) The life history of *Paryphostomum segregatum* Dietz, 1909. *J. Parasitol.*, **53**, 280–86.

Lie, K. J., Basch, P. F. and Heyneman, D. (1968) Direct and indirect antagonism between *Paryphostomum segregatum* and *Echinostoma paraensei* in the snail *Biomphalaria glabrata. Z. Parasitenk.*, **31**, 101–7.

Lie, K. J., Basch, P. F. and Umathevy, T. (1965) Antagonism between two species of larval trematodes in the same snail. *Nature (Lond.)*, **206**, 422–3.

Lie, K. J., Basch, P. F. and Umathevy, T. (1966) Studies on Echinostomatidae (Trematoda) in Malaya. XII. Antagonism between two species of echinostome trematodes in the same lymnaeid snail. *J. Parasitol.*, **52**, 454–7.

Lie, K. J., Heyneman, D. and Jeong, K. H. (1976) Studies on resistance in snails. 7. Evidence of interference with the defense reaction in *Biomphalaria glabrata* by trematode larvae. *J. Parasitol.*, **62**, 608–15.

Lie, K. J., Lim, H. K. and Ow-Yang, C. K. (1973a) Synergism and antagonism between two trematode species in the snail *Lymnaea rubiginosa. Intern. J. Parasitol.*, **3**, 719–33.

Lie, K. J., Lim, H. K. and Ow-Yang, C. K. (1973b) Antagonism between *Echinostoma audyi* and *Echinostoma hystricosum* in the snail *Lymnaea rubiginosa* with a discussion on patterns of trematode interaction. *Southeast Asian J. Trop. Med. Pub. Hlth.*, **4**, 504–8.

Lie, K. J., Lim, H. K. and Ow-Yang, C. K. (1973c) The pattern of antagonism between *Echinostoma hystricosum* and *Hypoderaeum dingeri* in the snail host. *Southeast Asian J. Trop. Med. Pub. Health.*, **4**, 596–7.

Lie, K. J. and Umathevy, T. (1965a) Studies on Echinostomatidae (Trematoda) in Malaya. VIII. The life history of *Echinostoma audyi*, sp.n. *J. Parasitol.*, **51**, 781–8.

Lie, K. J. and Umathevy, T. (1965b) Studies on Echinostomatidae (Trematoda) in Malaya. X. The life history of *Echinoparyphium dunni* sp.n. *J. Parasitol.*, **51**, 793–9.

Lim, H. K. and Heyneman, D. (1972) Intramolluscan intertrematode antagonism: a review of factors influencing the host–parasite system and its possible role in biological control. *Adv. Parasitol.*, **10**, 191–268.

Loker, E. S., Moyo, H. G. and Gardner, S. L. (1981) Trematode–gastropod associations in nine non-lacustrine habitats in the Mwanze region of Tanzania. *Parasitology*, **83**, 381–99.

Margolis, L., Esch, G. W., Holmes, J. C. *et al.* (1982) The use of ecological terms in parasitology (report of an *ad hoc* committee of the American Society of Parasitiologists). *J. Parasitol.*, **68**, 131–3.

Martin, W. E. (1950a) *Euhaplorchis californiensis* n.g., n.sp., Heterophyidae, Trematoda, with notes on its life cycle. *Trans. Amer. Micros. Soc.*, **69**, 194–209.

Martin, W. E. (1950b) *Parastictodora hancocki*, n.gen., n.sp. (Trematoda: Heterophyidae), with observations on its life cycle. *J. Parasitol.*, **36**, 360–70.

Martin, W. E. (1950c) *Phocitremoides ovale* n.g., n.sp., (Trematoda: Opisthorchiidae), with observations on its life cycle. *J. Parasitol.*, **36**, 552–8.

Martin, W. E. (1955) Seasonal infections of the snail, *Cerithidea californica* Haldeman, with larval trematodes. *Essays Nat. Sci. Honor of Capt. A. Hancock*, 203–10.

Martin, W. E. (1956) The life cycle of *Catatropis johnstoni* n.sp. (Trematoda: Notocotylidae). *Trans. Amer. Micros. Soc.*, **75**, 117–28.

Martin, W. E. (1972) An annotated key to the cercariae that develop in the snail *Cerithidea californica. Bull. South. Calif. Acad. Sci.*, **71**, 39–43.

Martin, W. E. and Adams, J. E. (1961) Life cycle of *Acanthoparyphium spinulosum* Johnston, 1917 (Echinostomatidae: Trematoda). *J. Parasitol.*, **47**, 777–82.

Nassi, H., Pointer, J. P. and Golvan, Y. J. (1979) Bilan d'un essai de contrôle de *Biomphalaria glabrata* en Guadeloupe à l'aide d'un Trématode stérilisant. *Ann. Parasitol.*, **54**, 185–92.

Pointier, J.-P., Théron, A. and de Vathaire, F. (1985) Correlations entre la présence du Mollusque *Biomphalaria glabrata* et quelques paramètres du milieu dans les mares de la Grand Terre de Guadeloupe (Antilles Françaises). *Malacol. Rev.*, **18**, 37–49.

Price, P. W. (1973) Parasitoid strategies and community organization. *Environ. Entomol.*, **2**, 623–6.

Robson, E. M. and Williams, I. C. (1970) Relationships of some species of Digenea with the marine prosobranch *Littorina littorea* (L.) I. The occurrence of larval Digenea in *L. littorea* on the North Yorkshire Coast. *J. Helminthol.*, **44**, 153–68.

Rohde, K. (1981) Population dynamics of the two snail species. *Planaxis sulcatus* and *Cerithium moniliferum*, and their trematode species at Heron Island, Great Barrier Reef. *Oecologia*, **49**, 344–52.

Root, R. B. (1973) Organization of a plant–arthropod association in simple and diverse habitats: the fauna of collards (*Brassica oleracea*). *Ecol. Monogr.*, **43**, 95–124.

Roughgarden, J., Gaines, S. and Possingham, H. (1988) Recruitment dynamics in complex life cycles. *Science*, **241**, 1460–66.

Sarkisian, L. N. (1957) *Maritrema uca*, new species (Trematoda: Microphallidae), from the fiddler crab *Uca crenulata* (Lockington). *Wasmann J. Biol.*, **15**, 35–48.

Schmidt, G. D. and Roberts, L. S. (1985) *Foundations of Parasitology*, Times Mirror/Mosby, St. Louis.

Sousa, W. P. (1983) Host life history and the effect of parasitic castration on growth: a field study of *Cerithidea californica* Haldeman (Gatropoda: Prosobranchia) and its trematode parasites. *J. Exp. Mar. Biol. Ecol.*, **73**, 273–96.

Stunkard, H. W. and Cable, R. M. (1932) The life history of *Parorchis avitus* (Linton) a trematode from the cloaca of the gull. *Biol. Bull.*, **62**, 328–38.
Vaes, M. (1979) Multiple infection of *Hydrobia stagnorum* with larval trematodes. *Ann. Parasitol.*, **54**, 303–12.
Vernberg, B., Vernberg, F. and Beckerdite, F. Jr (1969) Larval trematodes: double infections in the common mud-flat snail. *Science*, **164**, 1287–8.
Walker, J. C. (1979) *Austrobilharzia terrigalensis*: a schistosome dominant in interspecific interactions with the molluscan host. *Intern. J. Parasitol.*, **9**, 137–40.
Werding, B. (1969) Morphologie, Entwicklung und Ökologie digener Trematoden–Larven der Strandschnecke, *Littorina littorea*. *Mar. Biol.*, **3**, 306–33.
Yoshino, T. P. (1975) A seasonal and historical study of larval Digenea infecting *Cerithidea californica* (Gastropoda: Prosobranchia) from Goleta Slough, Santa Barbara County, California. *Veliger*, **18**, 156–61.

5

Helminth communities in marine fishes

John C. Holmes

5.1 INTRODUCTION

The literature on parasites of marine fishes is voluminous, diverse, and scattered. Although a substantial portion of that literature is descriptive (either taxonomic or survey), many surveys are restricted to one taxon (or few taxa), and the parasite fauna of only a small fraction of the marine fishes can be regarded as well-known. Most fishes with well-known faunas are economically important species of north temperate seas; parasites have been studied largely to elucidate their effects on fish populations (Lester, 1984; Sindermann, 1987), their zoonotic potential (Williams and Jones, 1976; Oshima and Kliks, 1987), or (especially recently) their usefulness as biological tags (MacKenzie, 1983, 1987; Sindermann, 1983). Studies on life histories of marine parasites have progressed to the point that basic patterns of the life cycles of most major groups have been determined, but there is relatively little information on the variation in details of those patterns, and specific life cycles are known for only a minute fraction of parasitic species. Knowledge of population dynamics, and especially processes which determine those dynamics, is rudimentary.

Despite the imperfections in our knowledge, ecological studies of marine parasites have progressed enough to be the topic of an excellent book (Rohde, 1982). Rohde provides reviews of three subjects particularly important to community ecology: the significance of cleaning symbioses, ecological niches and factors restricting niches, and zoogeographic patterns. This book, along with earlier ones by Dogiel *et al.* (1961) and Dogiel (1964), should be required reading for anyone working in parasite ecology.

One feature is abundantly clear from a perusal of the literature: there are essentially no studies presenting data at the infracommunity level. There are many studies (e.g. MacKenzie and Gibson, 1970; Holmes, 1971; Shotter, 1976; Rohde, 1977a; McVicar, 1979) which address questions appropriate to that level (such as microhabitat [site] occupation, microhabitat restriction, and complementarity of microhabitats in related species), but all are approached using data summed across individual hosts. Kennedy (1985) has clearly demonstrated that patterns apparent in summed data may have little relevance to patterns at the individual infracommunity level, and patterns apparent from analyses at the infracommunity level in several studies from our laboratory (e.g. Bush and Holmes, 1986b; Stock and Holmes, 1988) are not necessarily apparent in summed data.

Numerous studies (including those cited above) have demonstrated that marine parasites are limited in their distributions within individual fish. Factors implicated have been morphological specificity to the substrate (Llewellyn, 1956; Williams, 1960); physiological specificity (Uglem and Beck, 1972 [in freshwater fishes]); response to physical forces, such as water currents (Kabata, 1959, as an active selection of suitable currents; Wootten, 1974, [in freshwater fishes] as a passive source of infective agents) or disturbance by solid food particles (McVicar, 1979); concentration of individuals to facilitate cross-fertilization (Rohde, 1977b); mutualism, or habitat modification by another species (Shostak and Dick, 1986 [in freshwater fishes]); and niche segregation to avoid hybridization (Sogandares-Bernal, 1959) or to avoid competition (MacKenzie and Gibson, 1970). Only the last four directly invoke selection pressures associated with other species, thus infracommunity-level processes.

Several studies have investigated relationships between population density and habitat occupation. In some cases, densities appear to be regulated; these cases involve both restricted habitats (e.g. gall bladder, Burn, 1980; Beverley-Burton and Early, 1982) and those with extensive space (spiral intestine of holocephalans, Williams *et al.*, 1987). In most others, the site occupied broadens with increases in the parasite's population (see review in Rohde, 1979). These differences would obviously affect other infracommunity processes.

A second feature of the literature is that there are relatively few studies at the compound community level. Most of these have investigated parasites of deep-sea fishes, and have contrasted the parasites of benthic and pelagic species. Several patterns have emerged (see reviews by Noble, 1973; Campbell, 1983): parasite abundance is correlated with the abundance of the free-living fauna, parasite abundance and diversity in deep faunas are not very different from those in shallow faunas (except for major decreases at great depths), benthic feeders generally have more parasites than pelagic or benthopelagic forms, and the composition of the parasite community in different species of fishes depends largely on their diet. Houston and Haedrich (1986) have used parasite data to demonstrate that a high proportion of deep demersal fishes feed extensively on pelagic and benthopelagic prey, as well as on benthic forms.

The communities of parasites in coastal fishes have also received some attention. Polyanski (1961a) noted that littoral fishes have a characteristic fauna; a number of species are limited to that zone, and other species are precluded from that zone, by the distributions of their intermediate hosts.

Component communities have received by far the most attention. There appear to be two major thrusts: attempts to elucidate significant relationships between some aspect of the community (e.g. richness) and host or environmental factors, and attempts to determine spatial patterns in component communities (and to use those patterns as biological markers).

The first comprehensive review of the influences of host factors on parasite communities in marine fishes was Polyanski (1961a). He suggested that the 'main factors determining the variety of the parasite fauna, as well as the intensity and incidence of infestation' were: (a) the diet of the host, (b) the life span of the host, (c) the mobility of the host throughout its life, including the variety of habitats it encounters in its life, (d) its population density (or 'gregariousness'), and (e) the size attained (large hosts provide more habitats suitable for parasites than do small ones). Polyanski provided examples demonstrating the significance of each of these factors (and the fact that they are often correlated – fishes that attain large size often have long life spans; gregarious fish are often migratory, moving through a variety of habitats, and feeding on different foods in each). Other workers have since added many other examples for (and against) each of these patterns, but the only quantitative tests I am aware of are those of George-Nascimento (in preparation), who finds general support for some, but not all, of them.

Currently, one of the major reasons for survey work on parasites of marine fishes is the search for biological markers – parasites that can reveal whether or not local stocks exist (e.g. Kabata *et al.*, 1988), from what stock a particular sample has been taken (Margolis, 1963, 1982), or the migration routes taken by particular stocks (Bailey *et al.*, 1988). Most of the attention has been directed toward the use of the presence (or abundance) of individual species of parasites (see review by MacKenzie, 1987; Sindermann, 1987), but the population structure of specified species has also been used (Leaman and Kabata, 1987). Community characteristics have been used much less frequently (but see Arthur and Arai, 1980; Bailey *et al.*, 1988). Almost all of these studies have demonstrated geographic differences in at least some elements of the parasite communities.

Current questions

In a recent paper contrasting helminth communities in freshwater fishes with those in aquatic birds and mammals, Kennedy *et al.* (1986) concluded that the latter were richer and more diverse. They suggested that the greater complexity of the gastrointestinal tract, the greater energy requirements of endothermy, the greater vagility, and the broader diets of the birds and mammals may be responsible. They suggested that communities in marine fishes would be particularly interesting due to the greater vagility of marine fishes (as compared to freshwater fishes), and the greater diversity of invertebrates in the sea. That paper led me to reactivate an earlier study of the parasites of shallow water rockfishes (genus *Sebastes*) of the Pacific coast. In the remainder of this chapter, I will present data on gastrointestinal helminths of one common species, *Sebastes nebulosus*, then examine data from the literature on the helminths of other species of *Sebastes* (and other coastal marine fishes) to test whether or not patterns found in the *S. nebulosus* data are representative of those in data from other marine fish.

Sebastes nebulosus was chosen because adults are abundant on isolated rocky reefs in shallow coastal waters, where they occupy territories on the reefs, generally in water less than 30 m deep. They feed predominantly on crabs, brittle stars and other benthic prey with limited vagility, but include a wide variety of other organisms, including fishes, in their diet (personal observation). These characteristics would be expected to facilitate cycling of helminths within a reef (leading to high populations of individual species of helminths), to reduce the probability of movement of parasites among reefs (leading to differentiation of helminth communities on different reefs), and to facilitate the occasional ingestion of a wide variety of larval helminths.

Although *S. nebulosus* have a wide variety of parasites located in varied sites in the body, this analysis is limited to helminths of the gastrointestinal tract, the portion of the parasite community that was discussed by Kennedy *et al.* (1986). My analysis will focus primarily on the component community level, using infracommunities as replicate samples, as in Kennedy *et al.* I will focus on an analysis of the richness, diversity, and predictability of the communities, and on the extent to which they differ on different reefs.

5.2 MATERIALS AND METHODS

Sebastes nebulosus were collected from six rock reefs (Miller Reef, Folger Island, Seapool Rocks, Sandford Reef, Adamson Rocks, and Tyler Rock) in Barkley Sound, on the west coast of Vancouver Island, in the vicinity of the Bamfield Marine Station. Fish were caught by angling. Immediately on capture, each fish was killed by a blow on the head, weighed, and its fork length measured; it was then eviscerated and the gastrointestinal tract quick frozen in absolute ethanol, cooled with solid carbon dioxide. Frozen tracts were returned to the laboratory and examined using modifications of the methods of Bush and Holmes (1986a). The stomach was divided into three segments, each pyloric caecum was examined separately, and the intestinal tract (including rectum) was divided into ten segments. (Data on the microhabitat distributions of the parasites will be presented elsewhere; these methods are used herein only to define 'empty segments'.) All parasites were individually identified and enumerated. Only those helminths living within the gastrointestinal tract were used; larval stages were found encysted on the serosa, but were not included. Because this analysis of the data is a preliminary one, I have not presented the data as such, only the results of specified analyses. The complete data set will be published in the periodical literature.

Parasites from each fish are considered to be an infracommunity. All fish taken from the same reef in the same year are considered to be a sample of the population of fish there at that time. Infracommunities in the sampled fish are considered to be samples of the component community of parasites found in that population.

I have used the number of species (*S*), the number of individual worms (*N*), the proportion of worms belonging to the most numerous helminth species (dominance) and the number of unoccupied (empty) segments as measures of the richness of each infracommunity. To measure diversity, I have used Simpson's index (SI; expressed as $1 - \mathrm{SI}$, to make higher values indicate a more diverse community). This index emphasizes more common species and gives little weight to rare ones (Peet, 1974), attributes which I consider desirable for characterizing parasite communities. To examine predictability, I have used the number of core species (used as in Bush and Holmes, 1986a), and two measures of similarity: Jaccard's index (qualitative similarity, based on simple presence/absence) and per cent similarity (quantitative similarity, based on proportions of each species). Where warranted by the number of communities in the analysis, similarities have been analysed using ten randomly selected pairs.

Most data presented here (like most parasitological data) are non-normal in distribution; I have therefore summarized them using ranges, medians, and interquartile ranges. Because the intent is to examine data for broad patterns, rather than in detail, I have generally not presented results of statistical analyses. When statistics were needed, I used non-parametric methods.

Surveys of parasites of marine fishes are too abundant to cover in this presentation. As an (hopefully) unbiased sample, I have used data on the gastrointestinal helminths (defined as above) from recent surveys of fishes from the Pacific and Atlantic coasts of Canada (Margolis, 1963; Sekhar and

Threlfall, 1970; Sererak, 1975; Scott, 1981, 1985a, 1987, 1988; Appy and Burt, 1982; Sankurathri *et al.*, 1983; Arthur, 1984; Bourgeois and Ni, 1984; Kabata and Whitaker, 1984). In Figure 5.5, data from Kennedy *et al.* (1986) have been supplemented with data from Leong and Holmes (1981) and Shostak (1986), for freshwater fishes, and from Butterworth (1982) and Stock and Holmes (1987), for birds. Prevalence and intensity data presented in those papers have been used to calculate the mean number of species of helminths (sum of prevalences/100) and the total number of individuals of each species (prevalence × number examined × mean intensity). The latter have been used to calculate mean number of individuals (sum of number for each species/number of hosts examined), and to calculate diversity and quantitative similarity values.

5.3 RESULTS

Sebastes nebulosus

A total of 27 species of helminths was found in the 88 *S. nebulosus* examined (Table 5.1). On the basis of host records in Margolis and Arthur (1979) and Love and Moser (1983), four of these species are specialist parasites of rockfishes, 18 are wide-ranging generalists, and the rest are specialists in other host species (captured specialists), or of unknown specificity. Helminths varied markedly in prevalence; five core species (present in over 60% of the fish), 16 satellite species (found in less than 10% of the fish) and six species of intermediate prevalence are recognizable. Two of the core species and one of the intermediate species are specialists in rockfish; the other core and intermediate species are mostly generalists.

Individual fish harboured three to 13 species of helminths (with about two-thirds of the fish harbouring seven to nine species) and nine to 561 worms (with about two-thirds of the fish harbouring from 90 to 245 worms). The number of species of helminths, but not the number of individuals, was correlated with the size (fork length) of the fish (Spearman $r = 0.32$, $p < 0.01$ for S; $r = 0.18$, $p > 0.05$ for N). Despite the relative richness of the helminth fauna, 24–86% of the worms belong to the numerically dominant species of helminth in each fish; the dominant species accounted for over 50% of the worms in 47% of the fish, and over 75% of the worms in 7% of the fish. In over 90% of the fish, the numerically dominant helminth was one of the core species.

All but one fish contained segments without worms. These unoccupied segments comprised less than 25% of the total number of segments in about half of the fish, but over half of the segments in four fish. The number of

Table 5.1 Occurrence and dominance of gastrointestinal helminths in 88 individuals from ten populations of *Sebastes nebulosus*

Species	*Status*	*Populations*			*Fish*	
		Number where occurs	*Number where dominant*	*Number where core*	*Number infected*	*Number where dominant*
(*Deretrema cholaeum*)	S	1	0	0	2	0
Dollfustrema sp.	?	1	0	0	1	0
Genolinea laticauda	G	2	0	0	2	0
Helicometra sebastis	S	10	2	10	80	14.5†
Helicometrina nimia	G	1	0	0	1	0
Hemiurus levenseni	G	1	0	0	1	0
Lecithochirium exodicum	G	9	0	9	80	1
Lecithaster gibbosus	G	1	0	1	3	0
Neolepidapedon pugetense	S	10	6	10	87	46.5†
Opechona occidentalis	S	8	0	4	46	0
Parahemiurus merus	G	1	0	0	1	0
(*Phyllodistomum* sp.)	?	2	0	0	3	0
Podocotyle radifistuli	C	3	0	0	5	0
Prosorhynchus crucibulum	G	3	0	0	4	0
Stephanostomum californicum	G	9	0	3	42	0
Tubulovesicula lindbergi	G	2	0	0	2	0
Zoogonus dextrocirrus	G	7	0	4	36	6
Trematode A	?	1	0	0	1	0
Trematode B	?	2	0	0	3	0
Trematode C	?	1	0	0	1	0
Bothriocephalus sp.*	C	3	0	0	6	0
Echinorhynchus gadi	G	6	0	3	23	0
Ascarophis sebastodis	G	10	2	10	86	17
Cucullanus elongatus	C	6	0	3	34	0
Hysterothylacium aduncum	G	10	0	9	69	3
*Caballeronema wardlei**	C	9	0	6	51	0
Nematode A*	?	1	0	0	1	0

Key: S = specialist in rockfishes, G = generalist, C = captured specialist, * = did not mature, † = two co-dominants in one fish; helminth species enclosed in parentheses are normally inhabitants of other organs, only accidentally in the intestine; taxa identified by a letter could not be identified, but could be differentiated from other taxa in this study.

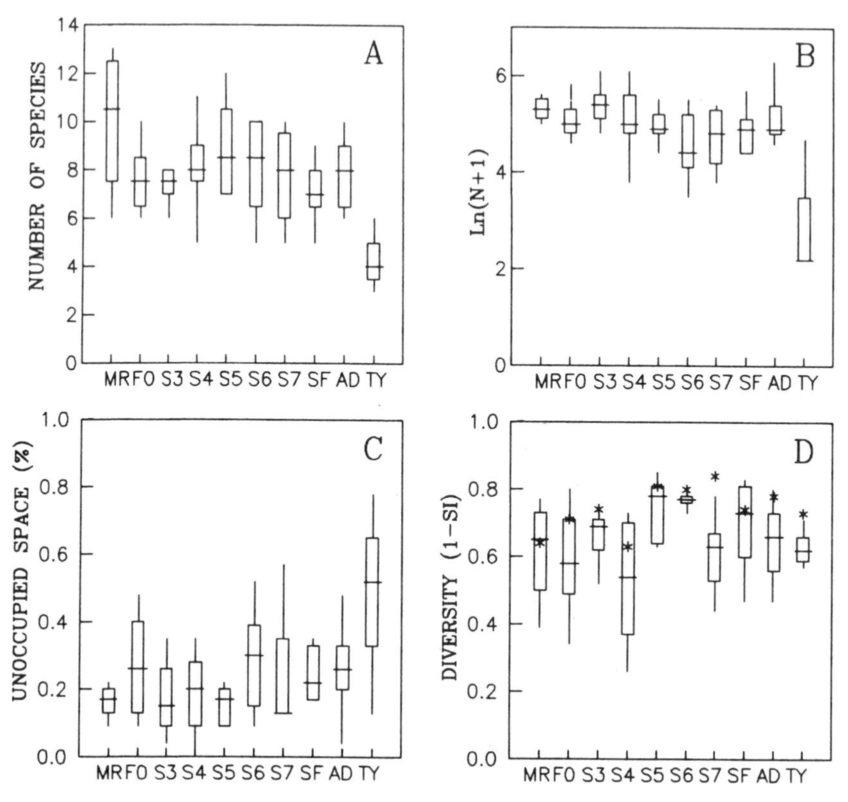

Fig. 5.1 Distribution of the number of helminth species (A) and individuals (B), unoccupied gut segments (C), and diversity (D) in ten populations of *Sebastes nebulosus*.

MR = Miller Reef, FO = Folger Island, S = Seapool Rocks (number following is date, 7 = 1987, etc), SF = Sandford Reef, AD = Adamson Rocks, TY = Tyler Rock. Data are medians (horizontal line), 25% and 75% quartiles (lower and upper limits to the box), and range (vertical lines). Asterisks in (D) indicate diversity calculated from summed data.

unoccupied segments was negatively related to the size (fork length) of the fish (Spearman $r = -0.29, p < 0.01$).

Variability in each of these parameters among samples of each component community was high, and there was no significant variation among component communities with respect to helminth species number (Kruskal–Wallis $H = 17.16$, $p > 0.3$), number of worms ($H = 23.65$, $p > 0.1$), or amount of unoccupied space ($H = 18.39$, $p > 0.1$) (although samples from Tyler Rock did appear to have fewer species and worms, and more unoccupied space, than those from elsewhere) (Fig. 5.1). There was significant variation among component communities ($H = 29.463, p < 0.05$)

PERCENT SIMILARITY

	MR	SF	FO	AD	SP7	SP5	SP3	SP4	SP6	TY
MR	.64 ±.03 / .52 ±.03	.52 ±.06	.45 ±.07	.45 ±.04	.42 ±.04	.49 ±.05	.43 ±.04	.61 ±.05	.50 ±.07	.35 ±.06
SF	.52 ±.02	.55 ±.07 / .78 ±.03	.48 ±.05	.47 ±.06	.42 ±.04	.49 ±.06	.31 ±.06	.40 ±.07	.46 ±.08	.30 ±.05
FO	.42 ±.02	.66 ±.03	.60 ±.07 / .69 ±.02	.55 ±.06	.39 ±.04	.41 ±.04	.36 ±.07	.41 ±.07	.49 ±.07	.33 ±.05
AD	.45 ±.01	.65 ±.04	.68 ±.03	.51 ±.07 / .70 ±.02	.47 ±.05	.42 ±.02	.43 ±.07	.45 ±.06	.42 ±.06	.30 ±.05
SP7	.50 ±.03	.57 ±.07	.53 ±.06	.57 ±.04	.53 ±.04 / .52 ±.05	.54 ±.04	.52 ±.03	.42 ±.06	.51 ±.03	.34 ±.02
SP5	.51 ±.03	.59 ±.04	.52 ±.03	.56 ±.04	.64 ±.04	.58 ±.04 / .71 ±.04	.47 ±.05	.51 ±.05	.58 ±.04	.47 ±.07
SP3	.37 ±.03	.48 ±.04	.52 ±.03	.52 ±.01	.54 ±.05	.62 ±.04	.53 ±.04 / .67 ±.02	.56 ±.05	.57 ±.04	.25 ±.05
SP4	.45 ±.02	.50 ±.02	.51 ±.03	.50 ±.03	.58 ±.04	.58 ±.04	.60 ±.04	.56 ±.03 / .57 ±.02	.53 ±.05	.34 ±.05
SP6	.45 ±.04	.55 ±.03	.52 ±.03	.45 ±.03	.54 ±.03	.56 ±.04	.57 ±.03	.56 ±.05	.62 ±.06 / .60 ±.04	.34 ±.04
TY	.30 ±.02	.37 ±.04	.37 ±.05	.33 ±.04	.28 ±.03	.36 ±.03	.36 ±.06	.37 ±.05	.31 ±.05	.50 ±.08 / .52 ±.08

JACCARD

Fig. 5.2 Similarity values (mean ± s.e.) between helminth communities in *Sebastes nebulosus*. Values above the diagonal represent quantitative similarities, those below represent qualitative similarities. Values on the diagonal are within-population similarities. All values (except the within-population values for MR [$n = 6$] or TY [$n = 3$], all possible) are based on ten randomly-selected pairs of infracommunities. Populations labelled as in Fig. 5.1

in diversity, with communities from Seapool Rocks in 1985 and 1986 having relatively high diversities. The overall impression, however, is one of considerable variation among the samples of each component community, with relatively little variation among communities on different reefs (with the possible exception of a depauperate community in fish from Tyler Rock).

Similarities between randomly-selected pairs of infracommunities taken from the same population of fish (i.e. samples of the same component community) can be regarded as measures of 'within-community similarity', those between infracommunities taken from different populations of fish (i.e. samples of different component communities) can be regarded as measures of 'across-community similarity'. Within-community qualitative similarities were highly variable within each component community (see standard errors on the diagonal in Fig. 5.2), but differed significantly ($H = 37.30$, $p < 0.005$) among component communities. The overall distribution of within-community similarities did not differ from the distribution of across-community similarities between fish taken from Seapool Rocks in different years ($H = 0.83$, $p > 0.1$). The similarities in these two groups were higher than across-community similarities between fish taken from different reefs ($H = 20.27, p < 0.005$). A furthest-neighbour clustering algorithm (which tends to maintain discrete clusters) applied to mean similarities between component communities (the off-diagonal values in Fig. 5.2) groups the five Seapool communities, joins them to a cluster of three of the other communities, and maintains two isolated communities (Miller Reef and Tyler Rock; Fig. 5.3).

Quantitative within-community similarities also varied markedly within each component community (Fig. 5.2), but did not vary significantly among communities ($H = 8.13$, $p > 0.5$). Overall, within-community similarities did not differ from across-community similarities between fish taken from

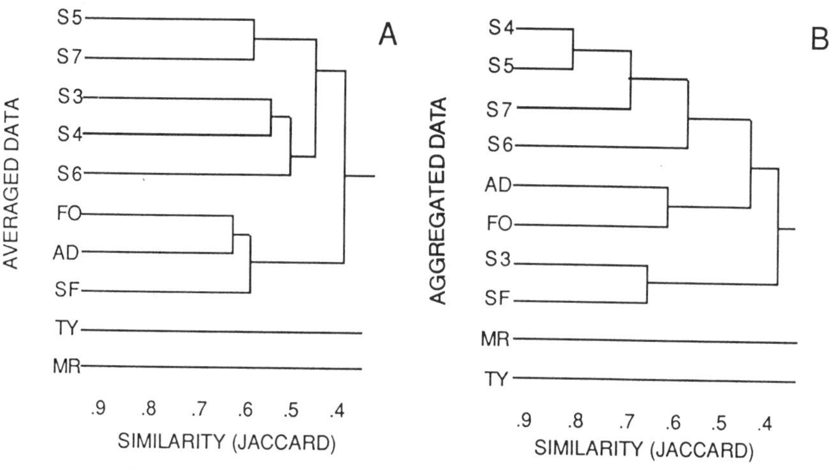

Fig. 5.3 Furthest-neighbour clustering patterns for qualitative similarities among helminth communities in *Sebastes nebulosus*. A: Clustering based on average similarities (data in Fig. 5.2). B: Clustering based on similarities between summed data. Populations labelled as in Fig. 5.1. Analysis terminated at similarity = 0.400.

Table 5.2 Distribution of core species in ten populations of *Sebastes nebulosus*

Parasite	*Site in gut*	*Geographic location*									
		MR	*SF*	*FO*	*AD*	*SP7*	*SP5*	*SP3*	*SP4*	*SP6*	*TY*
A. sebastodis	S	C	C*	C*	C*	C*	C*	C	C	C	C
L. exodicum	S	C	C*	C	C	C	C	C	C	C	–
H. aduncum	S/I	C	C	C	C	C	C	C	+	C	C*
C. wardlei	I	C	C	+	+	C	C	+	C	C	–
E. gadi	I	C	–	–	+	C	C	–	+	+	–
L. gibbosus	I	C	–	–	–	–	–	–	–	–	–
C. elongatus	I/PC	–	+	–	–	+	+	C	C	C	–
N. pugetense	PC/I	C*	C*	C*	C*	C*	C*	C*	C*	C*	C*
H. sebastis	PC/I	C	C	C	C*	C*	C*	C*	C*	C*	C
O. occidentalis	PC/I	+	C	C	C	+	C	+	+	+	–
S. californicus	R	–	+	C	C	+	+	C	+	+	+
Z. dextrocirrus	R	+	–	–	+	C*	C	C*	+	C*	–

Key: Sites: S = stomach, I = small intestine, PC = pyloric caeca, R = rectum; sites separated by a slash are both occupied, with the most commonly occupied site given first. C = core species, + = present, – = not present. Populations labelled as in Figure 5.1. Locations are listed in order determined by qualitative similarity.

* = numerical dominant in at least one infracommunity.

Seapool in different years ($H = 2.30$, $p > 0.1$); again, these values were significantly higher ($H = 26.91$, $p < 0.005$) than across-community similarities between fish taken from different reefs. A furthest-neighbour clustering algorithm grouped the Miller Reef community (one of those isolated in the Jaccard analysis) with the Seapool communities, and kept this group separate from a cluster of three other communities and the isolated Tyler Rock community (Fig. 5.4).

The number of core species in each component community varied from four to nine (Table 5.2). Three helminths were core species in all component communities and two others were core species in all but one of the component communities. Three helminths were core species on only one reef: *Lecithaster gibbosus* on Miller Reef and *Cucullanus elongatus* and *Zoogonus dextrocirrus* on Seapool (where they were core species for three and four of the five years, respectively). Several helminth species numerically dominated infracommunities from several component communities. The best example is *Neolepidapedon pugetense*, which dominated at least one infracommunity in each of the component communities.

The two measures of similarity appear to indicate that samples taken from the same reef, but in different years, are sampling the same component community. The existence of different core species in different years, however, suggests that the community does vary somewhat among years.

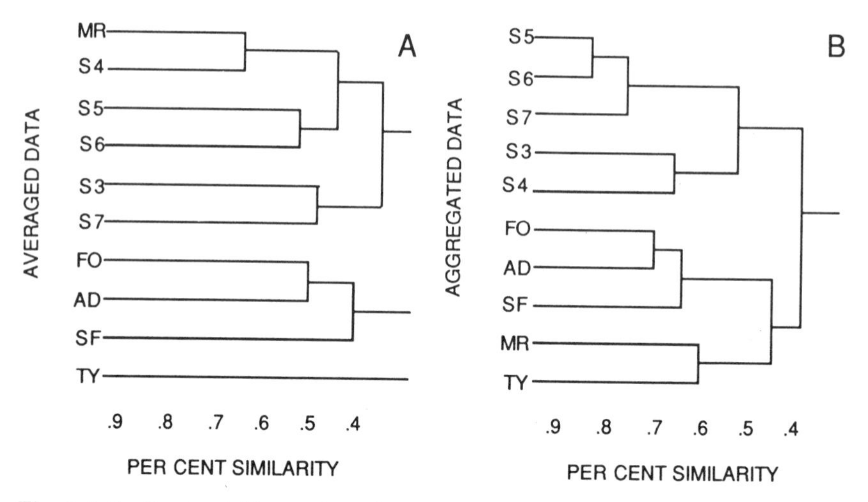

Fig. 5.4 Furthest-neighbour clustering patterns for quantitative similarities among helminth communities in *Sebastes nebulosus*. A: Clustering based on average similarities (data in Fig. 5.2). B: Clustering based on similarities between summed data. Populations labelled as in Fig. 5.1. Analysis terminated at similarity = 0.400.

Note that two or more species in the same feeding guild (Chapter 8), and occupying the same general part of the gastrointestinal tract, were frequently core species in the same community. The best example involves the digeneans (absorber–engulfers), *Neolepidapedon pugetense* and *Helicometra sebastis*, which were core species in all communities, and *Opechona occidentalis*, which joined them as a core species in four communities. All inhabited the pyloric caeca and the anterior small intestine, where they were frequently found together. All appear to be specialists of rockfish (Table 5.1).

The pattern that I see in these data is one of infracommunities that vary extensively, generally contain a moderate number of species and individuals, and contain some empty space. The presence of a suite of nearly-ubiquitous core species (less than half of which are rockfish specialists) confers a basic similarity to the infracommunities; this basic similarity is modified by the differential abundance of other species, which may be core species in particular locations, so that there are discernable differences in the component communities among locations.

To what extent is this pattern characteristic of communities of intestinal helminths in marine fishes? This question cannot be answered with data on infracommunities from other fishes because the data are not available in the literature. However, the literature is replete with surveys of parasites of marine fishes, presenting summed data on component communities which can be used to examine many of the same features. Before doing so, I should evaluate the effect of using summed data on the measures used to characterize component communities.

Mean numbers of species and individuals can readily be calculated from prevalence and intensity figures (section 5.2); within rounding error, they are identical to those calculated from infracommunity data. However diversity values are markedly higher (see asterisks on Fig. 5.1D); for three of the component communities examined here, the summed value is outside the range of values calculated for samples of that community; for four others, the summed value is at the upper limit of, or outside, the inter-quartile range. Jaccard values differ unpredictably from means of those calculated from infracommunity data; they varied from 0.183 lower to 0.287 higher, with an average difference of 0.065 ± 0.095. Cluster analysis indicated that these differences did have some effect on the patterns detected (compare Fig. 5.3A and B). Per cent similarity values from summed data are almost always higher than mean values from infracommunity data, varying from 0.013 lower to 0.456 higher, with an average difference of 0.179 ± 0.084. The differences in similarity values, plus some consequent differences in the patterns detected, can be seen in Figure 5.4 (compare A and B). Thus, using summed data will significantly increase diversity and per cent similarity values and unpredictably alter Jaccard values; this may alter patterns.

Other marine fishes

Community richness

Data on mean numbers of species and individuals per fish examined, from recent surveys of marine fishes from the Atlantic and Pacific coasts of Canada, are presented in Fig. 5.5. Maximum values are plotted when more than one population has been surveyed. The corresponding value for *S. nebulosus* is shown by a star. For comparison, ranges of values for freshwater fishes and aquatic birds and mammals have also been outlined. The data suggest that communities of gastrointestinal helminth in marine fishes are highly variable in richness, but less rich than communities in *S. nebulosus*, and that they have approximately the same range of numbers of individuals, but higher species richness (for a given number of worms), as communities in freshwater fishes. Maximum species richnesses (for a given *N*) increase with worm number at about the same rate as maximum richnesses of communities in birds and mammals. There is little support for the hypothesis that benthic feeders have more complex communities, or

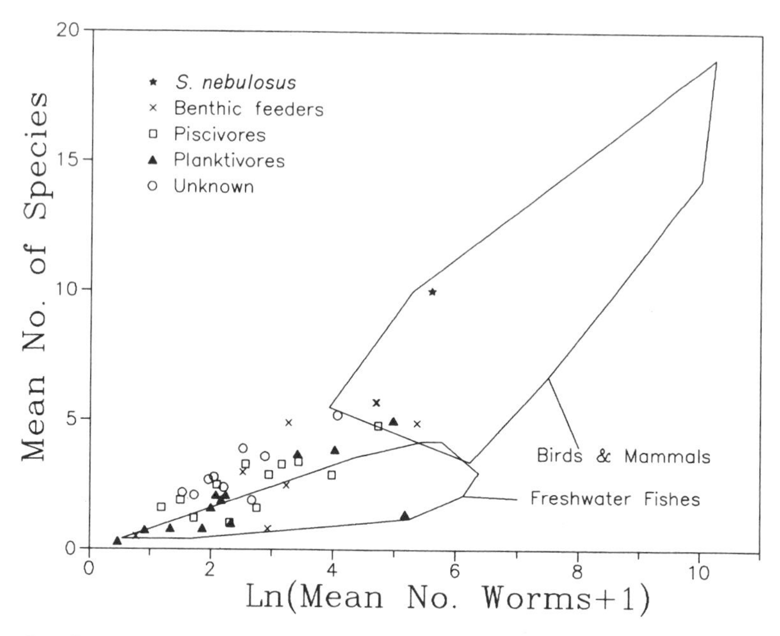

Fig. 5.5 Community richness of gastrointestinal helminths of marine fishes in comparison with those in freshwater fishes (polygon in lower left) and aquatic birds (polygon in upper right). See text for sources of data.

communities with more individual worms, than plankton-feeders or pelagic piscivores (Fig. 5.5).

Variation in total numbers of species, numbers of core species, and diversity are shown in Fig. 5.6 for those fishes for which at least five populations have been surveyed. (Of these fishes, *S. nebulosus* is a benthic feeder, *Merluccius productus* is a pelagic piscivore, and the rest are planktivores.) Generally, these other fishes have fewer species of helminths, fewer core species, and lower diversities than *S. nebulosus*. Variability, especially in the number of species and in diversity, is high.

Predictability

The only data available to assess the predictability of infracommunities are those on numbers of core species. The low numbers for most host species (Fig. 5.6B) suggest that infracommunities in most marine fishes are not as predictable as are those in *S. nebulosus*.

The predictability of component communities, as indicated by their similarities, is highly variable (Fig. 5.7). Both Jaccard and per cent similarity

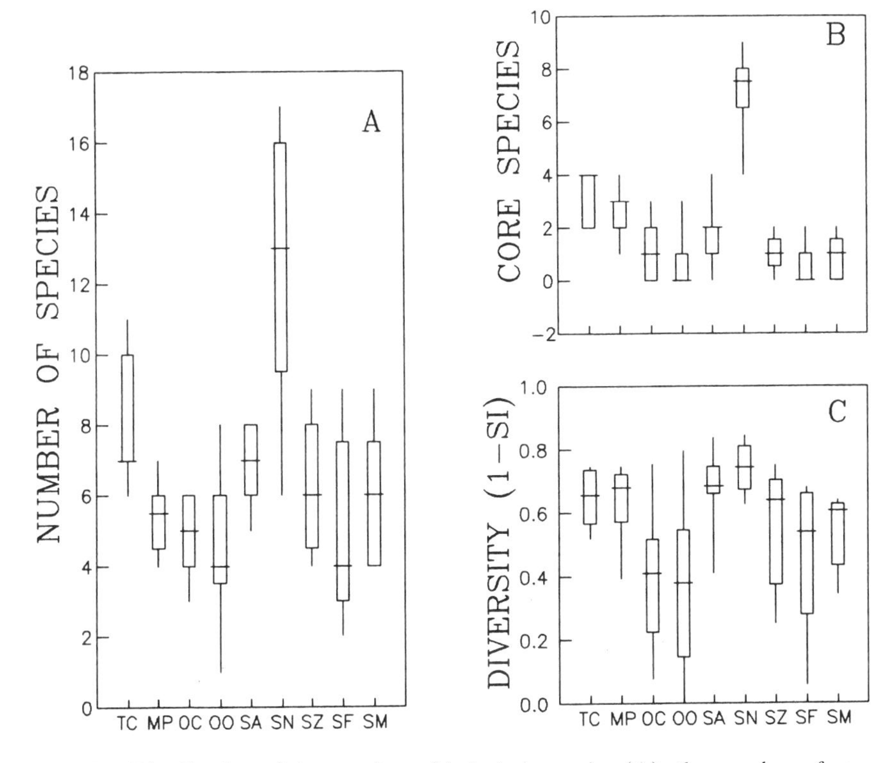

Fig. 5.6 Distribution of the number of helminth species (A), the number of core species (B), and diversity (C) in eight species of marine fishes for which at least five populations have been sampled. Data are shown as in Fig. 5.1.

TC = *Theragra chalcogramma*, MP = *Merluccius productus*, OC = *Oncorhynchus nerka*, coastal samples, OO = *O. nerka*, offshore samples, SA = *Sebastes alutus*, SN = *S. nebulosus*, SZ = *S. zacentrus*, SF = *S. fasciatus*, SM = *S. marinus*.

indexes can take values of 0.0 to 1.0. Observed Jaccard values for similarities between component communities in different populations of the same species of fish encompassed 27–88% of this possible range, and per cent similarity values encompassed 40–90% of the possible range. Median values for both measures also varied extensively across fish species. High median values (for per cent similarity) were associated with both complex communities (with high median numbers of all species and of core species, such as those in *Sebastes nebulosus* and *Theragra chalcogramma*) and depauperate communities (such as those in coastal *Oncorhynchus nerka* and in *Sebastes marinus*). Component communities in *Sebastes alutus* from the same area

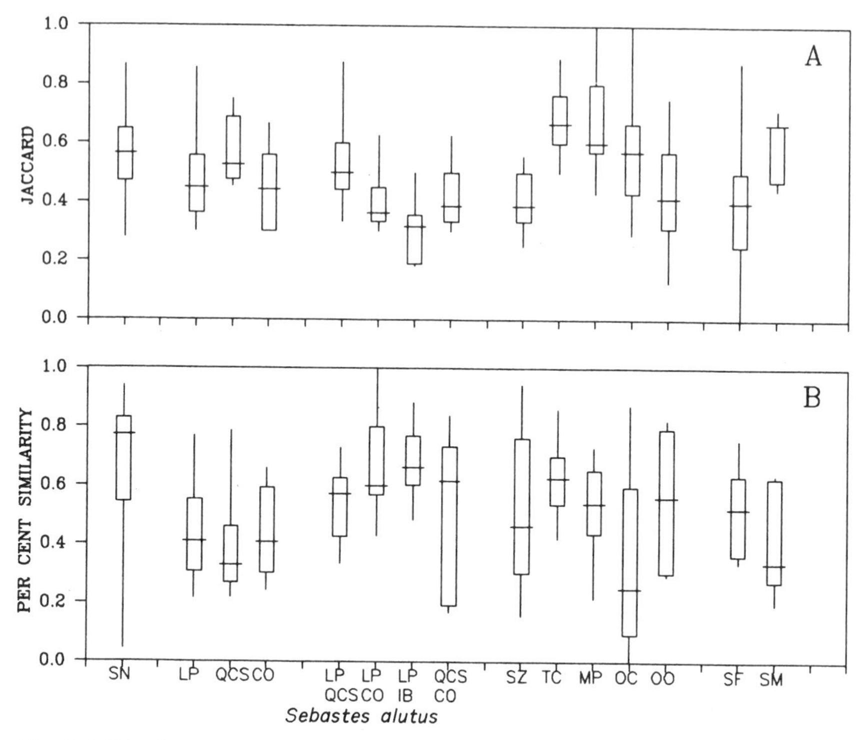

Fig. 5.7 Distribution of similarity values among populations of the eight species of marine fishes from Fig. 5.6. A: Qualitative similarities. B: Quantitative similarities. Data are shown as in Fig. 5.1. Codes for fishes as in Fig. 5.6.

Within- and between-area comparisons are shown for *S. alutus*: LP = LaPerouse Bank, QCS = Queen Charlotte Sound, CO = Cape Ommaney, IB = Icy Bay.

(but which differed in size, or depth of date of capture) appeared to be only slightly more predictable (and only using the Jaccard index) than those in *S. alutus* taken from different areas.

Another way of assessing the predictability of component communities is to determine the extent to which core species are shared amongst them. The data in Table 5.3 indicate that all host species with a median of two or more core species per component community had at least two individual helminth species which attained core status in at least 50% of those component communities. Even some depauperate communities (such as those in coastal *O. nerka* and *S. zacentrus*) had the same core species in over half of the component communities examined. This sharing of core species is obviously one of the major reasons for the high similarities among component communities in these host species.

Table 5.3 Shared core species within and across host species. Entries for each species of helminth are the number of populations in that host species in which the helminth was a core species

	Host species															
	Mrp	*Mcp*	*Onc*	*Ono*	*Sa*	*Sn*	*Sz*	*Sspp*	*Tc*	*Gm*	*Mea*	*Mra*	*Sf*	*Sma*	*Sme*	*Uc*
No. pop'ns exam.	10	3	22	55	18	10	5	24(19)	7	1	3	1	11	5	4	2
Median no. core spp.	3	1	1	0	2	7.5	1	–	4	–	4	–	0	1	0	1
No pop'ns where core																
Brachyphallus crenatus			4		9											
D. varicus	9				9	1	1	10		1	1		1	1		
H. sebastis						10										
H. levenseni				3					2	1	2					
L. gibbosus			15			1										
L. exodicum						9										
Lecithophyllum botryophorum								1							1	
Lepidapedon rachion											1					
N. pugetense						10										
N. sebastici								3(3)								
Opechona alaskensis					6											
O. occidentalis						4		2(1)								
Podocotyle reflexa													3	2		
Podocotyle spp.			1		15		4	7(6)	6							
Prosorhynchus crucibulum								6(6)								
S. californicum						3		1								
Z. dextrocirrus						4										
Abothrium gadi									3		1					
Bothriocephalus scorpii								1								
Clestobothrium scapiceps	10															
Bolbosoma caeniforme			4	13												
E. gadi		2		1		3		2(2)	6		2					
E. vancleavei										1						
A. sebastis						10		1								
Ascarophis sp.	1															
C. wardlei						6										
Capillaria spp.								2(2)								
C. elongatus						3										
Cucullanus sp.								2(2)								
H. aduncum	6	1			3	9		7(7)	7	1	3	1	3			2

Mrp = *Merluccius productus*; Mcp = *Microstomus pacificus*; Onc = *Oncorhynchus nerka*, coastal; Ono = *O. nerka*, offshore; Sa = *S. alutus*; Sn = *S. nebulosus*; Sz = *S. zacentrus*; Sspp = other species of *Sebastes* (values in parentheses are numbers of species); Tc = *Theragra chalcogramma*; Gm = Gadus morhua; Mea = *Melanogrammus aeglefinus*; Mra = *Merluccius albida*; Sf = *S. fasciatus*; Sma = *S. marinus*; Sme = *S. mentella*; Uc = *Urophycis chuss*.

Host specificity

Twenty-six helminth species attained core status in one or more populations (Table 5.3); seven of them attained core status in more than a single genus of host (seven of the eight host genera for *H. aduncum*). The records in Margolis and Arthur (1979) and Love and Moser (1983) indicate that 20 of these helminths are generalists found in many genera of marine fishes. The only host specialists appeared to be a few species of digeneans limited to *Sebastes* spp., and *Caballeronema wardlei* in *Scorpaenichthys*.

5.4 DISCUSSION

Community richness

If the parasite communities shown in Fig. 5.5 are representative, marine fishes have infracommunities of gastrointestinal helminths that are highly variable in numbers of individuals and species, but frequently are more complex (contain more species for a given number of individuals) than communities in freshwater fishes. Two interacting factors are probably responsible for this difference. First, marine fishes (or the organisms on which they feed) exhibit greater vagility. The isolated reefs inhabited by *S. nebulosus* are not completely cut off from invasion by occasional migrating rockfish, and are regularly traversed by swarms of plankton or schools of forage fishes, such as herring, both of which may carry larval parasites. As a result, helminth species that are eliminated from a particular habitat may be reintroduced (as in the 'rescue phenomenon' of island biogeography [Brown and Kodric-Brown, 1977]), keeping local helminth communities more complex.

Second, most gastrointestinal helminths of marine teleosts have a relatively broad host specificity. For territorial species of fish, like *S. nebulosus*, which probably do not move much as adults, reintroduction through parasites carried in more vagile hosts would probably be more important than that through parasites carried by settling juveniles or the occasional migrating adult. It is not surprising, then, that most of the core species of helminths in *S. nebulosus* are host generalists; only *Helicometra sebastis*, *Neolepidapedon pugetense*, and *Opechona occidentalis* are host specialists. The first two were more abundant in *S. nebulosus* than in any other species of *Sebastes* studied by Sekerak (1975), and both were found in all populations examined. *Opechona occidentalis* appears to be more abundant in *S. caurinus* (Sekerak, 1975, personal observation). Perhaps periodic local extinctions mitigate the development of strict host specificity in the gastrointestinal helminths of marine fishes.

The similarity in mean numbers of individuals from infracommunities in marine and freshwater fishes, and the differences between fish and birds, suggest that a common factor may be limiting worm numbers in fishes. Ectotherms require (and can digest) considerably less food than similarly-sized endotherms; reduced food intake should translate into reduced exposure to parasites (or, perhaps, reduced ability to support parasites). Thus, low numbers of worms in fishes may be a function of the rate of energy flow through the individual host.

Sebastes nebulosus had the richest gastrointestinal helminth community of any marine fish covered in this study. Five core species were found in almost all populations, and in a high proportion of the individual fish. These features are suggestive of an interactive community (*sensu* Holmes and Price, 1986). However, data on infracommunities emphasize the influence of stochastic factors. Virtually all fish had substantial amounts of empty space, infracommunities (even those in fish from the same population) were often dominated by different core species, and every measure of infracommunity structure showed considerable variation. In addition, populations differed in the number, and identity, of other core species; in some populations, three core species (all digeneans) appeared to share the same microhabitat (caeca and anterior small intestine). These latter features suggest that most infracommunities, and most component communities, are probably invasible. If such is the case for the rich parasite communities in *S. nebulosus*, it probably applies to the less rich parasite communities in most other marine fishes as well.

The evidence for interactions among gastrointestinal helminths of *S. nebulosus* will be examined elsewhere. However, it should be apparent that invasibility (or unoccupied space) does not preclude important interactions amongst species that regularly encounter each other (Bush and Holmes, 1986b; Goater and Bush, 1988; Stock and Holmes, 1988). Wellington and Victor (1988) have presented a more cogent argument: the potential significance of processes (such as competition) as selective (i.e. evolutionary) agents, acting at the level of the individual, may be independent of their significance (if any) in determining population (or, by extension, community) dynamics. That is, interactions among helminth species may determine the reproductive success of individuals, thereby selecting for niche differentiation, even if those interactions are unable to affect species richness (diversity, or other community characteristics). Parasitologists would be wise to heed their plea to recognize these different potentials, and keep them clearly in mind when investigating communities.

Predictability

Data on *S. nebulosus* infracommunities clearly demonstrated that each fish was essentially an independent sampler of the parasites potentially available to it. Presumably, their territories differed in depth, topography, relationship to prevailing currents, and other features that influenced the kinds of organisms that live there and were available as potential food (or intermediate hosts). Individual fish may also specialize on particular foods, as has been shown for some freshwater fishes (Curtis and Stenzel, 1984). Despite this independence, similarities between samples of the same component community were high (means of 0.52–0.78 for qualitative similaritity, 0.50–0.64 for quantitative similarity), and numbers of core species were high. Although there were some differences in numbers and identities of core species among infracommunities collected from the same reef in different years, there were no significant differences in either measure of similarity. All these features show the basic predictability (stability) of the system.

All three measures of predicability dropped when analyses were extended to cover infracommunities from different reefs. This differentiation of component communities on different reefs was predicted from the territorial behaviour of *S. nebulosus*. However, differences among component communities are not primarily due to differences in abundances of host specialists, but to differences in the presence (or abundances) of host generalists, suggesting that the mechanism responsible is not the lack of vagility of *S. nebulosus*, but something that determines the abundance of host generalists.

Stochasticity affecting infracommunities can be effectively wiped out by using summed data at the component community level. Doing so increases the diversity calculated for the community, increases its similarity to other communities, and may modify the patterns of similarity indicated by clustering analyses. Parasitologists traditionally use summed data to examine component communities, as I have in the latter part of the chapter. However, some recent studies (e.g. Kennedy *et al.*, 1986; Chapter 8) advocate the use of averaged data, as I have in the earlier part of this chapter. The use of averaged data has two obvious procedural advantages: (a) variation within component communities can be estimated and used to test statistically for variation among component communities, and (b) values such as the mean number of species (or individuals) are not critically dependent on the number of infracommunities sampled, as total numbers of species (or individuals) may be. However, neglecting procedural considerations, which set of data, averaged or summed, best represents the component community? Whenever data are summed, information is lost. If the information lost by using summed data was essentially a measure of stochastic variation, as would be the case if individual hosts acted essentially as random samplers of the available parasites, summed data could provide a good estimate of the diversity and similarity patterns among component communities. However, if individual hosts (or processes acting on or within them) actively filtered out parasites to which they were exposed, so that the information lost through summation included the results of infracommunity processes, summed data may provide misleading estimates of diversity and similarity patterns. Thus, where a choice exists, the use of averaged or summed information may reflect the bias of the research worker as to the relative importance of infracommunity processes, or may reflect the nature of the questions asked.

The only infracommunity-level measure of predictability that can be calculated from summed data is the number of core species. Communities in other marine fishes have fewer core species than *S. nebulosus*, suggesting less predictable infracommunities (and, by inference, component communities). However, similarity values between summed component communities in other marine fishes indicate that these component communities are not necessarily less predictable than those in *S. nebulosus*. Obviously, we need more information on the predictability of helminth communities in marine fishes, especially information comparing infracommunity and component community levels.

The extensive variation in similarity values for some species (such as *S. fasciatus* and coastal *Oncorhynchus nerka*) apparently reflects variation on larger spatial scales (see discussions by original authors), suggesting that community patterns may be useful as biological markers. However, data on *S. alutus* suggest that there may be considerable variation among samples taken from the same area. This variation stresses the need for caution in using community-level patterns as markers. The need to evaluate seasonal, host size-related, or annual variation before using individual parasites as tags has been stressed by MacKenzie (1987); such a need would be even more important (and harder to meet) for community-level patterns (but see Bailey *et al.*, 1988, for an example).

Host specificity

Several factors suggest that host specificity plays a relatively minor role in determining gastrointestinal helminth communities in marine fishes: (a) the relatively few specialists and the prominent role played by host generalists in communities in *S. nebulosus*, (b) the few host specialists that are core species in the fishes surveyed, and (c) the number of helminth species that are core species in several host genera (and in both Atlantic and Pacific oceans). This essentially agrees with conclusions reached much earlier by Polyanski (1961a).

This does not mean that host specificity is lacking. Parasites still have restricted host ranges, and are more abundant in some host species than in others. Abundances (or host ranges) may be correlated with the phylogeny of the hosts (as in the specialists on rockfish), but may not be (as in *Derogenes varicus*). However, under the right ecological conditions, a given helminth may occur in high abundance in an unusual host (as in *Lecithochirium exodicum*, normally a parasite of *Sebastes* or *Ophiodon*, in *Anoplopoma* on two seamounts [Kabata *et al.*, 1988]).

The major groups of marine parasites differ markedly in host specificity. Rohde (1978b) has shown that marine digeneans are markedly less host specific than monogeneans, and show decreasing specificity in colder oceans. Judging from host lists in Margolis and Arthur (1979) and Love and Moser (1983), acanthocephalans, nematodes, and cestodes of marine teleosts appear to be similar to digeneans in this respect; copepods and cestodes of elasmobranchs appear to be similar to monogeneans. The analysis in this chapter is focused on gastrointestinal parasites of cold-ocean teleosts, dominated by parasites of low host specificity. Host specificity is likely to play a much more important role in communities of helminths of tropical marine fishes or elasmobranchs, or in those on gills or body surfaces.

Determinants of helminth community structure in marine teleosts

In previous papers (Holmes, 1987, 1988), I have suggested that the infracommunities (and component communities) we study are results of actions of a number of screens (or filters) acting on the potential parasite fauna. My current concept of those screens and how they interact is shown in Fig. 5.8. That figure is derived from earlier work, modified by independent formulations of similar screens (or filter systems) acting on fish communities (Tonn *et al.*, in preparation; P. Moyles, personal communication) and discussions on determinants of parasite communities with D. R. Brooks, C. R. Kennedy, V. Kontrimavichus and O. Pugachev. In the remainder of this chapter, I will try to evaluate the significance of various screens in determining communities of gastrointestinal helminths in marine teleosts. I will focus on screens directly affecting component communities and infracommunities.

Historical and zoogeographic factors essentially determine the potentially available regional parasite (and host) fauna. These factors are obviously important, but an analysis of them is beyond the scope of this chapter (see Polyanski, 1961b; Rohde, 1982 for reviews). (Introductions by humans do

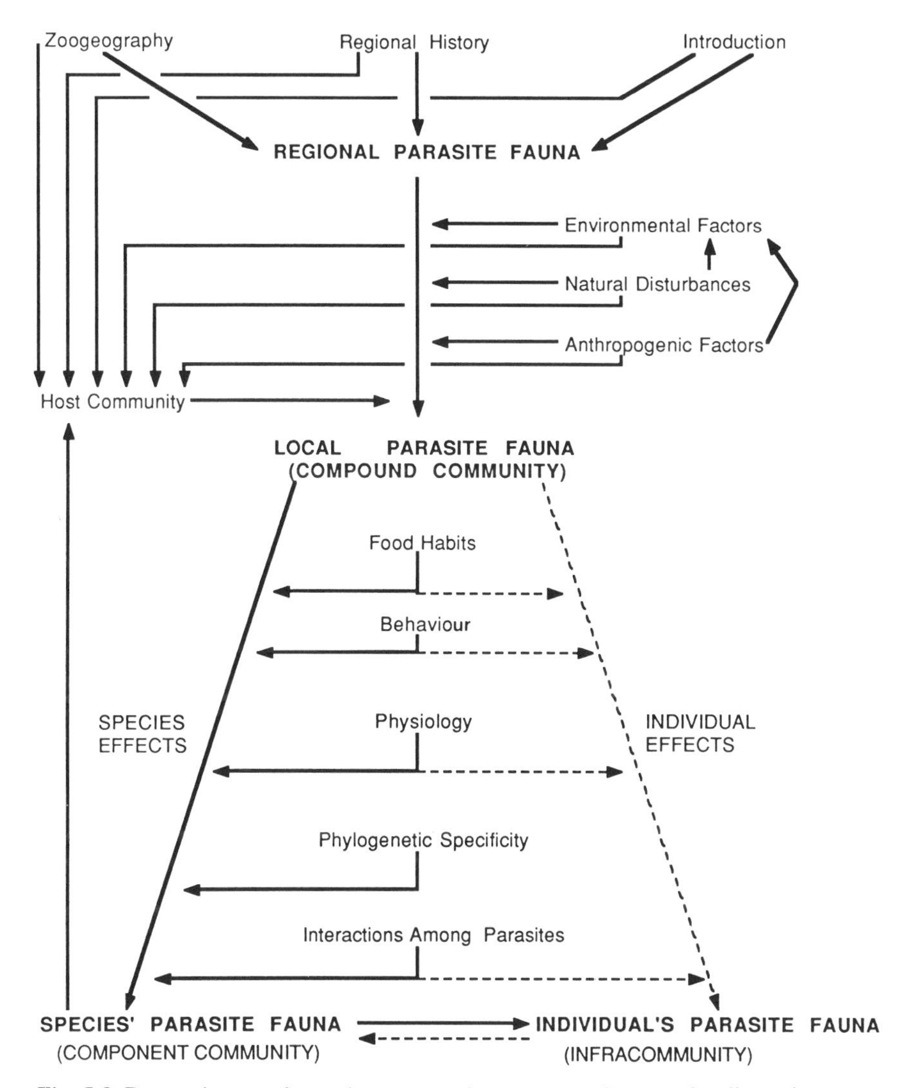

Fig. 5.8 Determinants of parasite community structure. See text for discussion.

not appear to be a major factor as yet in the marine environment; such introductions are very important in terrestrial and freshwater environments, and will probably increase in importance in marine systems with increasing mariculture.)

Environmental factors are very important in determining the local parasite fauna. For the parasites of marine fishes, most attention has been paid to depth, pelagic versus benthic conditions, coastal versus offshore conditions, temperature, and water chemistry; for more details see reviews by Polyanski (1961a), Rohde (1978a), Rohde (1982), and Campbell (1983). The general productivity of the local ecosystem is obviously important (Campbell, 1983), and deserves much more attention. The influences of natural disturbances (such as El Niño effects or catastrophic epidemic disease affecting keystone species) have not been studied in great detail but must be important, considering their potentially long-lasting effects on the host community (Dayton and Tegner, 1984; Johnson and Mann, 1988). The influences of anthropogenic factors such as pollution (Moller, 1987; Sulgostowska *et al.*, 1987) or fish farming (Paperna, 1987) are just beginning to be recognized (and deserve more attention in the future). The extent to which these factors operate directly on parasites, or operate indirectly through their effects on the host community, deserves more attention.

The composition of the host community (including both fish and invertebrates) must be important, but has received relatively little attention. For digenean-dominated communities, such as those in marine teleosts, the molluscan fauna, which determines what kinds of life cycles are possible, is crucial (Køie, 1983; see also Chapters 3 and 4). For similar reasons, the fauna of other invertebrate intermediate hosts may also be important. Food webs, and particularly the extent to which members of the fish community concentrate on temporally (or locally) dominant food items, should also be very important, but their roles in determining local parasite communities have not been investigated extensively.

The preceding factors determine the composition of the local parasite fauna, or compound community. Component communities and infracommunities may be either restricted subsets or random samples of that compound community. If the component community is determined by phylogenetic specificity of the parasites, or by ecological or physiological specificity dependent upon behavioural or physiological features of the host species, then that component community will be essentially a restricted subset of the compound community, and infracommunities are likely to be random samples of that subset. (This is the pattern shown by the solid lines in Figure 5.8.) However, if the infracommunity is dependent upon ecological, behavioural, or physiological factors operating at the level of the individual fish, then infracommunities will be random samples of the compound community, and the component community will be essentially an artificial construct. (This pattern is shown by the dashed lines in Fig. 5.8.) Obviously, these are opposite ends of a continuum, not strict alternatives. The extent to which component communities are real (and infracommunities are samples of component communities) is a function of the importance of specificity, or the extent to which frequent interactions amongst parasites influence their population dynamics. The extent to which infracommunities are samples of the compound community (and component communities are merely aggregates without much ecological significance) is a function of the importance of differences among individual hosts, stochastic factors, and generalized adaptations of helminth species.

Where along this continuum do communities of gastrointestinal helminths of marine teleosts lie? The evidence seems contradictory. Support for host specificity, or for the importance of interactions between parasites, is weak. The evidence does suggest that stochastic factors are important, and that most helminth species are relative generalists. However, for component communities in many species of fish, predictability seems extraordinarily high for stochastic systems, and the successful use of some parasites as biological tags indicates that at least those species are reasonably consistent from year to year (i.e. stable). In the data on *S. nebulosus*, the lack of any significant difference between similarities between samples from the same component community and similarities between infracommunities in fish taken from Seapool in different years also suggests a stable system. (Note, however, that if individual parasites persist for long periods of time, either as adults in the intestine of the fish, or as larval stages in long-lived intermediate hosts, the apparent stability may be an illustration of the 'storage effect', i.e. a function of the inertia in the system.)

My conclusion is that component communities of gastrointestinal helminths in marine teleosts occupy a broad band along the continuum from predictive subsets to stochastic samples of the compound community. Factors acting at the host species level are important determinants for some helminths, and stochastic factors are important for others. In addition, it appears that host lists taken from checklists such as those of Margolis and Arthur (1979) or Love and Moser (1983) are not adequate indicators of the extent of host specificity, if the latter is taken to mean the distribution of egg-producing adult parasites among host species. As has been pointed out previously (Holmes *et al.*, 1977; Rohde, 1980, 1982; Freeland, 1983), a parasite may have been found in a relatively long list of hosts, but still have most of its reproduction occurring in one (or few) hosts; it is this host (or hosts) that is (are) important in determining how abundant the parasite will be (and where).

If host specificity is more important than it appears from host lists (and therefore, component communities are meaningful), then one factor that should be an important determinant of the compound community existing, for example, on a reef, would be the composition of the fish community on that reef. Each of the common fish would presumably contribute its own major parasites to the compound community, and exchange of parasites within the fish community would be an important mechanism leading to the observed higher species richnesses of the communities in marine fishes. Data of Sekerak (1975), on parasites of *Sebastes*, and of Scott (1980, 1985b), on digenean parasites of flatfishes, agree with this interpretation, but the materials on which they are based were taken over too large areas to constitute good tests. I have collected many other species of rockfishes (and some other common fish species) on the same reefs from which the *S. nebulosus* were taken; when examinations of those fish are complete, I should have a better picture of quantitative host specificity, and the importance of exchange of parasites, on those reefs.

ACKNOWLEDGEMENTS

Data on helminths from *S. nebulosus* from Seapool Rocks in 1983 are from a class report by R. E. Bailey. I thank colleagues at the Bamfield Marine Station for assistance with fishing; E. Street and T. M. Stock for technical assistance; D. Welch and T. M. Stock for preparing figures; D. R. Brooks, A. O. Bush, C. R. Kennedy, V. Kontrimavichus, P. Moyles, O. Pugachev, T. M. Stock and W. M. Tonn for discussions on the ideas expressed here; A. O. Bush, G. W. Esch, L. Measures and T. M. Stock for comments on the manuscript. This study was supported by a research grant from the National Science and Engineering Research Council of Canada.

REFERENCES

Appy, R. G. and Burt, M. D. B. (1982) Metazoan parasites of cod, *Gadus morhua* L., in Canadian Atlantic waters. *Can. J. Zool.*, **60**, 1573–9.

Arthur, J. R. (1984) A survey of the parasites of Walleye Pollack (*Theragra chalcogramma*) from the Northeastern Pacific Ocean off Canada and a zoogeographical analysis of the parasite fauna of this fish throughout its range. *Can. J. Zool.*, **62**, 675–84.

Arthur, J. R. and Arai, H. P. (1980) Studies on the parasites of Pacific herring (*Clupea harengus pallasi* Valenciennes): a preliminary evaluation of parasites as indicators of geographical origin for spawning herring. *Can. J. Zool.*, **58**, 521–7.

Bailey, R. E., Margolis, L. and Groot, C. (1988) Estimating stock composition of migrating juvenile Fraser River (British Columbia) sockeye salmon, *Oncorhynchus nerka*, using parasites as natural tags. *Can J. Fish. Aquat. Sci.*, **45**. 586–91.

Beverley-Burton, M. and Early, G. (1982) *Deretrema philippinensis* n. sp. (Digenea: Zoogonidae) from *Anomalops katoptron* (Perciformes: Anomalopidae) from the Philippines. *Can. J. Zool.*, **60**, 2403–8.

Bourgeois, C. E. and Ni, I. H. (1984) Metazoan parasites of northwest Atlantic redfishes (*Sebastes* spp.). *Can. J. Zool.*, **62**, 1879–85.

Brown, J. H. and Kodric-Brown, A. (1977) Turnover rates in insular biogeography: effect of immigration on extinction. *Ecology*, **58**, 445–9.

Burn, P. R. (1980) Density dependent regulation of a fish trematode population. *J. Parasitol.*, **66**, 173–4.

Bush, A. O. and Holmes, J. C. (1986a) Intestinal helminths of lesser scaup ducks: patterns of association. *Can. J. Zool.*, **64**, 132–41.

Bush, A. O. and Holmes, J. C. (1986b) Intestinal helminths of lesser scaup ducks: an interactive community. *Can. J. Zool.*, **64**, 142–52.

Butterworth, E. W. (1982) *A Study of the Structure and Organization of Intestinal Helminth Communities in Ten Species of Waterfowl (Anatinae)* PhD Thesis, University of Alberta, Edmonton, Alberta.

Campbell, R. A. (1983) Parasitism in the deep sea. In *The Sea* (ed. G. T. Rowe), John Wiley and Sons, New York, pp. 473–552.

Curtis, M. A. and Stenzel, A. (1984) Parasitological evidence for food specialization within Arctic char populations. (abstr.). *Parasitology*, **89**, xxxii.

Dayton, P. K. and Tegner, M. J. (1984) Catastrophic storms, El Niño, and patch stability in a Southern California kelp community. *Science*, **224**, 283–5.

Dogiel, V. A. (1964) *General Parasitology*, Oliver and Boyd, Edinburgh.

Dogiel, V. A., Petrushevski, G. K. and Polyanski, Yu. I. (1961) *Parasitology of Fishes*, Oliver and Boyd, Edinburgh.

Freeland, W. J. (1983) Parasites and the coexistence of host animal species. *Am. Nat.*, **121**, 223–36.

Goater, C. P. and Bush, A. O. (1988) Intestinal helminth communities in long-billed curlews: the importance of congeneric host-specialists. *Holarctic. Ecol.*, **11**, 140–5.

Holmes, J. C. (1971) Habitat segregation in sanguinicolid blood flukes (Digenea) of scorpaenid rockfishes (Perciformes) on the Pacific Coast of North America. *J. Fish. Res. Bd. Can.*, **28**, 903–9.

Holmes, J. C. (1987) The structure of helminth communities. *Intern. J. Parasitol.*, **17**, 203–8.

Holmes, J. C. (1988) Progress in ecological parasitology – parasite communities. *Parazitologiya*, **22**, 113–22 (in Russian).

Holmes, J. C., Hobbs, R. P. and Leong, T. S. (1977) Populations in perspective: community organization and regulation of parasite populations. In *Regulation of Parasite Populations*, (ed. G. W. Esch), Academic Press, New York, pp. 209–45.

Holmes, J. C. and Price, P. W. (1986) Communities of parasites. In *Community Ecology: Pattern and Process* (eds, J. Kikkawa and D. J. Anderson), Blackwell Scientific Publications, Oxford, pp. 187–213.

Houston, K. A. and Haedrich, R. L. (1986) Food habits and intestinal parasites of deep demersal fishes from the upper continental slope east of Newfoundland, Northwest Atlantic Ocean. *Mar. Biol.*, **92**, 563–74.

Johnson, C. R. and Mann, K. H. (1988) Diversity, patterns of adaptation, and stability of Nova Scotian kelp beds. *Ecol. Monogr.*, **58**, 129–54.

Kabata, Z. (1959) Ecology of the genus *Acanthochondria* Oakley (Copepoda Parasitica). *J. Mar. Biol. Assoc. U.K.*, **38**, 249–61.

Kabata, Z., McFarlane, G. A. and Whitaker, D. J. (1988) Trematoda of sablefish, *Anoplopoma fimbria* (Pallas, 1811), as possible biological tags for stock identification. *Can. J. Zool.*, **66**, 195–200.

Kabata, Z. and Whitaker, D. J. (1984) Results of three investigations of the parasite fauna of several marine fishes of British Columbia. *Can. Tech. Rep. Fish. Aquat. Sci.*, **1303**, 1–19.

Kennedy, C. R. (1985) Site segregation by species of Acanthocephala in fish, with special reference to eels, *Anguilla anguilla*. *Parasitology*, **90**, 375–90.

Kennedy, C. R., Bush, A. O. and Aho, J. M. (1986) Patterns in helminth communities: why are birds and fish different? *Parasitology*, **93**, 205–15.

Køie, M. (1983) Digenetic trematodes from *Limanda limanda* (L.) (Osteichthyes, Pleuronectidae) from Danish and adjacent waters, with special reference to their life histories. *Ophelia*, **22**, 201–28.

Leaman, B. M. and Kabata, Z. (1987) *Neobrachiella robusta* (Wilson, 1912) (Copepoda: Lernaeopodidae) as a tag for identification of stocks of its host, *Sebastes alutus* (Gilbert, 1890) (Pisces: Teleostei). *Can. J. Zool.*, **65**, 2579–82.

Leong, T. S. and Holmes, J. C. (1981) Communities of Metazoan parasites in open water fishes of Cold Lake, Alberta. *J. Fish. Biol.*, **18**, 693–713.

Lester, R. J. G. (1984) A review of methods for estimating mortality due to parasites in wild fish populations. *Helgol. Meeresunters.*, **37**, 53–64.

Llewellyn, J. (1956) The host-specificity, micro-ecology, adhesive attitudes and comparative morphology of some trematode gill parasites. *J. Mar. Biol. Assoc. U.K.*, **35**, 113–27.

Love, M. S. and Moser, M. (1983) A checklist of parasites of California, Oregon and Washington marine and estuarine fishes. *NOAA Tech. Rep. NMFS SSRF-777*, NOAA, Washington, D.C., pp. 1–576.

MacKenzie, K. (1983) Parasites as biological tags in fish population studies. In *Advances in Applied Biology*, (ed. R. H. Coaker), Academic Press, Cambridge, pp. 251–3.

MacKenzie, K. (1987) Parasites as indicators of host populations. *Intern. J. Parasitol.*, **17**, 345–52.

MacKenzie, K. and Gibson, D. (1970) Ecological studies of some parasites of plaice, *Pleuronectes platessa* (L.), and flounder, *Platichthys flesus* (L.). In *Aspects of Fish Parasitology*, (eds, A. E. R. Taylor and R. Muller), *Symp. Brit. Soc. Parasitology*, Blackwell Scientific Publications, Oxford, pp. 1–42.

Margolis, L. (1963) Parasites as indicators of the geographical origin of sockeye salmon, *Oncorhynchus nerka* (Walbaum) occurring in the North Pacific Ocean and adjacent seas. *Bull. Intern. N. Pac. Fish Comm.*, **11**, 101–56.

Margolis, L. (1982) Parasitology of Pacific salmon – an overview. In *Aspects of Parasitology – A Festschrift dedicated to the fiftieth anniversary of the Institute of Parasitology of McGill University*, (ed. E. Meerovitch), McGill University, Montreal, pp. 135–226.

Margolis, L. and Arthur, J. R. (1979) Synopsis of the parasites of fishes of Canada. *Fish. Res. Bd. Can., Bull.*, **199**, Canada Fisheries and Oceans, Ottawa.

McVicar, A. H. (1979) The distribution of cestodes within the spiral intestine of *Raja naevus* Muller and Henle. *Intern. J. Parasitol.*, **9**, 165–76.

Moller, H. (1987) Pollution and parasitism in the aquatic environment. *Intern. J. Parasitol.*, **17**, 353–61.

Noble, E. R. (1973) Parasites and fishes in a deep-sea environment. *Adv. Mar. Biol.*, **11**, 121–95.

Oshima, T. and Kliks, M. (1987) Effects of marine parasites on human health. *Intern. J. Parasitol.*, **17**, 415–21.

Paperna, I. (1987) Solving parasite-related problems in cultured marine fish. *Intern. J. Parasitol.*, **17**, 327–36.

Peet, R. K. (1974) The measurement of species diversity. *Ann. Rev. Ecol. Syst.*, **5**, 285–307.

Polyanski, Yu. I. (1961a) Ecology of parasites of marine fishes. In *Parasitology of Fishes*, (eds, V. A. Dogiel, G. K. Petrushevski and Yu. I. Polyanski), Oliver and Boyd, Edinburgh, pp. 48–83.

Polyanski, Yu. I. (1961b) Zoogeography of the parasites of USSR marine fishes. In *Parasitology of Fishes*, (eds, V. A. Dogiel, G. K. Petrushevski and Yu. I. Polyanski), Oliver and Boyd, Edinburgh, pp. 230–45.

Rohde, K. (1977a) Habitat partitioning in Monogenea of marine fishes. *Heteromicrocotyla australiensis*, sp. nov. and *Heteromicrocotyloides mirabilis*, gen. et sp. nov. (Heteromicrocotylidae) on the gills of *Carangoides emburyi* (Carangidae) on the Great Barrier Reef, Australia. *Z. Parasitenk.*, **53**, 171–82.

Rohde, K. (1977b) A non-competitive mechanism responsible for restricting niches. *Zool. Anz.*, **199**, 164–72.

Rohde, K. (1978a) Latitudinal gradients in species diversity and their causes. II. Marine parasitological evidence for a time hypothesis. *Biol. Zentralbl.*, **97**, 405–18.

Rohde, K. (1978b) Latitudinal differences in host-specificity of marine Monogenea and Digenea. *Mar. Biol.*, **47**, 125–34.

Rohde, K. (1979) A critical evaluation of intrinsic and extrinsic factors responsible for niche restriction in parasites. *Am. Nat.*, **114**, 648–71.

Rohde, K. (1980) Host specificity indices of parasites and their application. *Experientia*, **36**, 1368–9.

Rohde, K. (1982) *Ecology of Marine Parasites*, University of Queensland Press, St. Lucia, Queensland.

Sankurathri, C. S., Kabata, Z. and Whitaker, D. J. (1983) Parasites of the Pacific hake, *Merluccius productus* (Ayres, 1855) in the Strait of Georgia, in 1974–1975. *Syesis*, **16**, 5–22.

Scott, J. S. (1980) Digenean parasite communities in flatfishes of the Scotian Shelf and Southern Gulf of St. Lawrence. *Can. J. Zool.*, **60**, 2804–11.

Scott, J. S. (1981) Alimentary tract parasites of haddock (*Melanogrammus aeglefinus* L.) on the Scotian Shelf. *Can. J. Zool.*, **59**, 2244–52.

Scott, J. S. (1985a) Occurrence of alimentary tract helminth parasites of pollack (*Pollachius virens* L.) on the Scotian Shelf. *Can. J. Zool.*, **63**, 1695–8.

Scott, J. S. (1985b) Digenean (Trematoda) populations in winter flounder (*Pseudopleuronectes americanus*) from Passamaquoddy Bay, New Brunswick, Canada. *Can. J. Zool.*, **63**, 1699–705.

Scott, J. S. (1987) Helminth parasites of the alimentary tract of the hakes (*Merluccius, Urophycis, Phycis*: Teleostei) of the Scotian Shelf. *Can. J. Zool.*, **65**, 304–11.

Scott, J. S. (1988) Helminth parasites of redfish (*Sebastes fasciatus*) from the Scotian Shelf, Bay of Fundy, and Eastern Gulf of Maine. *Can. J. Zool.*, **66**, 617–21.

Sekerak, A. D. (1975) *Parasites as Indicators of Populations and Species of Rockfishes (Sebastes: Scorpaenidae) of the Northeastern Pacific Ocean*, PhD Thesis, University of Calgary, Calgary, Alberta.

Sekhar, C. S. and Threlfall, W. (1970) Helminth parasites of the cunner, *Tautogolabrus adspersus* (Walbaum) in Newfoundland, *J. Helminthol.*, **44**, 169–88.

Shostak, A. W. (1986) *Sources of Variability in Life-history Characteristics of the Annual Phase of Triaenophorus crassus (Cestoda: Pseudophyllidea)*, PhD Thesis, University of Manitoba, Winnipeg, Manitoba.

Shostak, A. W. and Dick, T. A. (1986) Intestinal pathology in northern pike, *Esox lucius* L., infected with *Triaenophorus crassus* Forel, 1868 (Cestoda: Pseudophyllidea). *J. Fish. Dis.*, **9**, 35–45.

Shotter, R. A. (1976) The distribution of some helminth and copepod parasites in tissues of whiting, *Merlangius merlangus* L., from Manx Waters. *J. Fish. Biol.*, **8**, 101–17.

Sindermann, C. J. (1983) Parasites as natural tags for marine fish: a review. *NAFO Science Counc. Stud.*, **6**, 63–71.

Sindermann, C. J. (1987) Effects of parasites on fish populations: practical considerations. *Intern. J. Parasitol.*, **17**, 371–82.

Sogandares-Bernal, F. (1959) Digenetic trematodes of marine fishes from the Gulf of Panama and Bimini, British West Indies. *Tulane Stud. Zool.*, **7**, 70–117.

Stock, T. M. and Holmes, J. C. (1987) Host specificity and exchange of intestinal helminths among four species of grebes (Podicipedidae). *Can. J. Zool.*, **65**, 669–76.

Stock, T. M. and Holmes, J. C. (1988) Functional relationships and microhabitat distributions of enteric helminths of grebes (Podicipedidae): the evidence for interactive communities. *J. Parasitol.*, **74**, 214–27.

Sulgostowska, T., Banaczyk, G. and Grabda-Kazubska, B. (1987) Helminth fauna of flatfish (Pleuronectiformes) from Gdansk Bay and adjacent areas (South-east Baltic). *Acta Parasitol. Polon.*, **31**, 231–40.

Uglem, G. L. and Beck, S. M. (1972) Habitat specificity and correlated aminopeptidase activity in the acanthocephalans *Neoechinorhynchus cristatus* and *N. crassus*. *J. Parasitol.*, **58**, 911–20.

Wellington, G. M. and Victor, B. C. (1988) Variation in components of reproductive success in an undersaturated population of coral-reef damselfish: a field perspective. *Am. Nat.*, **131**, 588–601.

Williams, H. H. (1960) The intestine in members of the genus *Raja* and host specificity in the tetraphyllidea. *Nature*, **188**, 514–16.

Williams, H. H., Colin, J. A. and Halvorsen, O. (1987) Biology of Gyrocotylideans with emphasis on reproduction, population ecology and phylogeny. *Parasitology*, **95**, 173–207.

Williams, H. H. and Jones, A. (1976) Marine helminths and human health. Farnham Royal, UK Commonwealth Agricultural Bureaux, *CIH Misc. Publ., No. 3*.

Wootten, R. (1974) The spatial distribution of *Dactylogyrus amphibothrium* on the gills of ruffe *Gymnocephalus cernua* and its relation to the relative amounts of water passing over the parts of the gills. *J. Helminthol.*, **48**, 167–74.

6

Helminth communities in freshwater fish: structured communities or stochastic assemblages?

C. R. Kennedy

6.1 INTRODUCTION

Historical perspectives

The helminth parasites of freshwater fish have long proved an attractive field for parasitologists, and have spawned an extensive and voluminous literature. It was not until the seminal publications of Dogiel (1961, 1964), however, that attention was directed primarily to their ecology. Dogiel reviewed much of the existing literature in the light of ecological concepts, and examined the dependence of the parasite fauna as a whole, the parasitocoenosis, upon the environment. As well as laying down general principles, he discussed in detail (1961) the influence of physical factors, such as water chemistry and habitat size, and biological factors, such as host age, diet and migration, upon the composition of the parasite fauna of a host population. Dogiel's reviews came to form the foundations not only for the study of the ecology of fish parasites, but also for the new discipline of ecological parasitology.

In the following decades, there was something of an explosion of publications on fish parasites. Many investigators, for example Bauer and Stolyarov (1961) and Pugachev (1983), have continued to follow along the lines indicated by Dogiel and have concentrated on parasitocoenoses, but many others, e.g. Chubb (1964), Anderson (1974, 1978) and Kennedy (1977, 1985a) diversified into the field of population dynamics. Taxonomic studies have continued to appear (Amin, 1986; Gibson, 1987), as have surveys of fish parasites from all parts of the world (e.g. Hanek and Threlfall, 1970; Batra, 1984; Chen, 1984; Moravec, 1985). In the face of this extensive literature, it might therefore be thought that there is little or no need for an analytical review of helminth communities in freshwater fish. Unfortunately, despite the numerous surveys and the large quantity of data available, few studies actually provide information suitable for analyses of community structure and determinant processes and fewer still have actually attempted such analyses. Many studies were undertaken for other purposes, and are often incomplete in respect of quantitative coverage of all parasites and/or specific determinations of all species. Above all, there is a dearth of information in the literature on individual hosts, i.e. the infracommunity level, as the great majority of published accounts present only sample summaries of infection parameters. At the compound community level, the difficulties of surveying the entire parasite community in a locality may be virtually insuperable. Most of the published data are descriptive and refer to the component community level. It is only in the last few years, following the development of a theoretical framework for parasite community studies by Holmes (1973, 1983, 1986), Holmes and Price (1986) and Price 1984a, b, 1986), that some authors (Rohde, 1979; Kennedy, 1981; Price and Clancy, 1983) have actually started to address the fundamental problems of organization, replicability and determinant processes of helminth communities of fish.

Current perspectives

General

Amongst the most recent studies on helminth communities in freshwater fish are a few that take a more general, holistic approach. The most general study is that of Kennedy *et al.* (1986a), who examined the fundamental differences between communities of helminths in fish and bird hosts. They concluded that helminth communities in fish were significantly poorer in numbers of species, individuals and diversity than such communities in birds. They further identified a number of factors as being essential to the production of a diverse helminth community, including complexity of the host alimentary canal, endothermy, host vagility and the breadth and selectivity of the host diet. Fish, being ectothermic, less vagile than birds and having a simpler alimentary tract, would be expected to have less diverse intestinal helminth communities. The role of colonization in influencing helminth community structure was explored in a subsequent paper by Esch *et al.* (1988). Focusing on helminth colonization strategies, they recognized two categories of helminths: autogenic species which matured in fish and allogenic species which matured in vertebrates other than fish and thus had a greater colonization potential and ability. In their view, recognition and appreciation of the different colonization strategies of autogenic and allogenic species in respect of host vagility and ability to cross land and sea barriers, thus breaking down habitat isolation, provided an understanding of, and explanation for, the patchy spatial distribution and apparently stochastic nature of many freshwater fish helminth communities. They thus considered that colonization is a major determinant of helminth community structure.

Compound communities

Many recent studies have, by contrast, focused on habitats rather than processes *per se*. At the compound community level, there have been a number of attempts to predict the nature of a parasite community on the basis of knowledge of the characteristics of the locality. This approach was pioneered by Wisniewski (1958), who believed that the character of a water body, especially the trophic status of a lake, influenced and determined the composition of the parasite species present, i.e. the nature of the compound community. Whilst these views received some support from the studies of Chubb (1963, 1970), other authors believed that the same fish species had similar parasites in water bodies of widely differing trophic status, and that the composition of a compound community thus depended mainly upon the fish species present (Halvorsen, 1971; Wootten, 1973). The parasites of a dominant fish species or group of fish species could then come to dominate the whole lake (Leong and Holmes, 1981). In contrast, Esch (1971) concluded that the nature of the predator–prey relationships provided the best predictor of the structure of a parasite community in any given aquatic ecosystem. Kennedy (1978a), emphasized the differences between lakes associated with historical factors and colonization events, and stressed the individuality of each lake in respect of its parasite community. These somewhat conflicting conclusions serve to emphasize the importance of colonization as a community determinant, but otherwise reveal relatively little about the processes involved in structuring the community.

Component communities

At the level of the component community there have again been a number of different approaches and views. Brooks (1980) emphasized the importance of the phylogeny of parasites and hosts in determining the development and organization of helminth communities, and believed that most parasite communities can be considered as co-evolving units. By contrast, Price (1980) considered that most parasite communities are far more recent, and are rather more in the nature of chance assemblages that have seldom, if ever, co-evolved. Some support for this view comes from the several studies that have attempted to use island biogeographical theory (MacArthur and Wilson, 1967) as a predictor of helminth community structure in freshwater localities. Kennedy (1978a, b) examined helminth communities in *Salmo trutta* in British lakes, and later in *Salvelinus alpinus* on Arctic islands. He concluded that the theory was not a very good predictor of parasite community richness. In general terms, the larger the habitat, the richer was the parasite community but more distant habitats often had parasite communities as rich as, or richer than, nearer ones. It proved impossible, also, to predict the occurrence of any particular helminth species, let alone community, composition. A similar approach by Price and Clancy (1983) showed that within Britain, the wider the range of a fish species, the richer was its helminth fauna. In a later paper, Kennedy *et al.* (1986b) surveyed the entire fish parasite fauna of an island in the English Channel and, as a result of this and earlier studies, concluded that the presence of a helminth species in any particular habitat depended primarily on chance colonization events and that helminth communities in freshwater fish were, on the whole, chance assemblages rather than structured organizations.

Infracommunities

There have been very few attempts to study helminth infracommunity structure in fishes. Although information on site location by one or more particular species of helminth is given in many publications, very few workers have placed their findings in the context of niche theory, or have actually addressed the concept of the niche in studies of fish helminth communities. Rohde (1979) has used this approach, but was concerned primarily with marine mongeneans. Crompton (1973) and Holmes (1973) included freshwater helminths in their reviews of site location, and Holmes also attempted to relate the data to niche theory and to consider location in relation to selective or interactive site segregation by helminths. In a later detailed study of site selection by five species of acanthocephalans in fish, Kennedy (1985b) demonstrated that each species showed a degree of site preference, but, since niche overlap was high, all could be considered host and site generalists. Fidelity of site location was poor in individual fish, the concepts of site specificity and segregation were at best applicable only at the population (component) level, and mixed species infections were extremely rare. He concluded that acanthocephalan communities scarcely existed as such; the normal situation was chance assemblages, with vacant niches, little organization and few or no interactions between species. In a few localities, congeneric species of acanthocephalans can co-exist without any clear evidence of interspecific competition (Uglem and Beck, 1972; Kennedy and

Moriarty, 1987), although the possibility of such competition occurring in the future or having occurred in the past could not be excluded. Interactions, presumably competitive, between cestodes and acanthocephalans in fish intestines have been reported on rare occasions (Chappell, 1969; Grey and Hayunga, 1980), but their significance in structuring helminth communities is far from clear. In other localities (Kennedy, 1985a, b; Kennedy *et al.*, 1986b), there are clear indications of vacant niches in helminth communities. Overall, there appear to be both site specialists and generalists present amongst fish helminths and, amongst the latter, niche definition is often poor. Since the niche itself is often very broad, it is not surprising that the identity of the species occupying it tends to vary from locality to locality in an unpredictable and stochastic manner (Kennedy, 1985b). Interactions are scarce in the species poor communities.

Current problems

It should be clear from the foregoing remarks, that despite the considerable amount of information available, we are still far from understanding the processes involved in the structuring and organization of fish helminth communities. There are conflicting opinions on the role of phylogeny and the importance of co-evolutionary processes, on the role of habitat area and isolation, on the nature of the niche and its occupancy and on the significance of intra- and interspecific interactions. Indeed, it is still unclear whether helminth communities in freshwater fish are structured and organized, or are merely stochastic assemblages, or whether they are isolationist in nature or interactive (Holmes, 1986; Holmes and Price, 1986).

These basic general questions of how helminth communities in freshwater fish are structured and what factors or processes determine their composition and organization can be best addressed by posing a number of more specific questions. It is necessary to ask whether, and under what conditions, communities are diverse in character or are dominated by a few species only, and whether it is possible to recognize core and satellite species. We need to know whether communities are predictable from individual to individual host at the infracommunity level, and from locality to locality at the component level. We need to know whether communities are generally dominated by host specialists (confined almost exclusively to one species or genus of hosts) or generalists. We need to know whether there is a fixed number of niches available in a single host and, if so, how many. Then we need to know whether they are all filled, or whether some are vacant. Finally, we need to know who fills them under what circumstances, and whether communities can be invaded. Clearly it is not going to be possible to answer many of these questions until a great deal more information of the right quality becomes available, but an attempt will be made to answer some of them in this chapter.

Therefore instead of continuing to attempt to review the literature regarding all helminths in all freshwater fish, an alternative approach has been adopted here for the remainder of this chapter. I will focus intensively upon helminth communities in one species of fish in a restricted geographical area using, for the most part, new and original data. This has the advantage of ensuring taxonomic consistency, thorough spatial coverage in sampling, and uniformity of data treatment. For reasons discussed by Kennedy *et al.* (1986a), the investigations are concerned only with helminths from the alimentary tract.

The species of fish selected is the European eel, *Anguilla anguilla* (L), and the geographical area is the United Kingdom and Ireland. Eels were chosen for several reasons. They are common and abundant and their catadramous life history and physiological tolerance contribute to their ability to occur throughout the whole of the area and in all types of freshwater habitat. There are thus no problems arising from local or restricted distributions of the host. Eels are recognized as being good colonists, able to invade new localities by land or sea and are believed to have been one of the first species to colonize the British Isles post-glacially. They are thus not recent immigrants, and as an old, phylogenetically isolated, group of fish, harbour a suite of specialist parasites, specific to anguillids, as well as a number of generalist species. Finally, previous studies in Britain (Kennedy, 1985b; Esch *et al.*, 1988) have provided some background data on the composition and structure of their gastro-intestinal helminth communities and on niche segregation by helminths (Kennedy and Moriarty, 1987). The investigation will be restricted to the infra- and component community levels, especially the latter, and it is hoped that such a novel, intensive study of one species of freshwater fish in a limited area will enable at least some of the questions posed above to be answered.

6.2 MATERIALS AND METHODS

Samples of adult eels were collected by a variety of methods from 39 localities in Britain and Ireland, including one canal, ten lakes and 28 rivers or streams (Table 6.1). Extensive coverage of all regions of England, Wales and Ireland was achieved. Wherever possible a minimum of 15 eels of a similar size range was examined from each locality, all samples being taken between October and December. Eels and their parasites were treated in the manner described previously by Kennedy (1985b) and Kennedy *et al.*,

(1986a). Altogether, data from 842 eels from these 39 localities were available for infracommunity analysis. Published data from an additional 11 localities in the United Kingdom and Ireland were used for some aspects of component community analysis.

To describe helminth communities in each individual fish, the number of species of helminths and the total number of individuals in the alimentary canal were recorded. Preliminary studies had indicated species poor communities often dominated by a single species, so in order to focus attention on the dominant helminth species in each host in each locality, the non-parametric Berger–Parker dominance index was used (Southwood, 1978) as this measures the proportion of the total sample that is due to the dominant species. It is free of sample bias and can also serve as a measure of community diversity (Southwood, 1978). Site location and niche dimensions of helminths are not considered here, partly through limitations of space, and partly because some aspects have been considered previously (Kennedy, 1985b). Similarity of communities between individuals within and between localities has also been considered previously (Esch *et al.*, 1988) and so is not discussed in any detail here.

6.3 RESULTS

General

Altogether, 19 species of helminths were recorded from the intestines of eels in the survey (Table 6.2). Of these, five species were strict host specialists found only in eels, eight were fairly broad generalists that occur in several species of fish in addition to eels and the remaining species are specialists of other hosts that appear to be accidental occurrences in eels.

Preliminary investigations of the inter-relationships between the parameters used to describe the helminth communities revealed that over the range studied there were no significant relationships between the percentage occurrence of all helminth species and sample size, total number of helminth species in a sample and dominance index, nor between sample size and total number of helminth species and dominance index. There was, as expected, a relationship between dominance index and total number of helminth species, since in a species poor community containing only a single helminth species there must *de facto* be a dominance index of 1. Thus, with this one exception, none of the parameters studied were related to sample size or to each other.

Table 6.1 The occurrence and dominance of intestinal helminths in the localities studied

Locality	*Nature of region*	*Sample size*	*Prevalence of all species*	*Total no. of species*	*No. of specialist species*	*Dominance index*	*Dominant species*
Avon	River, S.England	24	66.6	3	1	0.96	*Pomphorhynchus laevis*
Clyst	River, S.W.England	32	56.2	6	4	0.3	*Paraquimperia tenerrima*
Crafnant	River, Wales	10	10.0	1	1	1.0	*Bothriocephalus claviceps*
Conway	River, Wales	25	24.0	3	2	0.67	*B. claviceps*
Culm	River, S.W.England	25	24.0	3	2	0.4	*Acanthocephalus lucii*
Dart	River, S.W.England	30	13.3	2	1	0.72	*P. tenerrima*
Driffield	Stream, Mid England	25	28.0	3	1	0.6	*B. claviceps*
Douglas	River, N.W.England	35	60.0	3	2	0.91	*A. lucii*
Erme	River, S.W.England	30	16.7	1	1	1.0	*P. tenerrima*
Exminister	Stream, S.W.England	18	100.0	5	2	0.49	*Acanthocephalus clavula*
Frome	River, S.England	10	60.0	3	2	0.84	*A. lucii*
Gara	River, S.W.England	20	5.0	1	1	1.0	*P. tenerrima*
Idle	River, Mid England	16	75.0	1	0	1.0	*Acanthocephalus anguillae*
Itchen	River, S.England	13	53.8	3	1	0.87	*Raphidascaris acus*
Jersey	River, Channel Isles	20	55.0	3	2	0.88	*Echinorhynchus truttae*

Lune	River, N.England	23	0.0	0	0	0	None
Mawn	River, Mid England	21	52.4	3	1	0.94	*A. lucii*
Otter	River, S.W.England	35	48.6	5	3	0.41	*P. tenerrima*
Ouse	River, E.England	10	40.0	1	0	1.0	*A. clavula*
Ribble	River, N.England	18	5.5	1	1	1.0	*Spinitectus inermis*
Shannon	River, Ireland	35	85.7	4	2	0.98	*A. lucii*
Sheme	River, N.England	24	37.5	3	1	0.68	*P. tenerrima*
Stour	River, E.England	30	63.3	4	2	0.61	*B. claviceps*
Teifi	River, Wales	25	16.0	2	2	0.73	*P. tenerrima*
Trent	River, Mid England	27	85.2	1	1	1.0	*E. truttae*
Tyne	River, N.England	21	19.0	3	2	0.5	*P. tenerrima*
Wear	N.England	19	21.0	3	3	0.57	*P. tenerrima*
Wych	River, N.England	20	75.0	1	0	1.0	*A. clavula*
Lancaster	Canal, N.England	18	44.4	3	2	0.8	*A. lucii*
Bofin	Lake, Ireland	8	100.0	1	0	1.0	*A. anguillae*
Bodey	Lake, Ireland	8	100.0	2	0	0.83	*A. lucii*
Bunny	Lake, Ireland	11	54.5	3	2	0.45	*A. lucii*
Coosan	Lake, Ireland	8	100.0	2	0	0.79	*A. lucii*
Dromore	Lake, Ireland	17	88.2	3	2	0.92	*A. lucii*
Hanningfield	Lake, E.England	21	42.8	2	1	0.96	*A. lucii*
Neagh	Lake, Ireland	30	63.3	3	2	0.97	*A. lucii*
Ree	Lake, Ireland	16	80.5	3	1	0.85	*A. lucii*
Shobrooke	Lake, S.W.England	55	61.8	4	2	0.65	*B. claviceps*
Slapton	Lake, S.W.England	35	55.0	3	2	0.96	*A. clavula*

Table 6.2 The occurrence and dominance of species in the helminth communities of eels

		Component community level			*Infracommunity level*		
Parasite species	*Status*	*No. of localities where occurs* ($n = 50$)	*Localities where dominant (%)*	*Localities in which it occurs where dominant (%)*	*No. of eels where occurs* ($n = 842$)	*Infected eels where dominant (%)*	*Eels in which it occurs where dominant (%)*
Sphaerostoma bramae	G	3	2.1	33.3	3	0.2	33.3
Credpidostomum farionis	A	1	0	0	NP		
Credpidostomum metoecus	A	1	0	0	NP		
Deropristis inflata	S	2	0	0	3	0.7	100.0
Bothriocephalus claviceps	S	33	12.7	18.2	94	15.8	72.3
Proteocephalus macrocephalus	S	10	0	0	10	1.4	60.0
Cyathocephalus truncata	A	1	0	0	NP		
Paraquimperia tenerrima	S	32	23.4	34.3	99	15.5	67.6
Spinitectus inermis	S	3	2.1	33.3	3	0.2	33.3
Raphidascaris acus	G	12	2.1	8.3	14	2.1	64.2
Camallanus lacustris	A	4	0	0	12	2.1	75.0
Cucullanus truttae	A	6	0	0	NP		
Capillaria sp.	A	2	0	0	9	1.4	66.7
Acanthocephalus clavula	G	13	10.6	38.5	101	21.1	90.1
Acanthocephalus lucii	G	20	31.9	75.0	127	22.9	77.9
Acanthocephalus anguillae	G	9	4.2	22.2	32	4.8	65.6
Neoechinorhynchus rutili	G	3	2.1	33.3	3	0.7	100.0
Pomphorhynchus laevis	G	5	4.2	40.0	17	3.9	100.0
Echinorhynchus truttae	G	2	4.2	100.0	31	6.9	96.7

Key: G = generalist, S = specialist, A = accidental, NP = not present in my samples. Infracommunity data original (39 localities): component community data original and from Chubb (1963); Conneelly and McCarthy (1986); Jeacock (1969); Lacey *et al.* (1982); Mishra and Chubb (1969); Thomas (1964); Wootten (1973).

Infracommunity level

Of the 842 eels examined, 409 (48.6%) were infected with one or more species of helminth. The proportion of infected eels varied considerably from locality to locality, with a minimum of 5% and a maximum of 100%. Of those infected, the great majority (77%) harboured only one species of helminth (Fig. 6.1). No eel was found to harbour more than three species, and only 13 (3.2%) had even three species. Most infracommunities thus consisted of only a single species of helminth. Where more than one species occurred in a locality, the species were normally found in different individuals. Where two species co-occurred in a single individual eel, the combination varied from locality to locality. The commonest combination was *Acanthocephalus lucii* and *Paraquimperia tenerrima* (22% of cases), followed by (14% of cases) *A. clavula* and *P. tenerrima*. Co-occurrences of *A. lucii* and *A. anguillae* were found in some Irish lakes. Almost all other possible pairs were found in at least one eel somewhere. In seven out of ten cases where three species co-occurred, one of the acanthocephalan species, the cestode, *Bothriocephalus claviceps*, and the nematode, *P. tenerrima*, were usually present; in each case, an acanthocephalan dominated the infracommunity.

No clear and consistent pattern of species occurrence and dominance is apparent (Table 6.2). This supports the previous studies of Esch *et al.* (1988) which have shown that the mean percentage similarity in parasite communities between individual eels varies greatly from locality to locality. In the present study, some species were found very seldom, whereas others were fairly common. There was no relationship between frequency of occurrence and the specialist or generalist nature of the helminth. For example, *Sphaerostoma bramae* and *Neoechinorhynchus rutili*, both generalists,

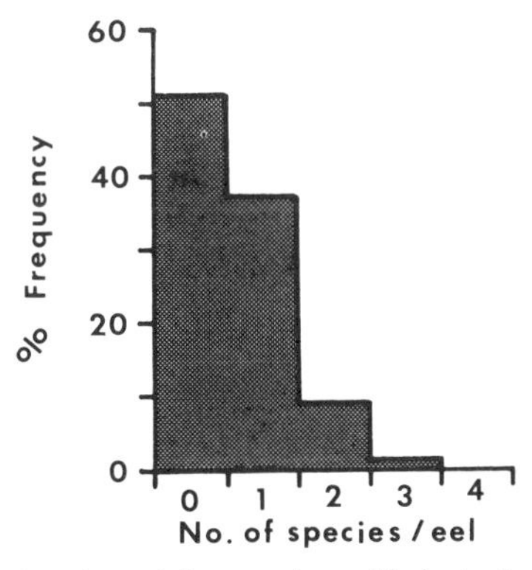

Fig. 6.1 Frequency distribution of the number of helminth species per individual eel (all localities combined).

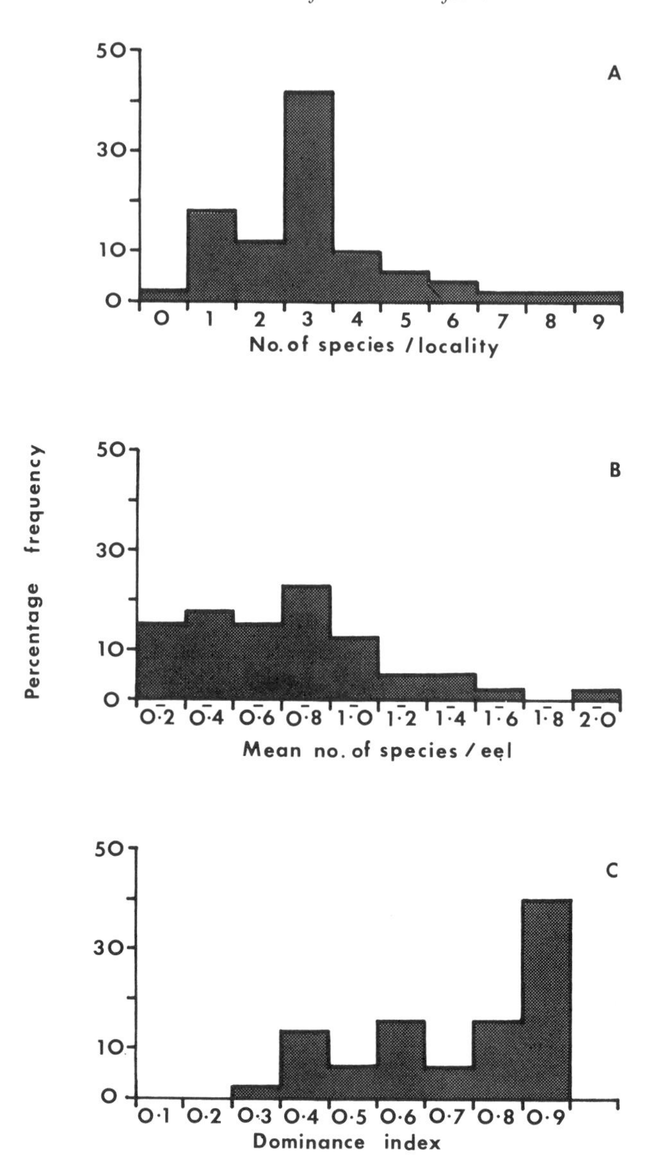

Fig. 6.2 Frequency distributions of selected parameters of helminth component communities in eels.

occurred in only three eels each. Low frequencies of occurrence were also evident in the specialists *Spinitectus inermis, Deropristis inflata* and *Proteocephalus macrocephalus*. The remaining two specialists *Bothriocephalus claviceps* and *P. tenerrima* were far commoner, being found in 94 and 99 eels respectively, but two generalist acanthocephalans, *A. clavula* and *A. lucii* occurred even more commonly. Both specialist and generalist species could dominate the infracommunities (Tables 6.1 and 6.2) and the four commonest species were also found to dominate the infracommunities most frequently. The generalist acanthocephalans were the most common dominants but, nevertheless, every species, even the rarest, dominated some infracommunities.

The pattern of dominance differed somewhat when only those eels in which the species occurred (as opposed to all infected eels) were considered, and the percentage of occurrences in which a species was dominant was examined. Consideration of these data (Table 6.2) confirmed the view that every species can dominate an infracommunity. Even those rare species that dominated less than 1% of all infected eels could nevertheless dominate the infracommunities of those eels in which they occurred far more frequently; for example, *N. rutili* dominated 0.7% of all infected eels, but 100% of those eels in which it occurred. The importance of the commoner species was also more apparent; the specialist *B. claviceps* dominated only 15.8% of all eels, but 72.3% of the eels in which it occurred, and the generalist *A. clavula* dominated 21.1 and 90.1%, respectively. These patterns of dominance to some extent reflect the fact that most infracommunities consist of only a single species (which is thus *de facto* dominant).

Thus, at the infracommunity level, the majority of helminth communities in eels show very low diversity indeed. Within the intestine, the site locations of the different species overlap considerably and it is extremely difficult to define a niche as such (Kennedy, 1985b), let alone determine if there is a change in the realised niche in the presence of a possible competitor when there are so few co-occurrences. Since individual eels can clearly harbour three helminth species, but generally only harbour one, it can be said that there are vacant niches in the majority of infracommunities. There is no discernible pattern or structure to the infracommunities. The overall impression, then, is one of independent assortment by each helminth species; if a species is present in a locality, it may occur in, and dominate some eels and, if it is common, it will dominate the majority. Infracommunities as such scarcely exist and may be dominated by any helminth species whether an eel specialist or a generalist.

Component community level

Of the 50 localities used in analyses of component communities, 49 harboured at least one species of helminth. The maximum number of species found in eels in any locality was nine, but 42% of component communities contained only three species (Fig. 6.2a), and a further 32% two or one.

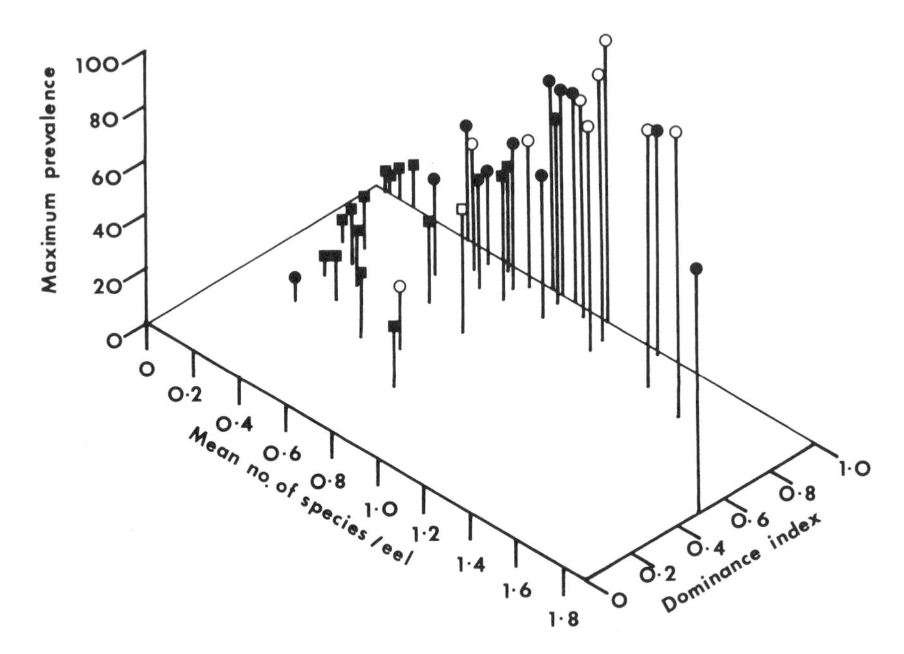

Fig. 6.3 Inter-relationships between the three summary parameters of helminth component communities illustrated in Fig. 6.2. Each point represents one locality. Solid symbols = river; hollow symbols = lake; circles = dominance by acanthocephalan generalists; squares = dominance by eel specialists.

There was no regional or habitat pattern evident (Table 6.1 and Fig. 6.3). One lake and eight rivers harboured only one species, and these included localities in Wales and Ireland as well as in the northeast and southwest of England. Similarly, of the richer communities (more than five species), two were in rivers and three in lakes. In some cases, but by no means all, the richness of the component community was due in part to the acquisition of specialists from other hosts (e.g. *Camallanus lacustris* from percids and *Cucullanus truttae* from salmonids).

The most characteristic feature of the component communities is the concentration of dominance resulting in a lack of diversity (Fig. 6.2c). The dominance index was greater than 0.7 in 62% of the communities and greater than 0.9 in 40%. Thus, the great majority of component communities were dominated by a single species of helminth. The low diversity of the communities is also borne out by the low mean number of species per eel (Fig. 6.2b); in very few localities was this greater than one. Although prevalence varied from locality to locality, there was no correlation between the maximum prevalence of the commonest species and the dominance index ($r = 0.329$, $n = 39$) or between the mean number of species per eel and the dominance index ($r = -0.0073$, $n = 39$). There is thus no indication that a high level of dominance is associated with a high or low prevalance or mean number of species per eel.

The relationship between these three variables (maximum prevalence of any species; mean number of species per eel; dominance index) and hence the general characteristics of the component communities, are shown in Fig. 6.3. The majority of component communities were characterized by high dominance of one species, high maximum prevalence of one species (generally, but not always, the dominant species) and a mean number of species per eel of greater than 0.5. A minority of communities formed a second cluster in which dominance was often still high, but both the maximum prevalence and mean number of species per eel was low. These tended to be in riverine localities in which overall helminth prevalence and abundance were very low and the number of uninfected eels high. There was generally only one (at most two) species of helminth present, most often an eel specialist, and so the community was, *de facto*, dominated by one species despite the paucity of individuals.

The occurrence and identity of the species comprising the component communities were variable (Tables 6.1 and 6.2). Some species occurred very infrequently and some very commonly, but there was no relationship between their frequency of occurrence and their host generalist or specialist nature. Amongst the rare species were the specialists, *D. inflata* and *S. inermis*, and the generalists, *S. bramae* and *E. truttae*. The four commonest species were two specialists, *B. claviceps* and *P. tenerrima*, and two generalists, *A. lucii* and *A. clavula*. Digeneans appeared to be relatively uncommon members of the community and acanthocephalans relatively common ones.

The pattern of dominance shows several features of interest (Table 6.2). There was a general tendency for the commoner species to dominate more often, and so as expected component communities were never dominated by accidentals but only by generalists or eel specialists. All generalists dominated some communities, but not all specialists; neither *D. inflata*, nor *P. macrocephalus* (despite its occurrence in 20% of localities) ever dominated, but *S. inermis*, despite its rarity (occurrence in 6% of localities) did in one community in which it was the only species present. Although the remaining two specialists, *B. claviceps* and *P. tenerrima*, dominated 36.1% of the communities collectively, most of these were in the riverine localities in which few or only one species of helminth occurred in eels and in which prevalence of acanthocephalans was very low (less than 10%) (the second cluster in Fig. 6.3). The majority of communities (54.7%), particularly those in which more than one species was present, were likely to be dominated by an acanthocephalan. The most frequent dominant species was the generalist, *A. lucii* (31.9% of cases). The importance of the acanthocephalans is even more apparent when only those localities in which the species occurs (as opposed to all localities) are considered. For example, *A. lucii* dominated 31.9% of all the communities studied, but 75% of those in which it occurs. *Acanthacephalus clavula* similarly dominated only 10.6% of all localities, but 38% of those communities in which it occurred. Even rare species, such as *E. truttae*, which only occurred in two localities, nevertheless dominated both of them. By contrast, the specialists, *B. claviceps* and *P. tenerrima*, dominated only 12.7% and 23.4% of all communities, respectively, and only 18.2% and 34.3% of those in which they occurred. Therefore, although almost any species may dominate any given component community, it is the generalist acanthocephalans that do so most often and the single most likely species to do so is *A. lucii*. Only if few or no acanthocephalans are present in the community, as for example in several rivers, is it likely to be dominated by an eel specialist, such as *B. claviceps* or *P. tenerrima*.

Component communities, like infracommunities, are thus characterized by their lack of diversity and by their dominance by a single species. The dominance of acanthocephalans, noted at the infracommunity level, is again evident at the component community level. The degree of similarity between component communities in different localities is quite variable (Esch *et al.*, 1988); it may reach 40%, but on average is far lower. However, because of the patchy distribution of each species it does not appear to be possible to predict which species of acanthocephalan or eel specialist will dominate any particular community and it seems simply to be a matter of which species happens to be present in the locality. In the face of such low diversity within each component community and such a concentration of dominance, application of the concept of core and satellite species appears to be of little relevance. In any single community, the dominant species is the only core species, but the satellites in that community may well be the dominants and cores in others.

6.4 DISCUSSION

Eels are an old group of fish, phylogentically and ecologically very distinct. Their catadromous habit, and ability to cross estuarine, and to a limited extent terrestrial barriers, makes them excellent colonizers and so very widespread in their distribution. Tolerance of variable and extreme physiochemical conditions in freshwater allows them to occur in all types of habitat. These features should mean that they and their parasites, which are also often equally tolerant (Kennedy and Lord, 1982), can be disseminated widely throughout a restricted geographical area such as Britain and

Ireland. Consequently, within this area, biogeographical factors should play little or no part in determining the composition of their helminth communities. Their broad diet also means that they have the potential to ingest a wide range of intermediate hosts and so of larval helminth parasites.

The helminth species found in eels fall into three quite distinct groups: there is a suite of specialists, very specific to eels and presumably having co-evolved with that host, a group of generalists, and a group of accidental species. Within the area covered by this study, 19 species, of which five were specialists, were recorded in eels. Despite this potential for a rich and diverse helminth community, the majority of component communities were comprised of only three species (maximum nine) and the majority of infracommunities of only one species (maximum three). The potential for a diverse community does not therefore appear to be realized. Both infra and component communities were similar in general characteristics; they exhibited low diversity, and were usually dominated by a single species. This could be a generalist or a specialist, but was most often a generalist, in particular a species of acanthocephalan.

This close similarity in the general characteristics of the component and infracommunities naturally directs attention to the relationship between these two levels of community organization and suggests two quite different explanations for the observed pattern. The species poor infracommunities could reflect the fact that there is a fixed number of niches available in the alimentary canal of an eel. No matter, therefore, how many species may be present in the component community, an individual eel cannot harbour more than a limited number of helminth species (three, based on the present study). Alternatively, the number of species present in an infracommunity may reflect the outcome of two different processes: (a) the number of species present in a particular locality, i.e. the richness of the compound community and hence colonization events, and (b) transmission and infection opportunities within that locality and, thus, the probability of being infected. In more general terms, this raises the questions of whether the infracommunity determines the component community or, vice versa, of whether the component community is, as defined, truly the sum of all the infracommunities, or whether the infracommunities are merely sub-sets of the component and compound communities. This latter situation would imply that under suitable conditions in a particular locality, i.e. if a large number of generalist species are present, the compound community is diverse and transmission levels are very high, then infracommunity richness could be far higher than observed in the present study. The answers to these questions may be sought initially by considering each community level separately.

At the component level, there are two features of particular interest in this context. The first is the observation that most component communities contain only three species, but may contain up to nine, and the second, the observation that most communities are dominated by the generalist acanthocephalans. The suite of eel specialists can, at best, comprise only five species, and in most localities is represented by only one or two species. Where component communities are richer, it is because eels are infected by species accidently acquired from other hosts or by generalists which are also circulating through other host species in the locality. Only when generalists are absent, or scarce, do specialists dominate communities. Whilst considerations of specificity clearly play some part in the ability of non-specialists to infect eels, it does appear that the composition of the component community reflects in large measure the species of generalists present, and their abundance. The occurrence of digeneans in eels illustrates this point particularly well. In freshwater localities in Britain and Ireland digeneans are a relatively uncommon part of the community, e.g. *S. bramae* occurred in only three localities. When this generalist species is very common in a locality, however, it may, as in Lough Corrib, in Ireland, dominate the eel component community (Conneely and McCarthy, 1986). In marine localities, where digeneans are often abundant and are more commonly found in eels, up to five species may occur in the component community (Whitfield, personal communication), and in freshwater and estuarine localities in continental Europe there may be three to five species of digenean in the component communities of eels (Seyda, 1973; Kazic *et al.*, 1982). Thus, the contribution digeneans make to eel helminth component communities appears to reflect their presence and abundance in the compound community. Indeed, in several localities in continental Europe, the component communities in eels are far richer than those in Britain, e.g. Kazic *et al.* (1982) reported ten species of helminths, Seyda (1973) twelve, and Køie (1988) eleven. In virtually all these examples the increased richness of the community is due to the acquisition of generalist and accidental species, and not to the presence of more species of eel specialists. In Britain and Ireland, the generalists are predominantly acanthocephalans; elsewhere, other species may be involved. Even in Britain, component communities may be enriched and change when habitat conditions change; in 1979, the component community in eels in a small stream in Devon contained only three species, (Kennedy, 1984) but, following a number of habitat changes e.g. eutrophication and embanking and the consequent decline of some invertebrate species, by 1987 it had increased to six species (Kennedy, personal observation), and the dominant species had changed from *A. clavula* to *P. tenerrima*.

The identity of the species comprising the component community thus seems to be largely a matter of chance and to reflect the presence and abundance of the species present in the compound community. This would also appear to hold for the dominant species. Where the compound community is poor and eel specialists are present, as in some small rivers where only salmonid fish are present in addition, the eel specialists may dominate the component community (the minor cluster in Fig. 6.3). When a variety of other fish species are present, the compound community is richer, acanthocephalans or other generalists are present, and they dominate in preference to the specialists (the majority of communities in Fig. 6.3). The specialist suite is clearly an important part of the component community, but generalists dominate more often. The identity of the dominant species appears again to reflect the presence and abundance of the species present in the compound community. Any species of acanthocephalan, for example, may dominate and, if present and abundant, is likely to do so. In this respect, component communities in eels exhibit features comparable to the helminth communities of fish in Cold Lake, Alberta, Canada (Leong and Holmes, 1981). There, helminth communities in all species of fish are characterized and dominated by the most abundant acanthocephalan in the habitat. What is clear is that, in contrast to the views of Brooks (1980), phylogenetic factors are not of major importance as community determinants. Specialists are responsible for some of the similarity between communities in different localities (Esch *et al.*, 1988), but even though at least one species is present in most, though not all, localities (Table 6.1) they tend only to dominate in the absence of generalists and are often reduced to a subservient status in their presence. The question of whether core species are specialists or generalists seems therefore largely irrelevant.

At the infracommunity level, the same two features are of particular interest. First is the fact that the great majority of infected eels contain only one species of helminth although they can contain more, and, second, that individual eels can be dominated by any species of helminth, but are more frequently dominated by generalist acanthocephalans. Even when the component community is fairly rich, e.g. five or six species, the majority of individual eels may be uninfected or harbour only a single species. Since most localities harboured an appreciable number of uninfected eels, it seems likely that the species poor infracommunities merely reflect the low probabilities of transmission for each helminth. If more than one species is present in the component community, but abundance levels and transmission rates are low for each, then the probability of two species occurring in a single eel will be very low. There was no suggestion from the data that any one species regularly excluded any other. On the contrary, almost all possible combinations of two or three species occurred somewhere in some eels.

Nevertheless, co-occurrence, especially of species in the same guild, was rare, and when two or more species did occur in a single eel, they usually included a cestode and/or an acanthocephalan and an nematode. This could be held to suggest a fixed number of niches, each of which is filled stochastically by a member of the same guild. There is some evidence in support of this suggestion. Although acanthocephalans can occupy almost the entire gut length, suggesting very broad niches in eels, mixed species infections are very rare indeed (Kennedy, 1985b) and when they do occur it is under circumstances that do not exclude the possibility of exploitation or interference competition occurring between the species (Kennedy and Moriarty, 1987). The several examples indicative of interactive segregation between acanthocephalans and cestodes in other species of fish (Chappell, 1969; Grey and Hayunga, 1980) also point to the possibility of such interspecific interactions for limited resources occurring in eels.

It is of course never an easy matter to determine how many niches are available in a host; the only measure that can be used here is a comparison of the same host in different localities (Price, 1984b, 1986). On this basis, there are at least three niches available in the eel intestine, this being the maximum number of species found here in any single infracommunity. There may be more, but data on infracommunities from the richer continental localities are not available. If, however, an eel in Britain or Ireland can harbour three species of helminth but normally only harbours one, then it would seem that at least two niches are generally unfilled or vacant, and that most infracommunities are unsaturated. Since this is the commonest situation in the observed infracommunities, it would seem unlikely that it was a fixed number of niches that determined the richness of the infracommunity although this may ultimately determine maximum richness. This does not exclude the possibility of interactions occurring between members of the same guild, e.g. between species of acanthocephalans, but the significance of this as a determinant of community structure of helminths in eels has yet to be demonstrated unequivocally. Based on the data presented here, it appears far more likely that each infracommunity represents a sub-set of the component community, and reflects both the abundance of different species in the locality and chance transmission events to individual eels.

The overall impression given by this study is that helminth communities in eels are potentially rich, but that this richness is seldom approached and never achieved. Both infra and component communities are generally species poor and exhibit low diversity. They are dominated by any species that happens to be present and abundant in the locality. Its presence is a consequence of stochastic colonization events affecting both eel specialists and generalist species, and its abundance is a consequence of favourable condition for its transmission within the locality. The major determinant of helminth community structure and pattern in eels is thus the environmental and habitat conditions that determine the compound community. There is no evidence of interactions between species playing a major determinant role at either level, but this possibility cannot yet be excluded. The evidence for a fixed number of niches in individual eels is equivocal, as it may equally reflect a lack of data from the right locality e.g. one in which the component community is very species rich and the mean number of the species per eel is greater than three. Such communities do occur in some localities and would suggest detailed investigation.

It is very difficult to know just how far these conclusions apply to helminth communities of eels in general. Despite the many studies on selected parasites of eels in other countries, there are few that report on the entire component community and none, to the author's knowledge, that contain data on infracommunities or compound communities. Some localities in continental Europe harbour eels with far richer component communities, e.g. in Denmark (Køie, 1988), in Yugoslavia (Kazic *et al.*, 1982) and in Poland (Seyda, 1973). Others, e.g. in Czechoslovakia (Moravec, 1985), in North America (Hanek and Threlfall, 1970), and in New Zealand (Hine, 1978), harbour component communities of a richness similar to those in the British Isles. When a given community is richer, it is often so because of the acquisition of more species of generalists, including digeneans and nematodes, and accidentals, and it is usually dominated by a generalist, even if not so frequently by an acanthocephalan. It tentatively appears as if the dominance of so many eel communities by acanthocephalans may be primarily a British and Irish characteristic, and may thus reflect the predominance of this group in the British and Irish fish parasite fauna and the absence from these islands of many other species of fish parasite present in continental localities.

It is even more difficult to know how far these conclusions apply to helminth communities in other species of freshwater fish.The present study is the most detailed and concentrated investigation of communities in one species of fish in a defined geographical area, and there are no comparable studies known to the author. However, many features of eel helminth communities are common to those in other groups of fish. For example, salmonids also have a recognizable suite of host specialist helminths but can additionally acquire generalist acanthocephalans (Kennedy, 1978a). Salmonid helminth communities are in general not very rich, diverse or organized when compared to those of aquatic birds. (Kennedy *et al.*, 1986a). The composition of the helminth community in salmonids in any locality appears to be a result of stochastic colonization events (Kennedy, 1978a, 1981), but most component communities are also characterized by the dominance of a single species (Esch *et al.*, 1988). Apparent vacant niches exist at the infracommunity level in some localities (Kennedy *et al.*, 1986b). Following migration to sea, the identity of the occupants of the niches changes (Konovalov, 1975). Such data as do exist on community structure in other species of freshwater fish (Kennedy, 1978a, b, 1985; Kennedy *et al.*, 1986a, b) suggest that, as in eels, helminth communities are isolationist in nature, their composition being determined by chance colonization events.

In many respects these conclusions should not be too surprising, as Kennedy *et al.* (1986a) and Esch *et al.* (1988) have demonstrated the importance of host vagility and parasite colonization processes as determinants of helminth community structure. In freshwater, where habitats are discontinuous and isolated from each other (Kennedy, 1981), ecological factors promoting colonization and subsequent high transmission rates within a locality are likely to be of overwhelming importance as community determinants in comparison to, for example, phylogenetic factors. As Holmes (1986) and Holmes and Price (1986) pointed out, interactive communities are likely to be found only where helminths regularly co-occur at substantial population densities. Since such conditions seldom occur in freshwater habitats, fish helminth communities are likely to be isolationist in nature. Nevertheless, there is some evidence for interactive site segregation amongst fish parasites (Holmes, 1973, and references therein). Moreover, interactions between some pairs of species do occur (Chappell, 1969; Grey and Hayunga, 1980) and the scarcity of mixed species of infections of acanthocephalans in many British fish is still compatible with competition (Kennedy and Moriarty, 1987). Thus the potential for competitive interactions to influence helminth community structure in fish appears to exist, even if seldom realized.

There is a striking parallel here to some of the conclusions reached from studies on fish helminth population regulation. Kennedy (1977, 1985a) considers that the majority of fish parasite populations in natural waters are non-equilibrial, and that their population levels are normally determined by transmission events. Both the potential and mechanisms for population regulation have been recognized, but such regulation is seldom apparent. In both cases, the potential may only be realized in a locality where a large number of species are present at high abundance. There is an urgent need to study the ecology of fish parasites in all aspects in such a locality. Interactive helminth communities may in fact occur in some fishes under certain conditions, but in the localities studied to date the screens identified by Holmes (1986) are too effective. It must also be emphasized that my conclusions apply only to intestinal helminth communities. Communities on gills may be very different, as Rohde (1979) has shown for marine fish. There, and in some freshwater localities (Chen, 1984), up to nine species of congeneric monogeneans may occur in a single component community, several of them in relative abundance (Chen, 1984). If a comparable locality for intestinal helminths can ever be found, where several congeners co-exist, where transmission rates are high, and where the possibility of species packing exists, then it may be that intestinal helminth communities in fish will be more comparable in structure and dynamics to the interactive communities found in birds. The evidence to date, however, suggests that most fish helminth communities are isolationist in nature, and rather more in the way of being stochastic assemblages than structured communities. The infracommunities are indeed, as Holmes (1986) suggested, samples of the specialists in the host component community and generalists in the compound community. Interactions appear to be uncommon if indeed they occur at all.

ACKNOWLEDGEMENTS

I wish to thank all those who obtained eel samples for me, and in particular Eric Hudson of the MAFF Fish Diseases Laboratory, Weymouth; Phil Shears for his assistance with eel capture and examination; Al Bush, John Aho and John Holmes for their helpful and critical comments on drafts of this manuscript; and the British Council, the University of Exeter and Wake Forest University for making it possible for me to attend the Symposium.

REFERENCES

Amin, O. (1986) On the species and populations of the genus *Acanthocephalus* (Acanthocephala: Echinorhynchidae) from North American freshwater fishes: a cladistic analysis. *Proc. Helminthol. Soc. Wash.*, **99**, 574–9.

Anderson, R. M. (1974) Population dynamics of the cestode *Caryophyllaeus laticeps* (Pallas, 1781) in the bream (*Abramis brama* L.). *J. Anim. Ecol.*, **43**, 305–21.

Anderson, R. M. (1978) The regulation of host population growth by parasitic species. *Parasitology*, **76**, 119–57.

Batra, V. (1984) Prevalence of helminth parasites in three species of cichlids from a man-made lake in Zambia. *Zool. J. Linn. Soc.*, **82**, 319–33.

Bauer, O. N. and Stolyarov, V. P. (1961) Formation of the parasite fauna and parasitic diseases of fish in hydro-electric reservoirs. In *Parasitology of Fishes*, (eds, V. A. Dogiel, G. K. Petrushevski and Y. I. Polyanski), Oliver and Boyd, London, pp. 246–54.

Brooks, D. R. (1980) Allopatric speciation and non-interactive parasite community structure. *Syst. Zool.*, **29**, 192–203.

Chappell, L. H. (1969) Competitive exclusion between two intestinal parasites of the three-spined stickleback, *Gasterosteus aculeatus* L. *J. Parasitol.*, **55**, 775–8.

Chen, Chih-leu (1984) Parasitical fauna of fishes from Liao He (Liaoho River) of China. In *Parasitic organisms of freshwater fish of China*, Institute of Hydrobiology Academia Sinica, Beijing, Agricultural Publishing House, pp. 41–81.

Chubb, J. C. (1963) On the characterization of the parasite fauna of the fish of Llyn Tegid. *Proc. Zool. Soc. Lond.*, **141**, 609–21.

Chubb, J. C. (1964) Observations on the occurrence of the plerocercoids of *Triaenophorus nodulosus* (Pallas, 1781) (Cestoda: Pseudophyllidea) in the perch *Perca fluviatilis* L. of Llyn Tegid (Bala Lake) Merioneth. *Parasitology*, **54**, 481–91.

Chubb, J. C. (1970) The parasite fauna of British freshwater fish. *Symp. Brit. Soc. Parasitol.*, **8**, 119–44.

Conneely, J. J. and McCarthy, T. K. (1986) Ecological factors influencing the composition of the parasite fauna of the European eel, *Anguilla anguilla*, (L.), in Ireland. *J. Fish Biol.*, **28**, 207–19.

Crompton, D. W. T. (1973) The sites occupied by some parasitic helminths in the alimentary tract of vertebrates. *Biol. Rev.*, **48**, 27–83.

Dogiel, V. A. (1961) Ecology of the parasites of freshwater fishes. In *Parasitology of Fishes*, (eds, V. A. Dogiel, G. K. Petrushevski and Yu. I. Polyanski), Oliver and Boyd, London, pp. 1–47.

Dogiel, V. A. (1964) *General Parasitology*, Oliver and Boyd, London.

Esch, G. W. (1971) Impact of ecological succession on the parasite fauna in centrachids from oligotrophic and eutrophic ecosystems. *Am. Midl. Nat.*, **86**, 160–8.

Esch, G. W., Kennedy, C. R., Bush, A. O. and Aho, J. M. (1988) Patterns in helminth communities in freshwater fish in Great Britain: alternative strategies for colonisation. *Parasitology*, **96**, 519–32.

Gibson, D. I. (1987) Questions in digenean systematics and evolution. *Parasitology*, **95**, 429–60.

Grey, A. J. and Hayunga, E. G. (1980) Evidence for alternative site selection by *Glaridacris laruei* (Cestoidea: Caryophyllidea) as a result of interspecific competition. *J. Parasitol.*, **66**, 371–2.

Halvorsen, O. (1971) Studies on the helminth fauna of Norway XVIII: on the composition of the parasite fauna of coarse fish in the River Glomma, Southeastern Norway. *Norw. J. Zool.*, **19**, 181–92.

Hanek, G. and Threlfall, W. (1970) Metazoan parasites of the American eel (*Anguilla rostrata* (le Sueur)) in Newfoundland and Labrador. *Can. J. Zool.*, **48**, 597–600.

Hine, P. M. (1978) Distribution of some parasites of freshwater eels in New Zealand. *N.Z. J. Mar. Freshwater Res.*, **12**, 179–87.

Holmes, J. C. (1973) Site selection by parasitic helminths: interspecific interactions, site segregation, and their importance to the development of helminth communities. *Can. J. Zool.*, **51**, 333–47.

Holmes, J. C. (1983) Evolutionary relationships between parasitic helminths and their hosts. In *Coevolution*, (eds, D. J. Futuyama and M. Slatkin), Sinauer Associates, Sunderland, Mass., pp. 161–85.

Holmes, J. C. (1986) The structure of helminth communities. In *Parasitology – Quo Vadit? Proc. 6th Int. Cong. Parasitol.*, (ed. M. J. Howell), Australian Academy of Sciences, Brisbane, pp. 203–8.

Holmes, J. C. and Price, P. W. (1986) Communities of parasites. In *Community Ecology: Pattern & Process*, (eds, J. Kikkawa and D. J. Anderson), Blackwell Scientific Publications, pp. 187–213.

Jeacock, A. (1969) The parasites of the eel, *Anguilla anguilla*, (L.) *Parasitology*, **59**, 16P.

Kazic, D., Ubelaker, J. F. and Canovic, M. (1982) The endohelminths of eel (*Anguilla anguilla* Linne, 1758) of Lake Skadar and some tributaries. *Acta Biol. Yugoslavia, Ichthyol.*, **13**, 41–53.

Kennedy, C. R. (1977) The regulation of fish parasite populations. In *Regulation of Parasite Populations*, (ed. G. W. Esch), Academic Press, New York, pp. 63–109.

Kennedy, C. R. (1978a) An analysis of the metazoan parasitocoenoses of brown trout *Salmo trutta* from British Lakes. *J. Fish Biol.*, **13**, 255–63.

Kennedy, C. R. (1978b) The parasite fauna of resident char *Salvelinus alpinus* from Arctic Islands, with special reference to Bear Island. *J. Fish Biol.*, **13**, 457–66.

Kennedy, C. R. (1981) Parasitocoenoses dynamics in freshwater ecosystems in Britain. *Trudy Zool. Inst. AN USSR*, **108**, 9–22.

Kennedy, C. R. (1984) The dynamics of a declining population of the acanthocephalans *Acanthocephalus clavula* in eels *Anguilla anguilla* in a small river. *J. Fish Biol.*, **25**, 665–77.

Kennedy, C. R. (1985a) Interactions of fish and parasite populations: to perpetuate or pioneer? In *Ecology and Genetics of Host–Parasite Interactions*, (eds, D. Rollinson and R. M. Anderson), *Linnean Society Symposium Series* 11, Academic Press, London, pp. 1–20.

Kennedy, C. R. (1985b) Site segregation by species of Acanthocephala in fish, with special reference to eels, *Anguilla anguilla. Parasitology*, **90**, 375–90.

Kennedy, C. R. and Lord, D. (1982) Habitat specificity of the acanthocephalan *Acanthocephalus clavula* (Dujardin, 1845) in eels *Anguilla anguilla* (L.). *J. Helminthol*, **56**, 121–9.

Kennedy, C. R. and Moriarty, C. (1987) Co-existence of congeneric species of Acanthocephalans: *Acanthocephalus lucii* and *A. anguillae* in eels *Anguilla anguilla* in Ireland. *Parasitology*, **95**, 301–10.

Kennedy, C. R., Bush, A. O. and Aho, J. M. (1986a) Patterns in helminth communities: why are birds and fish different? *Parasitology*, **93**, 205–15.

Kennedy, C. R., Laffoley, D.d'A., Bishop, G. *et al.* (1986b) Communities of parasites of freshwater fish of Jersey, Channel Islands. *J. Fish Biol.*, **29**, 215–26.

Køie, M. (1988) Parasites in eels, *Anguilla anguilla* (L.), from eutrophic Lake Esrum (Denmark). *Acta Parasitol. Pol.*, **33**, 89–100.

Konovalov, S. M. (1975) Differentiation of local populations of sockeye salmon *Oncorhynchus nerka* (Walbaum). Trans. L. V. Sagen, *University of Washington Publications in Fisheries New Series VI*, pp. 1–290.

Lacey, S. M., Williams, I. C. and Carpenter, A. C. (1982) A note on the occurrence of the digenetic trematode *Sphaerostoma bramae* (Muller) in the intestine of the European eel, *Anguilla anguilla*, (L.). *J. Fish Biol.*, **20**, 593–6.

Leong, T. S. and Holmes, J. C. (1981) Communities of metazoan parasites in open water fishes of Cold Lake, Alberta. *J. Fish Biol.*, **18**, 693–713.

MacArthur, R. H. and Wilson, E. O. (1967) *The Theory of Island Biogeography*, Princeton University Press, Princeton.

Mishra, T. N. and Chubb, J. C. (1969) The parasite fauna of the fish of the Shropshire Union Canal, Cheshire. *J. Zool. Lond.*, **157**, 213–24.

Moravec, F. (1985) Occurrence of endoparasitic helminths in eels (*Anguilla anguilla* (L.)) from the Macha Lake fishpond system, Czechoslovakia. *Folia Parasitol. Praha*, **32**, 113–25.

Price, P. W. (1980) *Evolutionary Biology of Parasites*, Princeton University Press, Princeton.

Price, P. W. (1984a) Alternative paradigms in community ecology. In *A New Ecology: Novel Approaches to Interactive Systems*, (eds, P. W. Price, C. N. Slobodchikoff and W. S. Gaud), John Wiley and Sons, New York, pp. 353–83.

Price, P. W. (1984b) Communities of specialists: vacant niches in ecological and evolutionary time. In *Ecological Communities* (eds, D. R. Strong Jr, D. Simberloff, L. G. Abele and A. B. Thistle), Princeton University Press, Princeton, pp. 510–23.

Price, P. W. (1986) Evolution in parasite communities. In *Parasitology – Quo Vadit?* (ed. M. J. Howell), *Proc. 6th. Int. Cong. Parasitol.*, Australian Academy of Sciences, Brisbane, pp. 209–14.

Price, P. W. and Clancy, K. M. (1983) Patterns in number of helminth parasite species in freshwater fishes. *J. Parasitol.*, **69**, 449–54.

Pugachev, O. N. (1983) Helminths of freshwater fishes of Northeast Asia. *Trudy Zool. Inst. AN USSR Leningrad.*, **181**, 90–113.

Rohde, K. (1979) A critical evaluation of intrinsic and extrinsic factors responsible for niche restriction in parasites. *Am. Nat.*, **114**, 648–71.

Seyda, M. (1973) Parasites of eel *Anguilla anguilla* (L.) from the Szczecin Firth and adjacent waters, *Acta Icthyol. Pisc.*, **3**, 67–76.

Southwood, T. R. E. (1978) *Ecological Methods*, 2nd edn, Chapman and Hall, London.

Thomas, J. D. (1964) Studies on populations of helminth parasites in brown trout (*Salmo trutta*) (L.) *J. Anim. Ecol.*, **33**, 83–95.

Uglem, G. L. and Beck, S. M. (1972) Habitat specificity and correlated aminopeptidase activity in the acanthocephalans *Neoechinorhynchus cristatus* and *N. crassus*. *J. Parasitol.*, **58**, 911–20.

Wisniewski, W. L. (1958) Characterisation of the parasitofauna of an eutrophic lake (Parasitofauna of the biocoenosis of Druznno Lake – part 1.) *Acta Parasitol. Pol.*, **6**, 1–64.

Wootten, R. (1973) The metazoan parasite fauna of fish from Hanningfield Reservoir, Essex in relation to features of the habitat and host populations, *J. Zool. London*, **171**, 323–31.

7

Helminth communities of amphibians and reptiles: comparative approaches to understanding patterns and processes

John M. Aho

7.1 INTRODUCTION

General introduction

Interest in the ecology of parasite communities has increased over the last decade. The increase in attention appears to reflect a shift in emphasis away from descriptive studies toward more quantitative approaches identifying processes responsible for creating community patterns. Recognition of three general features of parasite communities have contributed to the change in focus. First, helminth communities have discrete boundaries with each host individual representing replicate communities for statistical analysis of patterns. Second, by distinguishing either between different guilds (*sensu* Root, 1973) or among dominant or rare species (core/satellite or centrifugal species: Hanski, 1982; Rosenzweig and Abramsky, 1986), structure within a restricted group of species may be more apparent than for the community as a whole. Finally, communities of specialist organisms, such as parasites, might be structured in fundamentally different ways from generalist organisms (Price, 1980, 1984a).

Empirical and theoretical views of community organization of parasitic helminths have recently been presented by several authors (Rohde, 1979; Holmes and Price, 1980, 1986; Price, 1980, 1984a, b; Holmes, 1983). Most attempt to assess the relative roles of species interactions among parasites, concluding that communities of parasites occur in two distinguishable forms, isolationist and interactive. Other investigations emphasize that communities exist at several different hierarchical scales (e.g. Holmes, 1987; Holmes and Price, 1986; Kennedy *et al.*, 1986), and that processes operating on larger scales may determine what happens at smaller scales. A pluralistic view is, therefore, a necessity if cogent interpretations of ecological and evolutionary mechanisms underlying community organization are to be made. Comparative studies on freshwater fish and birds have figured prominantly in the development of recent predictions; needed now are new studies on a broader range of host systems to disentangle the many mechanisms that can affect helminth community structure.

As a contribution to that endeavour, this chapter examines aspects of the helminth fauna in two broad groups of hosts, amphibians and reptiles. As hosts, herps (a collective term for the Amphibia and Reptilia) represent excellent systems to explore patterns and processes influencing helminth community organization. They have invaded a multitude of habitats and exhibit a striking diversity of life history patterns, reproductive modes, body sizes, foraging modes, and trophic relations. Consequently, herp-parasite systems provide a valuable comparative analyses of approach to understanding ecological and evolutionary relationships determining helminth species distribution and abundance.

Historical perspective

To date, no attempt has been made to synthesize what is known about the organization of helminth communities in this collective group. The literature on helminths of herps is extensive, diverse, and diffuse, but overall seems to have historically been focused on two major areas.

Fundamental questions concerning taxonomic relationships and life histories have received considerable attention. From a community perspective, these studies have made an invaluable contribution to understanding systematic/historical relationships of both parasite and host (e.g. Brooks, 1979a, b, 1981a, b; Ernst and Ernst, 1980; Anderson, 1984; Baker, 1984). The importance of host and parasite phylogenetic relationships as determinants of helminth community structure for amphibians and reptiles, as well as other classes of vertebrate hosts, however, has been controversial. In general, Brooks (1980) advocates a co-evolutionary view with host and parasite having a long evolutionary association, while others consider most parasite communities as recent, chance assemblages (e.g. Rohde, 1979; Holmes and Price, 1980; Price, 1980). Studies on life histories have also now progressed to where basic patterns for most helminth groups are known, but relatively little information exists to evaluate variability in these patterns. For only a relatively small number of parasite species are specific life cycles known. As a consequence, knowledge of population dynamics is limited to a few studies which have examined transmission dynamics and circulation patterns for specific species of parasites (Dronen, 1978; Jarroll, 1980; Hardin and Janovy, 1988).

Faunistic surveys have also received considerable attention. Two features, however, are clear from the literature: many investigations present information only for select helminth taxa or simply provide species lists with unequal representation among the different host groups. Studies on amphibian parasites are relatively common, but even amongst these host groups, there are few comprehensive surveys of the helminth fauna of either salamanders (e.g. Harwood, 1932; Rankin, 1937, 1945; Fischthal, 1955a, b: Dunbar and Moore, 1979; Coggins and Sajdak, 1982; Goater *et al.*, 1987) or anurans (Harwood, 1932; Brandt, 1936; Campbell, 1968). These studies still exceed the number of detailed surveys on lizards (e.g. Harwood, 1932; Telford, 1970; Pearce and Tanner, 1973; Benes, 1985; Bundy *et al* , 1987), and snakes (e.g. Harwood, 1932; Collins, 1969; Rau and Gordon, 1980; Detterline *et al.*, 1984). Turtles have been examined in greater detail than other reptiles (e.g. Rausch, 1947; Schad, 1963; Martin, 1972; Platt, 1977; Limsuwan and Dunn, 1978; Esch *et al.*, 1979a, b), but species representation and number of populations are limited relative to amphibians. Consequently, the parasite fauna of only a small fraction of the amphibians and reptiles can be regarded as well known.

Despite these limitations, the extant data derived from helminth surveys have been invaluable for understanding community structure, particularly patterns at the component community level. The nature of the questions vary between studies, but most focus on some aspect of the community and host or environmental factors. Three features have been consistently identified as important determinants of helminth distribution and abundance: geographic or habitat, temporal, and host demography.

Many studies have attempted to correlate the structure of helminth communities with geographic location or habitat conditions. Esch and Gibbons (1967) and Esch *et al.* (1979a) note compositional differences in the helminth faunas of emydid turtles from Wintergreen Lake in Michigan (*Chrysemys picta* dominated by trematodes) and several lentic habitats in South Carolina (*Trachemys scripta* dominated by several species of acanthocephalans) that parallel changes in the nature and quality of the habitat. Since trematodes require a molluscan intermediate host, the higher pH and alkalinity in Michigan waters presumably contributes to a richer molluscan fauna. The absence of acanthocephalans in Wintergreen Lake may be due to a lack of colonization by either the parasite, an appropriate intermediate host, or both. Variability in community structure has also been ascribed to the effects of environmental change. Esch *et al.* (1979b) collected turtles (*Trachemys scripta*) from lakes differing in their degree of perturbation and observed community complexity to be greatest in more benign habitats than in fluctuating, harsh environments. Host habitat selection has also been reported to determine complexity and size of the helminth community in several species of amphibians (Brandt, 1936; Rankin, 1937; Campbell, 1968; Dunbar and Moore, 1979; Goater *et al.*, 1987). In general, hosts associated with aquatic habitats have richer and larger helminth communities than terrestrial counterparts. Ecomorphological differences among congeneric species (e.g. *Anolis* lizards: Bundy *et al.*, 1987) have also been shown to influence community composition.

Where examined, seasonal variability in infections of individual helminth species has been commonly observed (e.g. Brandt, 1936; Rankin, 1937; Esch *et al.*, 1979a; Telford, 1970), but its influence on community development and structure is currently equivocal. Temporal recruitment patterns have been attributed to changes in intermediate host abundance, dietary shifts, development rates of parasites for infection, and features such as the host's immune responsiveness which, in some poikilotherms, can be influenced by temperature (Esch *et al.*, 1975). Winter dormancy for most host species does not appear as an important source of parasite mortality, as many helminth species show no decline in prevalance or population size (Dubinina, 1949; Esch and Gibbons, 1967; Rau and Gordon, 1978). Alternatively seasonal changes might reflect shifts in habitat use associated with animals possessing complex life cycles. Most studies, therefore, emphasize the importance of abiotic factors influencing temporal variations in helminth infection.

The third pattern regularly identified as being important is the influence of host demographic structure. For both amphibians and reptiles, helminth communities of larvae or juveniles typically are species-poor compared to adults (e.g. Landewe, 1963; Jackson and Beaudoin, 1967; Avery, 1971; Price and Buttner, 1982; Vogel and Bundy, 1987). Changes associated with size or age can simply be a function of time; larvae and juveniles may not have been exposed long enough to acquire their typical adult complement of helminths. Differences between age classes could also reflect ontogentic shifts in diet, habitat, or behaviour. The influence of gender on aspects of community structure is variable. For example, Vogel and Bundy (1987) found significant differences in population size for several helminth species in male and female *Anolis lineatopus*, but no gender effects were observed in *Desmognathus fuscus* (Dyer *et al.*, 1980).

In contrast to the numerous studies at the component community, few investigations present data at the infracommunity level. Questions pertaining to site selection within or between closely related hosts, presence of vacant niches, and predictability of the helminth composition between individuals are appropriate at this level, but data are lacking for most amphibian and reptile hosts. Critical observations are available for two species of turtles, and support the view that many species occupy specific and predictable locations along the intestine (Crompton and Nesheim, 1976). The importance of interactive site selection determining the distribution of helminth species populations is, however, questioned. Based on linear and circumferential distributions of three species of acanthocephalans, Jacobson (1987) concluded that competitive interactions are not important in structuring infracommunities of *T. scripta*. Schad (1963) offers a contrary opinion for the co-occurrence of eight nematode species (*Tachygonetria* spp.) in the colon of the European tortoise (*Testudo graeca*). Species distribution patterns overlapped along the length of the colon; however, radial separation (lumenal/mucosal) was apparent in most species, and where species combinations overlapped in distribution, there were usually obvious differences in oral morphology and food habits. Schad (1963) attributed competitive interaction as a major selective force in determining infracommunity structure by promoting the development of resource partitioning and niche diversification.

The literature also indicates that there are no studies at the compound level. Where holistic approaches to investigating community patterns have been attempted, they usually contrast parasite faunas among a restricted suite of host species (e.g. Brandt, 1936; Campbell, 1968; Dunbar and Moore, 1979; Goater *et al.*, 1987). These studies emphasize the importance of host abundance and habitat preference as determinants of helminth community richness.

Current questions

It should be obvious from these general patterns that one cannot summarily dismiss the importance of any biotic or abiotic factor influencing the helminth community organization of amphibians and reptiles. One goal of this chapter is, therefore, to assemble data available on amphibian and reptile parasite communities, and through the use of comparative analyses, analyse interspecific patterns of community structure. The inclusion of host species from different families or orders presents a wider range of variation for any aspect of community structure than would otherwise be accessible within a single species. Although caution must be used in inferring process from patterns identified by descriptive studies (e.g. Salt, 1982; Connell, 1983; Schoener, 1983), such analyses can enable one to make sense out of the overall pattern of community traits. They can also suggest testable hypotheses for the underlying causes of these patterns and, with some qualifications, be used to evaluate these hypotheses.

Although herps have a wide variety of parasites located in varied sites in the body, attention is restricted to only gastrointestinal helminths. The analysis will focus primarily on the component community, and whenever possible, use infracommunities as replicate samples. Comparisons of species richness, community size, and predictability at local (the host population at a sampling site) and regional (over the entire geographical range of the host) spatial scales will be used to address the following three major, interrelated sets of questions.

1. Are helminth communities in amphibians and reptiles rich in species or individuals or are they dominated by a few species? Are core species evident? How do characteristics of herp helminth communities compare with other vertebrate host groups?
2. What are the correlates of species richness? In particular, do geographically widespread hosts support more species of parasites than hosts with narrower ranges? Are regional and local richness correlated, suggesting that processes determining richness at both spatial scales are strongly linked?
3. Are helminth communities in amphibians and reptiles more similar among samples of related species in the same habitat, or among samples of the same species from different habitats? If core species are present, are they generalists or specialists in host occupation?

Approach

Data used in these analyses come from two sources: published host surveys and collections of red-spotted newts (*Notophthalmus viridescens*) within the southeastern United States. The published literature provides a convenient source from which to examine broad scale patterns in community organization. I have attempted to include representatives of all orders (Squamata, Crocodylia, Salientia, Caudata, Testudines) from as many geographical regions as possible (a complete listing of the dataset is available upon request). Most analyses, however, were dominated by taxa from North America because of accessibility to the literature. Multiple host surveys from a single locality were desired, but most examined only one or two host species thereby limiting interspecific comparisons within and between locations.

Several requirements had to be met for inclusion of a host survey in these analyses: the intestinal helminth fauna had to be fully censused, some quantitative estimate of infection (prevalence and/or intensity) be given, and in most instances (> 95% of the studies), the species had to be represented by at least ten individuals in each survey. To increase breadth of coverage for certain species (e.g. pelobatid toads), the minimum sample size was lowered to seven individuals. Adherence to these criteria substantially reduced the number of host species included and, for geographic comparisons, potentially reduced the total number of helminth species reported from a host species across its distributional range.

Collections of red-spotted newts (*Notophthalmus viridescens*) from nine locations in the southeastern United States were made during March–April 1988. Newts were chosen for study because of their abundance in a diversity of lentic habitats, generalist feeding behaviour, and site fidelity to breeding ponds (Likens, 1983). These characteristics would be expected to facilitate cycling of helminths within a locality and lead to compositional variability between habitats by limiting movement and exchange of helminth species.

From each survey, the total number of helminth species found, mean number of species per host individual, and the mean abundance of helminths per host were determined for each host species as measures of community richness and abundance. To examine predictability, the number of core species (defined as occurring in > 50% of the hosts) per host survey as well as measures of qualitative (Jaccard's coefficient, based on presence/absence) and quantitative (per cent similarity, based on proportional abundance of a species) similarity were determined. Measurements of host size, host geographic range estimated from range maps, and habitat preferences (classified as either arboreal, fossorial, terrestrial, semi-aquatic, or aquatic) are from information presented in Conant (1975) and Behler and King (1979).

7.2 PATTERNS IN CONTEMPORARY COMMUNITIES

Review of the literature produced data for 16 families of reptiles (nine lizards, two snakes, four turtles and one crocodilian) and 14 families of amphibians (seven salamanders and seven frogs and toads). In all, 155 host species from 393 separate surveys were represented.

Helminth species richness and abundance in reptiles and amphibians

Composition of the helminth fauna varied markedly among the Amphibia and Reptilia, but class-level comparisons indicated few differences in community structure (Table 7.1). There were no significant differences in either measure of community richness or the number of core species. A host species harboured at most ten helminth species per population with a mean total richness of 3.0 ± 0.2; on average, each individual host harboured less than five species of helminths with a mean individual richness of 0.8 ± 0.1.

Table 7.1 Distributional patterns of mean total number of helminth species per host species, mean number of helminth species per host individual, mean relative abundance of helminths per host, and mean total number of common helminth species per host species for members of the Amphibia and Reptilia. Analysis of variance used to detect between class differences in four measures of community structure. All tests done using log $(x + 1)$ transformation

	Amphibia	*Reptilia*	*ANOVA*
Mean total number of species per host	3.04 (0.13)* 0–9† 231‡	2.77 (0.21) 0–10 162	$p > 0.65$
Mean species richness per host individual	0.80 (0.04) 0–2.50 231	0.84 (0.07) 0–4.48 162	$p > 0.09$
Mean abundance	7.83 (1.19) 0–83.1 143	59.67 (19.30) 0–572.10 35	$p < 0.0001$
Mean total number of core species	0.55 (0.05) 0–3 231	0.67 (0.08) 0–7 162	$p > 0.15$

* $\bar{x}$ (± 1 SE).
† Range of values.
‡ Number of studies where information is available.

Core species were only a small component of the helminth fauna of each host population. Values ranged from zero to seven (mean = 0.6 ± 0.1), and compared to the total number of species per population, core species comprised approximately 17 and 24% of the reptilian and amphibian helminth fauna, respectively. Only the mean abundance of helminths differed between classes, with larger population sizes recorded from reptilian hosts.

Despite the superficial appearance of convergence, considerable variability in community structure was found within and among the Amphibia and Reptilia (Table 7.2). Excluding the single study on crocodilians, significant differences were noted in species richness, abundance, and number of core species among the five host groups (taxonomic order, except suborders for the Squamata) (Table 7.2). Ranking the four measures describing community structure, complexity of the helminth community was highly concordant across the five host groups (Kendall's rank concordance test, $W = 0.71$, $X^2 = 11.4$, $p < 0.05$). Turtles characteristically possessed the

Table 7.2 Distributional patterns of mean total number of helminth species per host species, mean number of helminth species per host individual, mean relative abundance of helminths per host, and mean total number of core helminth species per host species for five host groups of amphibians and reptiles. Analysis of variance used to detect between group differences in four measures of community structure; *a. posteriori* comparisons use Duncan's Multiple Range Test. Values with same letters are not significantly different. All tests done using log $(x + 1)$ transformation

	Amphibia		*Reptilia*			*ANOVA*
	Caudata	*Salientia*	*Lacertilia*	*Serpentes*	*Testudines*	
Mean total number of species per host	2.76 (0.15)*[bc] 0–7† 148‡	3.54 (0.24)[b] 0–9 83	2.06 (0.13)[cd] 0–5 100	1.88 (0.35)[d] 0–8 25	4.92 (0.46)[a] 0–10 36	$p < 0.0001$
Mean species richness per host individual	0.70 (0.04)[c] 0–2.5 148	0.98 (0.07)[b] 0–2.08 83	0.63 (0.06)[c] 0–2.50 100	0.56 (0.09)[c] 0–1.29 25	1.52 (0.20)[a] 0–4.48 36	$p < 0.0001$
Mean abundance	5.2 (1.38)[c] 0–83.1 98	11.55 (1.86)[c] 0–44.68 45	38.21 (18.93)[b] 0–303.50 17	13.49 (6.44)[bc] 0–35.0 5	66.59 (21.94)[a] 0–211.1 12	$p < 0.0001$
Mean total number of core species	0.45 (0.05)[c] 0–3 148	0.73 (0.08)[b] 0–3 83	0.48 (0.07)[c] 0–4 100	0.40 (0.10)[c] 0–1 25	1.36 (0.25)[a] 0–7 36	$p < 0.0001$

* $\bar{x}$ (± 1 SE).
† Range of values.
‡ Number of studies where information is available.

richest and largest helminth communities; complexity then progressively declined, particularly the measures of richness, from anurans to salamanders, lizards, and snakes.

Variability in helminth community complexity was pronounced within and between families of each host group (Fig. 7.1a). Among the turtles, members of the Kinosternidae typically had less complex helminth communities than the Emydidae. There were also gradients in complexity within the emydids; host species such as *Clemmys* had depauperate communities (1–2 species with < 10 individuals per host) while the helminth communities of *Trachemys* were larger and more speciose (3–7 species with > 100 individuals per host). Similar patterns depicting within family differences in species richness and abundance have also been found among plethodontid salamanders. Sympatric species of desmognathine salamanders differed in species richness and abundance (Goater *et al.*, 1987), which when compared to other members of the Plethodontidae, generally also had the richest and largest helminth communities. While comparable patterns were noted in the remaining host groups, the observed intra- and interfamily variability in helminth community complexity were minor compared to differences between the five host groups. There were no significant differences in the relationship of mean species richness and community abundance per host between families within a host group (Analysis of Covariance, test of heterogeneity of slopes: $F_{10,\ 139} = 1.30$, $p > 0.20$). Significant differences were only found between host groups (ANCOVA, test of heterogeneity of slopes: $F_{4,\ 139} = 9.58$, $p < 0.001$). Using orthogonal contrasts to distinguish among host groups, turtles differed significantly from all other host groups. There were no significant differences in the regression coefficients between anurans and caudataes or snakes and lizards. Host groups would appear, therefore, to have underlying differences in the development of their community richness and abundance.

Habitat preference of the host accounts for some of the variability in community structure observed within a host group (Fig. 7.2). Positioning host species along a terrestrial–aquatic environmental transition, there was a trend for helminth species richness (mean total per host population and mean per host individual) to peak in anurans and turtles with semi-aquatic life histories; terrestrial or aquatic hosts had comparable values. Salamanders and squamate reptiles exhibited slightly different patterns with either progressive increases (salamanders) along the terrestrial to aquatic gradient or having the fewest number of species in semi-aquatic hosts (squamate reptiles). Differences in habitat-associated patterns of species richness probably reflect inclusion in the semi-aquatic category, species that ecologically are more similar to terrestrial hosts. In particular, several species of mole salamanders (*Ambystoma*) spend most of their adult life underground, while several species of garter snakes (*Thamnophis*) are much

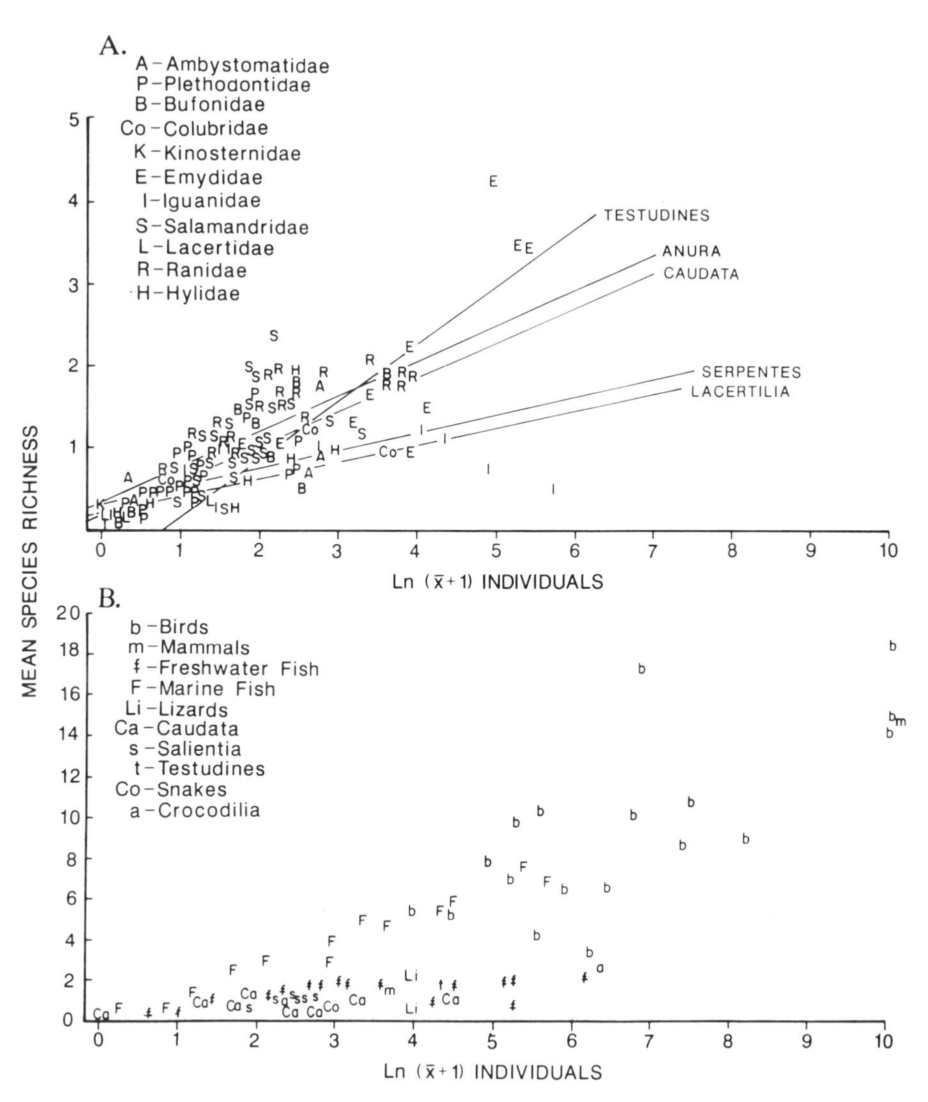

Fig. 7.1 Comparison of relationships for mean number of helminth species (mean species richness) and mean community abundance per host taxa. A. Patterns for individual species of herps, presented by family, for five host groups. Values are presented only for the most numerous families with many points hidden. Lines represent linear regressions for the data. B. Patterns for different vertebrate host taxa. For amphibians and reptiles, points represent mean values for different herp families within a host group; other vertebrate classes are represented by mean values for selected species.

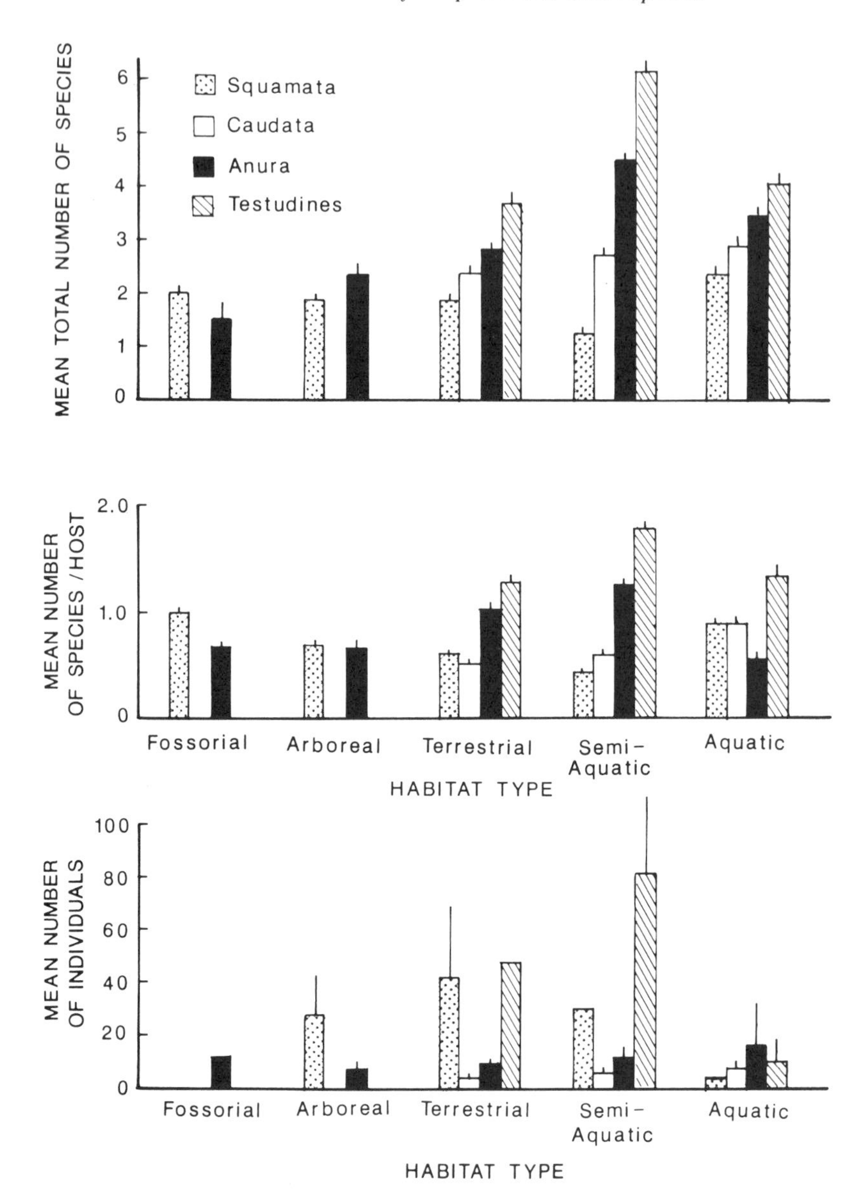

Fig. 7.2 Comparison of habitat associated changes in mean total number of helminth species per host species, mean number of helminth species per individual host, and mean community abundance for four orders of amphibians and reptiles. Host habitat preferences were assigned according to the habitat a species is primarily found throughout its lifespan. Vertical lines are 95% confidence interval.

more terrestrial than aquatic in habitat selection and food utilization (Seigel, 1986). Abundance patterns were variable, but in most instances followed similar habitat relations as species richness.

Comparison to helminth communities in other vertebrates

Helminth communities in reptiles and amphibians differed dramatically when compared to other classes of vertebrates (Fig. 7.1b). As previously shown by Kennedy *et al.* (1986), significant differences exist in patterns of helminth community richness and abundance between freshwater fish and certain aquatic birds. Holmes (Chapter 5) indicated (and further supported by data presented in Fig. 7.1b) that gastrointestinal helminth communities of marine fish were comparable to freshwater fish in the number of individuals, but had higher species richness (for a given number of worms). Species richness and abundance for marine fish also appeared to be increasing at about the same rate as communities in birds and mammals. Helminth communities of reptiles and amphibians were similar in abundance to other fishes, but were among the most depauperate hosts of all vertebrates.

7.3 LOCAL–REGIONAL PATTERNS OF SPECIES RICHNESS

Comparative approaches emphasizing differences in species richness at two spatial scales, regional and local, are relevant to identifying general mechanisms determining community structure (Cornell, 1985; Ricklefs, 1987). Specifically, it is necessary to quantify the true regional richness (total number of helminth species across a host's geographic range) from which patterns of local richness (total number of helminth species in a host at a particular site) for a host species are drawn. Many studies are founded on the premise that determinants of local and regional richness are independent processes; local richness is a deterministic outcome of local processes within the biological community and is not influenced by regional richness. Local factors undoubtedly influence local richness, but the larger issue is whether these factors can explain patterns in local diversity solely by local processes. The examination of the correlation between local/regional richness should be an important first step in understanding mechanisms by which hosts accumulate parasites in ecological and evolutionary time.

Some studies examine covariability in local and regional species richness for a single host species, but taken from different geographical regions (e.g. Lawton, 1982). An alternative approach, and the one taken here, compares several closely related host species possessing different regional helminth faunas and different regional richnesses. If host species with larger regional richnesses have larger average local richness values, then it would appear likely that regional and local richness are linked patterns. The use of closely related hosts should provide similar ecological conditions for the helminth faunas.

Size of geographic range vs regional richness

Correlation of host geographical range and regional helminth richness were done to determine whether more widespread host species supported richer helminth faunas than more narrowly distributed ones. Only species from North America where sufficient collections had been made over a major portion of the host distributional range were included. This reduced the likelihood that a major helminth species was omitted from the analysis. In total, data on eight families (Squamata – Iguanidae and Scincidae; Caudata – Salamandridae, Plethodontidae, Ambystomatidae; Anura – Ranidae, Bufonidae; and Testudines – Emydidae) representing 76 different species were included in the analysis.

Double logarithmic regression of regional helminth richness on host geographical range confirmed the dominant role played by distributional range in setting regional helminth species richness. More widespread host species harboured richer helminth faunas than more narrowly distributed ones. The proportion of the total variance explained by the relationship was, however, comparatively small, accounting for approximately 30%. Differences in helminth faunal composition between host families suggested that some residual variation of the regression might be explained along

Table 7.3 Species–area relationships for the number of intestinal helminth species within different families of amphibians and reptiles

Family	*c*	*z*	R^2	*p*
Scincidae	−3.31	0.67	0.91	<0.05
Ambystomatidae	−2.52	0.55	0.89	<0.02
Ranidae	−2.12	0.49	0.59	<0.05
Emydidae	−2.07	0.54	0.44	<0.05
Salamandridae	−1.15	0.39	0.81	<0.05
Plethodontidae	−0.86	0.31	0.47	<0.001
Bufonidae	−0.09	0.27	0.78	<0.05
Iguanidae	−0.01	0.11	0.19	<0.05

Relationship derived from function $(S + 1) = cA^z$ where S = number of helminth species, A = geographical range of host species, c and z are constants.

taxonomic lines or differences in body size. Each species was therefore coded by vertebrate class and family, and differences between taxonomic groups tested using ANCOVA heterogeneity of slopes. There was no difference in slope between the two host classes. Significant differences in species–area relationships were detected between orders within a class ($F_{1,60} = 1.11, p < 0.05$) and families within an order ($F_{4,60} = 2.69, p < 0.04$). Orthogonal contrasts comparing species–area relationships indicated the existence of three different groupings for the eight families (Table 7.3). No significant differences in regression coefficient were detected between the Ambystomatidae, Ranidae, Emydidae, and Scincidae or among the Bufonidae, Plethodontidae, and Salamandridae. The Iguanidae differed from all herp families having the lowest rate of change. In no comparison was host size correlated with parasite species richness.

Relationship between local and regional richness

Using hosts with different putative species pools, patterns relating local and regional richness can be recognized using a simple graphical approach

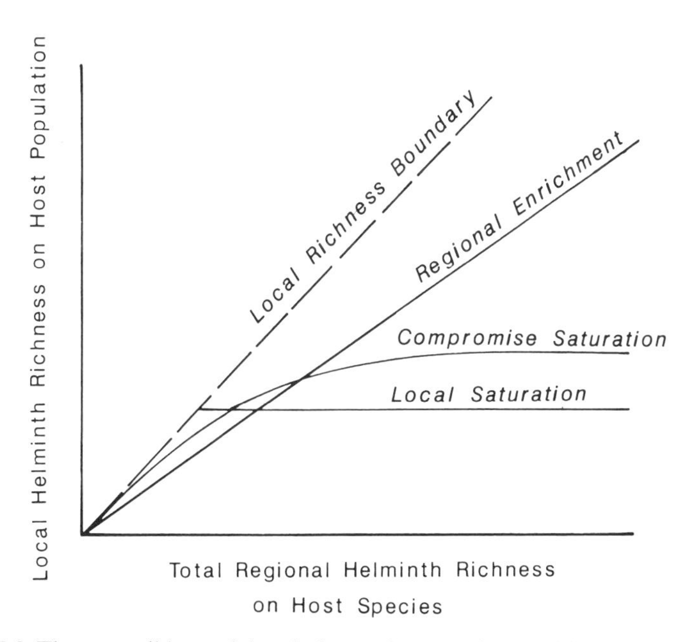

Fig. 7.3 Three possible models relating regional and local richness on a series of appropriately chosen host species representing saturated and unsaturated patterns. The local richness boundary indicates where local and regional richness are equal and cannot be exceeded.

(Fig. 7.3). Three idealized results are envisioned. The first represents an enrichment model whereby local richness is strongly influenced by regional species richness. In the extreme case, local richness is entirely determined by regional richness (all species found over the geographic range of the host occur at every local host population). Alternatively, some constant proportion of the regional pool is, on average, represented in the local population. A second pattern, a local saturation model, would be depicted by a straight line with a slope of 0 indicating a strict upper limit to local richness independent of the regional pool. In the third pattern, the compromise saturation model, local richness initially increases with regional richness, but eventually remains nearly constant or at least independent of regional richness in larger geographic areas.

Values of local helminth richness were plotted against regional helminth richness, and for the seven genera representing 23 host species presented, there was a strong similarity in the form of the relationship (Fig. 7.4). For species of *Rana, Ambystoma, Desmognathus* and *Plethodon*, there were no significant changes in local richness with increasing regional richness (ANOVA with linear regression: Sokal and Rohlf, 1981). Only for *Eurycea* did local richness change significantly with increasing regional richness ($F_{2,15} = 3.7$, $p < 0.05$). However, there was no evidence to suggest the relationship to be either linear (regression $F_{1,15} = 7.2$, $p > 0.05$) or curvilinear (deviation from regression, $F_{1,15} = 0.9$, $p > 0.05$). Though a limited number of observations for *Notophthalmus* and *Taricha* prevented statistical analysis of the local/regional values, the pattern appeared comparable to the other genera with no obvious change in local richness with increasing regional helminth richness. While not shown, similar local/regional richness patterns have also been observed in *Bufo* and *Sceloporus*. Thus, regional richness did not appear to make a strong contribution to local richness in any of the herp taxa examined.

Predictability of the helminth fauna for a host species at any local population, as indicated by their compositional similarities, was highly variable (Table 7.4). Values for Jaccard's coefficient can range from 0–100%. Observed values of qualitative similarity between component communities in different populations of the same species encompassed 15–100% of this possible range. Within host species variation in Jaccard values was not correlated with either regional richness (Spearman rank correlation, $r_s = 0.19$, $p > 0.25$) or the number of populations sampled ($r_s = 0.01$, $p > 0.50$). Mean similarity values varied extensively across host species, but were generally low (0–57%). There was a trend, however, for faunal similarity within each grouping of host taxa to increase with decreasing regional helminth richness. The implication is that parasite species are being replaced along ecological and geographical gradients.

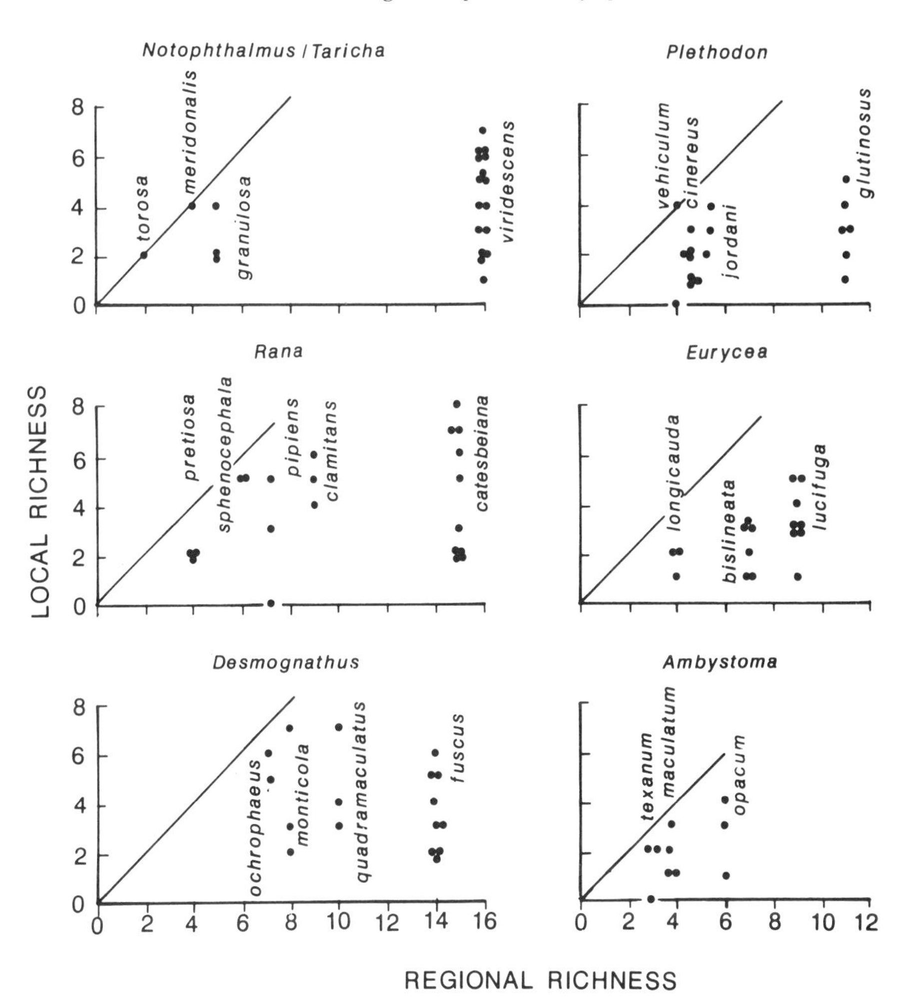

Fig. 7.4 Local–regional richness comparisons for intestinal helminth communities of six taxa groupings of amphibians. Each point represents a sampled local host population and its abscissa value represents the regional richness on each host species determined as the cumulative number of helminth species pooled over all local populations where that species was sampled. The straight line represents the local richness boundary.

An additional method of examining the local/regional richness relationship is based upon differences in the distribution patterns of individual helminth species found infecting each host species at specific locations. Examination of frequency distributions for each host tended to be bimodal and that species distribution and abundance were correlated in most hosts

Table 7.4 Comparison of helminth community similarity for a host species either among all populations where surveyed or between related species at locations when multiple hosts were examined. Values are mean and range of Jaccard's coefficient of similarity

Host species	*Within host species across locations (%)*		*Between related species within a location (%)*	
Desmognathus ochrophaeus	57	(57)	68	(33–86)
Desmognathus monticola	46	(43–67)	79	(50–100
Desmognathus quadramaculatus	29	(22–40)	71	(33–100)
Desmognathus fuscus	26	(0–67)	77	(67–83)
Plethodon jordani	40	(20–75)	25	(17–33)
Plethodon cinereus	38	(0–100)	43	(25–66)
Plethodon glutinosus	21	(0–50)	43	(17–66)
Plethodon vehiculum	0	(0)	0	(0)
Ambystoma maculatum	56	(33–100)	46	(0–100)
Ambystoma opacum	20	(0–60)	44	(0–100)
Ambystoma texanum	11	(0–33)	23	(20–25)
Eurycea lucifuga	36	(0–100)	67	(67)
Eurycea bislineata	33	(0–100)	*	
Eurycea longicauda	11	(0–33)	67	(67)
Rana sphenocephala	43	(43)	32	(27–38)
Rana clamitans	38	(25–50)	42	(25–67)
Rana pretiosa	33	(0–100)	0	(0)
Rana catesbeiana	29	(0–100)	27	(0–67)
Rana pipiens	5	(0–15)	8	(0–40)
Notophthalmus viridescens	20	(0–100)	*	
Taricha granulosa	34	(20–50)	*	

* No comparative material available

(Table 7.5). These features suggest that two distinct components of the helminth fauna can be recognized: common species that occurred at a high percentage of sites and uncommon species present in less than 50% of the local populations. The bimodal trend and species distribution–abundance correlations may represent general patterns having been reported in several helminth (e.g. Bush and Holmes, 1986; Stock and Holmes, 1987) and other animal communities (e.g. Cornell, 1985; Hanski, 1982; Fowler and Lawton, 1982). The relative proportion of these two faunal components, however, has important consequences for understanding the basic structure of, and interpreting spatial variability in, the helminth community.

Common and rare species were plotted separately against regional richness and then analyzed using ANOVA with linear regression (Fig. 7.5). For each genus of host examined, the number of rare species was highly variable within and between host species, but representation in local

Table 7.5 Spearman's rank correlation coefficient (r_s) between distribution and local abundance of helminth species on different host species. Distribution is the proportion of total sample sites where a particular species occurred on a specific host species. Abundance is the frequency of helminth species occurrence at each site.

Host species	*Number of helminth species*	r_s
Plethodon jordani	5	0.60 NS
Plethodon cinereus	5	0.90 *
Plethodon glutinosus	11	0.64 *
Eurycea longicauda	4	0.00 NS
Eurycea bislineata	7	0.83 *
Eurycea lucifuga	9	0.86 †
Desmognathus monticola	8	0.85 †
Desmognathus quadramaculatus	10	0.64 *
Desmognathus fuscus	14	0.88 ‡
Rana pretiosa	4	0.90 NS
Rana pipiens	7	0.45 NS
Rana clamitans	9	0.88 †
Rana catesbeiana	15	0.86 ‡
Ambystoma texanum	3	0.88 NS
Ambystoma maculatum	4	1.00 *
Ambystoma opacum	6	0.90 *

* $p < 0.05$; † $p < 0.01$; ‡ $p < 0.001$; NS = Not significant.

richness did not differ significantly with changes in regional richness. The number of common species was also variable within and between host taxa, but only among *Ambystoma* were there no significant changes in the local richness across regional richness. The remaining host taxa exhibited no consistent pattern in the results of the analysis. No significant linear or deviation from the regression was found for *Eurycea*. Significant deviations from a linear regression were detected for both *Rana* and *Plethodon* suggesting a possible curvilinear relationship, and the number of common species decreased linearly in *Desmognathus* (Fig. 7.5). Among *Notophthalmus* and *Taricha*, the number of rare species varied inconsistently between species while common species appeared to curvilinearly increase with increasing regional richness. The pattern suggests that for most host species, common species constituted only a relatively small component of the local richness at a site. Proportional representation in the community will generally be greater at higher values of regional richness, but turnover rates for common species must be high consistent with the observation of decreasing faunal similarity values with increasing regional richness.

For each host taxon, many of the same helminth species can be found in several congeneric host species (Table 7.6). The extent of faunal similarity,

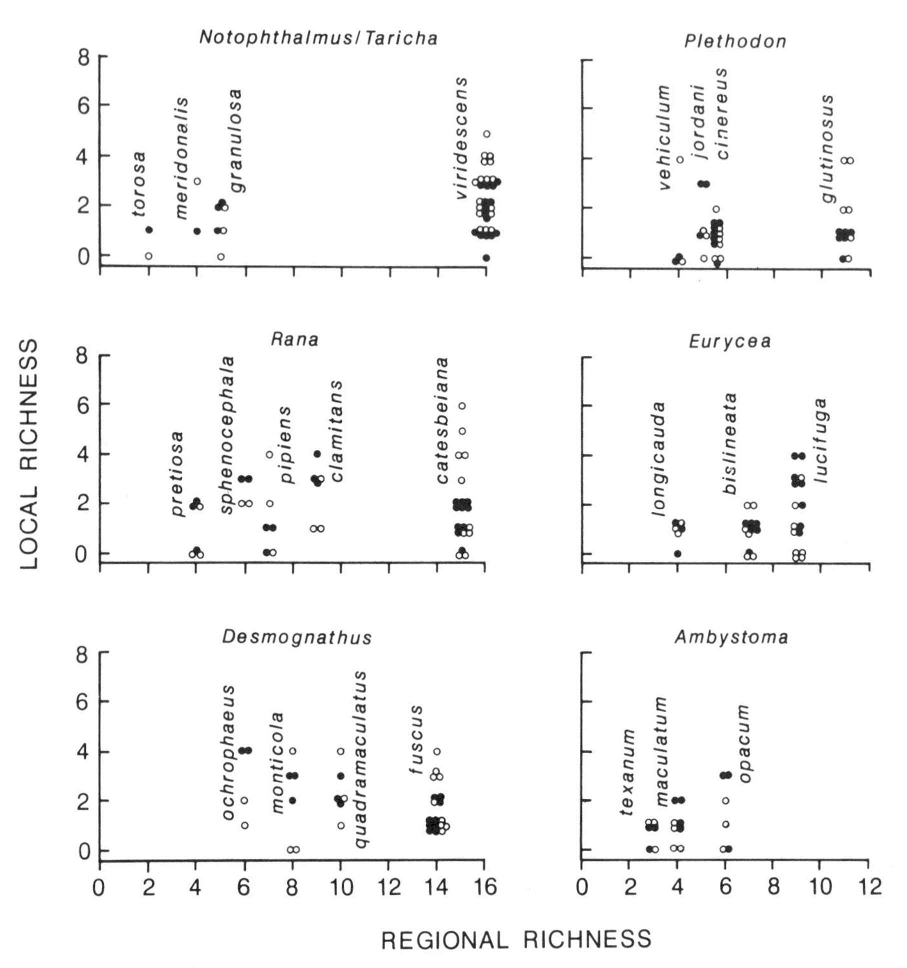

Fig. 7.5 Plots of the numbers of common (●) and rare (○) helminth species identified at each local population sampled against the total regional richness on each host species.

comparing related hosts taken from the same location, generally equalled, or exceeded, values obtained for the same host species comparing different populations (Table 7.4). Focusing on the hosts used in the previous analysis and three species of *Bufo*, many of the same helminth species were also shared among host species from different genera, families, or orders (Table 7.6). Eighteen helminth species were core species in one or more populations; five of them attained core status in more than a single genus of host. Some species such as *Cosmocercoides variabilis* and *Brachycoelium salamandrae* were cosmopolitan in host distribution. The composition of the

Table 7.6 Species composition of intestinal helminths harboured by congeneric host species and across host taxa groups

	Desmognathus	*Plethodon*	*Eurycea*	*Ambystoma*	*Notophthalmus/Taricha*	*Rana*	*Bufo*
Brachycoelium salamandrae	(2)	*(4)	*(3)	*(3)	(2)	(1)	(1)
Brachycoelium elongatum	(4)	(1)					
Brachycoelium meridonalis		(1)	(1)		(2)		
Megalodiscus temperatus	(1)			(2)	(1)	*(4)	(1)
Megalodiscus americanus					*(2)	(1)	
Megalodiscus intermedius	(1)					*(1)	
Plagitura salamandra			(2)	(1)	*(2)		
Glypthelmins quieta						*(5)	(1)
Cephalogonimus americanus						(2)	
Haplometrana intestinalis						*(1)	(1)
Cylindrotaenia americana	(4)	(3)				(1)	(1)
Ophiotaenia cryptobranchi	(2)	(1)	(1)				
Ophiotaenia saphenus						*(3)	
Bothriocephalus rarus	(1)		(1)		(1)		
Distoichometra bufonis							(2)
Cosmocercoides variabilis	(4)	*(4)	*(1)	(3)	*(4)	(3)	*(3)
Thelandros magnavulvaris	*(4)	(2)	*(2)		(1)		
Thelandros salamandrae	(1)	(2)	*(2)				
Capillaria inequalis	*(4)	(1)	(1)	*(2)	*(1)		
Capillaria tenua					*(1)	(1)	
Oswaldocruzia pipiens	(2)	(2)	(2)			*(4)	*(3)
Oswaldocruzia waltoni						(1)	*(1)
Oxysomatium variabilis						(2)	*(1)
Oxysomatium longicauda		(1)				(1)	
Oxysomatium americana	(1)		(1)			(1)	
Falcaustra plethodontis	(3)						
Desmognathinema nantahalaensis	(2)						
Falcaustra pretiosa						*(1)	
Falcaustra catesbeiana						*(1)	
Acanthocephalus aculeatus	(1)	(2)		(1)	*(1)		
Total number of helminth species by host genus	18	13	14	6	17	20	10
Number of host species examined	4	4	3	3	4	5	3

* = Presence of a particular parasite species being a core species in at least one species of congeneric host.
(n) = Number of host species from a taxa group that harboured a particular helminth species.

helminth communities in these different hosts appears to be largely composed of host generalists; no host species appears to possess a truly unique fauna.

7.4 HELMINTH COMMUNITY DYNAMICS OF *NOTOPHTHALMUS VIRIDESCENS*: A SECOND LOOK AT INFRA AND COMPONENT COMMUNITY STRUCTURE

Examination of the helminth fauna of red-spotted newts provides an additional, and complementary, view to the previous broad scale analyses of community structure. Selection of the nine different ponds emphasize differences in habitat conditions with four representing permanent (River Stix in Alachua County, FL and Flamingo Bay, Wood Duck Bay, and Dick's Pond in Aiken and Barnwell Counties, SC) and five being ephemeral (Sarracenia Bay, Canoe Bay, Little Robin's Bay, Bay 155, and Bay 23 in Aiken and Barnwell Counties, SC) water bodies. Comparing compositional variability within and between pond types of differing hydroperiods should stress the importance of local ecological conditions on helminth community structure.

Six species of helminths were found in the 127 newts examined (Table 7.7). Variability in the mean number of species and individuals was high within the different pond types, and there were no significant differences between permanent and ephemeral ponds (Nested ANOVA: richness, $F_{1, 118} = 1.7$; abundance, $F_{1, 118} = 1.6$, both $p > 0.2$). There were slight differences between the nine ponds in the mean number of species and individuals indicating that changes in community structure may reflect local environmental conditions (Table 7.7). In the two permanent ponds where richness and abundance was highest, both paedomorphic and metamorphic adults were found. Thus, ecological factors influencing variation in life history traits in local populations potentially play a major role in regulating helminth community structure.

Composition of the helminth fauna differed between locations. Similarities among newts from the same location were generally higher than similarities between populations (Fig. 7.6). There were no obvious within pond differences in either qualitative or quantitative similarity for either permanent or ephemeral habitats, but most values were low (except Wood Duck Bay) indicating an uneven distribution of helminth species among individual newts. The highest similarity values were found in those sites where there was at least one core species (Table 7.7). The patchy distribution pattern of species among ponds was reflected in high spatial variability in faunal similarity. Values were generally high when comparisons were

Table 7.7 Mean number of species and abundance (±1 SE) of helminths and distribution of core species in nine populations of *Notophthalmus viridescens* collected in the southeastern United States

	RS	*DP*	*WDB*	*FB*	*SB*	*CB*	*LRB*	*B155*	*B23*
Brachycoelium salmandrae	+	–	–	–	–	+*	–	+*	C*
Plagitura salamandra	C*	–	C*	–	–	+	–	–	–
Megalodiscus temperatus	+	+*	C	+*	+*	–	–	–	–
Cosmocercoides variabilis	–	+	–	–	–	+*	C*	+*	–
Capillaria inequalis	C*	–	–	–	–	–	–	–	–
Acanthocephalus aculeatus	–	–	–	–	–	–	+	–	–
Mean number of species	1.7 ± 0.2	0.4 ± 0.1	2.0 ± 0.0	0.4 ± 0.2	0.3 ± 0.2	0.5 ± 0.2	1.0 ± 0.3	0.5 ± 0.1	0.7 ± 0.2
Mean abundance	8.9 ± 2.7	1.8 ± 0.9	62.9 ± 8.2	2.4 ± 1.1	0.4 ± 0.3	0.7 ± 0.2	1.4 ± 0.4	1.1 ± 0.3	1.3 ± 0.5

Locations: RS = River Stix, DP = Dick's Pond, WDB = Wood Duck Bay, FB = Flamingo Bay, SB = Sarracenia Bay, CB =Canoe Bay, LRB = Little Robin's Bay, B155 = Bay 155, B23 = Bay 23.
C: core species, +: present, –: not present.
* Numerical dominant in at least one infracommunity.

JACCARD COEFFICIENT

	RS	DP	WDB	FB	SB	CB	LRB	B155	B23
RS	47.2 ±5.9 / 42.1 ±6.1	2.9 ±1.7	27.7 ±3.6	3.8 ±1.9	2.2 ±1.4	1.2 ±1.0	0.0	0.4 ±0.9	2.2 ±2.4
DP	1.8 ±1.2	9.1 ±8.3 / 9.1 ±8.3	16.7 ±4.9	14.8 ±7.8	8.3 ±6.3	2.5 ±2.6	5.2 ±5.3	3.1 ±3.3	0.0
WDB	33.5 ±5.4	3.6 ±1.6	100.0 ±0.0 / 90.5 ±2.9	22.2 ±5.5	12.5 ±4.9	1.7 ±1.3	0.0	0.0	0.0
FB	2.4 ±1.4	14.8 ±7.8	4.9 ±1.8	16.7 ±12.6 / 16.7 ±12.6	11.1 ±7.7	0.0	0.0	0.0	0.0
SB	1.3 ±1.1	8.3 ±6.3	2.7 ±1.5	11.1 ±7.7	3.6 ±6.8 / 3.6 ±6.8	0.0	0.0	0.0	0.0
CB	1.4 ±1.2	2.6 ±2.7	2.5 ±1.9	0.0	0.0	8.3 ±3.6 / 8.8 ±5.6	18.9 ±4.9	12.1 ±5.1	5.0 ±3.9
LRB	0.0	5.7 ±5.7	0.0	0.0	0.0	21.6 ±7.7	33.3 ±20.4 / 41.7 ±22.3	23.4 ±8.8	0.0
B155	0.4 ±0.9	3.1 ±3.3	0.0	0.0	0.0	12.4 ±5.2	25.8 ±9.3	13.3 ±8.5 / 13.3 ±8.5	8.3 ± 5.9
B23	2.2 ±2.4	0.0	0.0	0.0	0.0	4.4 ±3.7	0.0	8.3 ±5.9	40.0 ±20.8 / 40.0 ±20.8

PER CENT SIMILARITY

Fig. 7.6 Similarity values (mean ± 1 SE) between helminth communities in *Notophthalmus viridescens*. Values above the diagonal represent qualitative similarities; those below represent quantitative similarities. Values on the diagonal are within population similarities. Populations labelled as in Table 7.7).

made either among permanent or ephemeral sites; low similarities were found comparing helminth community composition among ephemeral and permanent sites.

The nested nature of changes in total species richness between ponds also presents a natural experiment to assess the importance of biotic interactions on community structure. By comparing community abundance at sites with differing numbers of species, one can ask: Do species present at species-poor sites increase in population size and compensate for the absence of species present at richer sites? Applied to the helminth fauna for the red-spotted newts, there were significant differences in the mean number of individuals across ponds from different species richness classes (ANOVA, $F_{3, 123} = 3.95$, $p < 0.01$; Fig. 7.7). There was no obvious pattern of change with increasing species richness; sites with one and three helminth species had the fewest individuals. The general pattern did not illustrate density

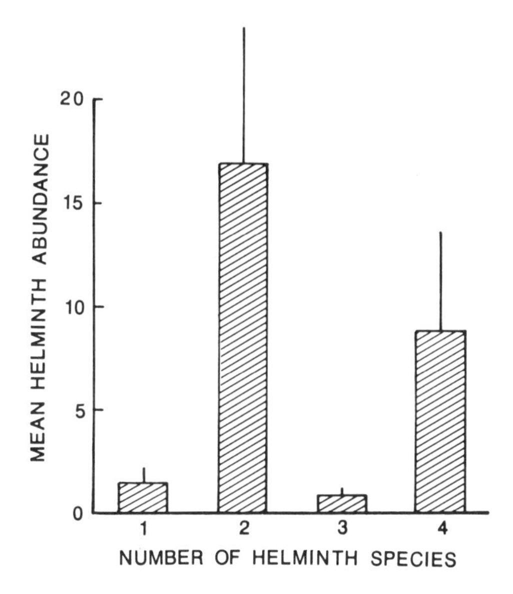

Fig. 7.7 Mean helminth community abundance for four species richness classes (ponds in each class combined) in red-spotted newts. Histograms represent mean values for all newts from specific ponds and 95% confidence intervals.

compensation, so there is no evidence supporting the importance of interspecific interactions in structuring newt helminth communities.

7.5 MECHANISMS

Community richness

Acquisition of parasites through food web linkages are ultimately set by ecological energy budgets with the assumption that energy available to an individual is finite. Within limits of resource availability, generalizations based on patterns of helminth community structure in birds and fish have identified several host attributes that contribute to establishment of communities of differing species complexity and size (Kennedy *et al.*, 1986; Goater *et al.*, 1987; Holmes and Price, 1986). These features include: breadth and selectivity of the diet, vagility of the host, structural complexity of the alimentary canal, and host physiology. Exposure to helminth species with direct life cycles is also recognized as contributing to species richness.

Although herps are gape-limited predators, they are known to consume a diversity of aquatic and terrestrial prey, and accordingly should lead to establishment of diverse helminth communities. However, in terms of species richness and community size, helminth communities of herps are among the most depauperate of all vertebrate groups. As ectotherms, environmental conditions limit helminth recruitment potentail and community development by affecting both feeding rates and foraging behaviour. Amphibians and reptiles are, for the most part, opportunistic, generalist feeders; where selective predation occurs, it is usually more a function of prey size rather than for a specific prey type (Toft, 1985). Consequently, there may be little selective pressure for helminths of amphibians and reptiles to adapt to utilizing specific prey as intermediate hosts. Changes in structural complexity (or habitat complexity) of the gut also appear to be unimportant to development of a diverse helminth community. Lizards and turtles possess true colic cecae, but are at opposite ends of the continuum in community complexity. Consequently, chances of developing a deterministic or 'core' suite of frequently co-occurring and abundant helminth species would be limited.

Low vagility of most hosts also precludes exposure to a wide range of parasites dependent on food web transfers. Aside from reproductive activity or movements associated with finding hibernacula, most herp species (exceptions would include species such as marine turtles) remain relatively sedentary, not moving great distances within or between habitats. The restricted movement may, however, contribute to the successful exploitation of hosts by parasites with direct life cycles (e.g. *Capillaria inequalis, Thelandros magnavulvaris, Cosmocercoides variabilis, Oswaldocruzia pipiens*). In certain hosts, such as desert lizards (e.g. *Sceloporus*), enrichment of the assemblage by direct life cycle nematodes is quite obvious. The number of nematode species and individuals varies between host species, but frequently can dominate the composition of the helminth community. The observed patterns of species richness and community size support the existence of an isolationist community (*sensu* Holmes and Price, 1986), in which biotic interactions should be limited because abundances are low and species diversity is restricted. The potential exists for hosts from either class to have a diverse helminth community, but it is rarely realized.

Consistent differentiation in community complexity among taxa of amphibians and reptiles stresses the generality of certain themes. Biophysical constraints (either due to temperature or hydric conditions) are probably the major feature controlling variability in helminth community structure by influencing rates of resource harvesting and processing. Temporal constraints on activity and foraging patterns effectively reduces exposure, both direct and indirect, of the host to helminths. Diel partitioning of temporal activity and foraging patterns is generally correlated with temperature and is most pronounced among lizards and snakes (Schoener, 1974; Toft, 1985), particularly terrestrial species, as they are less buffered against environmental change than aquatic hosts. Many species of lizards are active for only short periods of time in the morning or afternoon (e.g. Huey, 1982; Grant and Dunham, 1988), while diurnal or diel peaks in activity are seen for snakes (Gibbons and Semlitsch, 1986). Terrestrial amphibians are probably not constrained to the same degree, but are limited in distribution to moist microhabitats because they are subject to desiccation (Feder, 1983; Feder and Lynch, 1982; Duellman and Trueb, 1986). Aquatic amphibians and reptiles experience the least temporal constraints in activity periods and foraging. They may, however, face limits on rates of processing because digestion and gut clearance times are positively correlated with water temperature (Avery, 1987). As an important thermoregulatory behaviour, especially among turtles, aerial basking represents a means of increasing digestive efficiencies and gut clearance times to increase foraging rates (Crawford *et al.*, 1983). Variability in activity patterns coupled with differences in habitat preference and productivity result in differential exposure to potential prey used as intermediate hosts or transmission through direct penetration. The changes in helminth community structure among the different herp taxa, therefore, parallel the extent to which an organism is environmentally buffered.

Several additional features contribute to the variability in helminth community structure among herp taxa, but account more for where a species is positioned along the continuum of complexity typical of that host group. Two different adaptive peaks in foraging mode can be distinguished among the herps: sit-and-wait ambush predators and wide-ranging, searching predators (e.g. Huey and Pianka, 1981; Toft, 1981, 1985). Morphological and physiological commitments associated with each mode determines the type, size, and quantity of prey consumed. Widely foraging predators capitalize on small, sedentary prey that is locally abundant but unpredictably clumped. Sit-and-wait predators capture few individuals of large, mobile prey. Consistency seen in diets of many herp species suggests that microhabitat utilization, foraging mode, as well as anatomical and behavioural constraints exert a profound influence on trophic relations. Ambush predators (e.g. most lizards, pelobatid toads) tend to possess less complex communities while wide-ranging predators (e.g. turtles, ranid frogs) have richer, larger helminth communities. Thus, ecological correlates of foraging strategies could easily contribute to the differences in helminth community complexity within and between families of each order. The development of anti-predator defenses (e.g. skin toxicants, mimicking noxious insects) in wide-ranging foragers might also contribute to greater helminth community complexity in families such as the Salamandridae or Bufonidae by decreasing predation risk and subsequently increasing foraging time.

Comparing the helminth communities of herps to other vertebrate hosts, it is apparent that two interacting factors are responsible for the observed differences in structure. First, among the examples of marine fish, birds, and one mammal, host vagility (or the organisms on which they feed) is greater than herps. Movement of definitive hosts between habitats (within or between locations) or the immigration of potential intermediate hosts for territorial organisms exposes these hosts to more helminth species. These features may also aid the reintroduction of helminth species following local extinctions thereby contributing to increased helminth community complexity at specific locations (Chapter 5). Second, differences in helminth community size reflect the demands of homeothermy, or of high levels of activity, making high feeding rates necessary. Transmission of helminths through predator–prey interactions is simply not as important in herp systems as it is for certain birds or mammals. Consequently, differences in physiological processes controlling metabolic demands appear to be a fundamental feature in the development of isolationist (species-poor assemblages with low transmission rates) or interactive communities (species-rich assemblages with relatively high transmission rates). As emphasized by Kennedy *et al.* (1986) and Goater *et al.* (1987), other host attributes will then either amplify or negate these fundamental differences.

Community diversity: relative roles of local and regional processes

The total number of helminth species associated with several species of amphibians and reptiles is strongly linked to host geographic range. The generality of this pattern has been suggested in other studies by Price and Clancy (1983: geographic range), Freeland (1979: primate group size), and Kennedy (1978: lake area). Three classes of hypotheses (recently reviewed by Connor and McCoy, 1979) have been presented to account for the correlation. All are based in part on analogies with island biogeography (MacArthur and Wilson, 1967). These include: (a) the habitat heterogeneity hypothesis – diversity of habitat types occupied is greater in hosts with large geographic ranges, (b) the frequency encounter hypothesis – the host passively samples a larger number of parasites as its range (size) expands, and (c) the area *per se* hypothesis – area alone increases the probability of colonization and reduces the probability of extinctions. While the mechanism producing the pattern is not determined in the present study, the most likely explanation focuses on the habitat heterogeneity model with widespread hosts encountering a diversity of habitat types; support for this hypothesis is present in studies on a variety of organisms (e.g. Strong *et al.*, 1984). The lack of correlation of host size and species richness would argue against the hypothesis that larger hosts sample a greater range of potential helminth species. Although testing of species–area relationships has been questioned (Kuris *et al.*, 1980), many authors contend that such data are invaluable to understanding patterns of species richness (e.g. Price and Clancy, 1983; Lawton *et al.*, 1981).

Factors other than host range play important roles in influencing regional richness. The lack of consistent differences in regression coefficients within and between host taxonomic order, however, suggests that taxonomic affinity is unimportant. Rather, distinctions between the three groups appear to reflect ecological differences contrasting mesic/aquatic with xeric habitats. Why hosts from more mesic areas have different species–area relations is uncertain at this point. The differences may reflect variability in habitat primary productivity (being less in xeric habitats and determined water availability (Hadley and Szarek, 1981) which indirectly may affect secondary productivity (Dunham, 1981) as well as the degree of territoriality each species exhibits and home range size. While the importance of features accounting for the residual variation is stressed, ecological factors appear more important than phylogenetic factors in explaining the patterns in determinants of regional richness.

The asymptotic relationship of local to regional richness regardless of host taxa is, however, suggestive that an upper limit exists to the number of helminth species coexisting in a uniform habitat, the host population. When such a pattern, indicative of a saturated community, has been observed (e.g. terrestrial habitats – Terborgh and Faaborg, 1980; marine systems – Bohnsack and Talbot, 1980), the standard explanation has been synecological, stressing species interactions and/or size of the niche space in a given habitat as limiting factors. Local richness (alpha diversity) increases rapidly and reaches a plateau as niche space fills, and increases in regional richness (gamma diversity) can only occur through increases in beta (between habitat) diversity.

The existence of asymptotic recruitment curves alone, however, are weak inference for species interactions limiting local richness. An alternate explanation, a 'pool-exhaustion' hypothesis, proposes that the pool of potential colonists available in a habitat become exhausted, and not resources. Within the geographic range of the host, most helminth species will not be cosmopolitan in distribution. Many will be restricted to specific habitats. Thus, regional richness may not represent the true species pool from which the local population is drawn. Only a small subset of species may be adapted to local environmental conditions (hosts and/or habitat), and the level of saturation is maintained by the size of this group, not by niche size influencing colonization and extinction. Similar pool-exhaustion models have been proposed from plant–herbivore systems (Connor and McCoy, 1979; Lawton and Strong, 1981).

Helminth communities of amphibians and reptiles have several features that are consistent with a 'pool exhaustion' hypothesis of community development rather than a 'niche saturation' model. Although an asymptotic pattern exists, there is considerable variability in local richness within and among species, and different host taxa, suggesting there is no fixed limit to the number of helminth species a host can harbour. Regional richness sets the upper limit to local richness on a particular host species, while variation in local ecological factors probably accounts for most of the variation in richness among populations. Low values of local richness could reflect the location of the population (i.e. degree of isolation or at the periphery of the host range) while multi-species host assemblages could produce richer faunas on each host by increasing effective host density for host generalists. Low faunal similarity values among local populations for a given species emphasizes stochastic changes in the faunal components along ecological and geographical gradients, particularly when regional diversity is high. Although different local/regional richness relationships are evident for common species, the patterns seem to be less a result of interpretable differences than to unsystematic heterogeneity of mean local richness around the regression. Thus, comparatively few common species are present around which to assemble a basic core community and species turnover between sites is high. Where interactions have been presumed, or identified, to be important in determining community structure, they have usually been among species that are common, abundant, and regularly co-occurring (e.g. Holmes, 1987; Bush and Holmes, 1986; Stock and Holmes, 1988). From these patterns, it is inferred that the helminth communities of herps are invasible and many vacant niches exist.

The observations on helminth community dynamics in red-spotted newts provide additional support for a noninteractive structure and the presence of many vacant niches. The impoverished nature of the helminth communities of newts leads to no obvious density compensation. Data on the infracommunities of newts indicate that each individual is an independent sampler of the parasites potentially available to it. Similarities were greatest, but still relatively low, among samples from the same component community with no apparent differences between permanent or ephemeral ponds. The limited transmission potential within these ponds can be associated with either ontogenetic shifts in host habitat use or temporal changes in the habitat (e.g. Alford, 1986). The low predictability of helminth, community composition between ponds emphasizes the extent of stochastic variation in helminth composition between ponds because of the isolated nature of the habitats and philopatry to breeding site. Where species of parasites are abundant, it is because their life cycles are simple or possess adaptations ensuring transmission without dependence upon intermediate hosts. For example, transmission of the common digenean trematodes (*Glypthelmins* and *Megalodiscus* spp.) is accomplished by herps ingesting their moulted skin in which the cercariae penetrate. The environmental constraints (e.g. constancy of water level) or genetic characteristics that produce a shift in life history characteristics to paedomorphy can also markedly affect helminth community structure. Once again, we are left with the notion that predator–prey interactions are not a significant factor in influencing helminth distribution and abundance patterns in herps.

Host specificity

From the analyses presented here, and patterns of species distribution evident from surveys, host specificity seems unimportant in determining gastrointestinal communities in amphibians and reptiles. Several patterns are evident: (a) the relatively few specialists and dominance by host generalists and (b) the comparatively high number of helminth species that are core species in several host taxa at one or more local populations (component community). The lack of a unique helminth fauna among many herp species would appear to refute the cospeciation hypothesis (based on the ideas of Brooks, 1980) that predicts that parasite speciation events track host speciation events. These observations also indicate a noninteractive structure because the 'Ghost of Competition Past' argument (Connell, 1980) figures prominantly in the cospeciation argument. The occurrence of many phylogenetically unrelated helminth species of herps comprising the assemblage suggests that 'host-capture' (Chabaud, 1981) has been an important process in community enrichment, but that such communities will predominantly be isolationist in nature (Goater *et al.*, 1987). Thus, host-habitat selection for a helminth fauna overrides evolutionary trends in amphibian and reptile systems.

A generalist life history strategy, however, seems well-adapted to the life history patterns exhibited by most amphibians and reptiles. By exploiting temporally and spatially variable environments, helminths of amphibians and reptiles opt for spreading the risk among a large number of host species. An alternate strategy to ensure survival and successful recruitment would be to increase longevity; persistence of *Bothriocephalus rarus* in red-spotted newts for 2–7 years serves as a hedge against local extinctions (Jarroll, 1980). What is not known, and is of importance because of the variability seen in population dynamics of most amphibians and reptiles, is identification of which host species is responsible for circulating and maintaining the helminth within a system. Measures of fitness in parasitic species are essential if we are to understand circulation networks of helminths and local evolutionary events as parasites adapt to exploit different host complexes over time.

Conclusions

Helminth communities of amphibians and reptiles are highly variable, depauperate, and have traits characteristics of noninteractive community structure. Only among some species of turtles, and possibly crocodilians, does the complexity of the assemblage approximate conditions considered necessary to be interactive. Whether they are interactive in these cases, however, has yet to be convincingly shown. Phylogenetic differences between hosts are important in structuring herp helminth communities, but primarily from a physiological perspective determining patterns of resource acquisition and utilization. Unlike the structure of helminth communities in other vertebrate groups, development of amphibian and reptilian helminth faunas are not strictly responsive to food web dynamics. Direct life cycle strategies seem well adapted to exploiting life in temporally and spatially variable conditions, conditions for which herps are well-adapted.

Recognition of the importance of local environmental conditions in determining helminth community composition is well founded, as evidenced by observations as early as those of Dogiel (1964). Interpretation of the local/regional richness relationships for the amphibian helminths questions whether local diversity is solely a process of local conditions. In these cases, local communities appear not to be saturated by interactive processes. Local diversity may be prescribed by local conditions (biotic and abiotic), but the number of species found reflects such regional processes as geographic dispersal and the historical accumulation of parasites in different host animals.

Addressing these concepts necessitates consideration of determinants of pattern at the compound community level. Recent investigations have begun to focus on the importance of spatial and temporal variability in recruitment rates of marine invertebrates to resultant community structure (Connell, 1985; Gaines *et al.*, 1985). Specifically, if larval settlement rates are low, post-settlement mechanisms (including interactions) are almost of no importance. At high settlement rates, when vacant space is available, it is quickly saturated with newly arriving larvae. Biotic properties of the community are then primarily determined by post-settlement processes.

The role of transport processes, or 'supply-side' ecology, for helminths in a local population will initially be dependent on the structure of the potential host community in an environment. For many species of pond-breeding amphibians, suitability of the habitat depends on its hydrologic cycle, which determines the length of time available for larval growth of potential colonists and their risks of predation or competitive effects. The surrounding terrestrial habitats and life histories of individual species determine the suite of species that breed in a particular pond. The level of biological interaction among species then, is dependent on the structure of the food web and the level of disturbance (drying regime) (Wilbur, 1987). For species with complex life cycles, however, determination of where regulation of population size or community structure occurs (terrestrial, aquatic, or effects at one stage affect regulatory processes at another stage) will be extremely important to understanding helminth community structure. Similar parallels are present in evaluating the relative roles of habitat structural complexity, climate, and biotic interactions as determinants of lizard community structure (e.g. Pianka, 1986; Schiebe, 1987). Once the supply has been determined, the screens or filters determining exposure and establishment of parasites from the potential pool of species to a realized helminth community will take effect (see Chapter 5). Through the action of supply and screens, the actual structure of the infracommunity will be determined by either biotic interactions or stochastic processes.

Major strides have been made in the study of helminth community ecology. The focus has largely been on particular systems studied in isolation, and has formed the nucleus recognizing the importance of scale and the hierarchical nature of helminth community structure. The patterns observed for herps suggest that concepts of community processes must be broadened to incorporate regional and historical processes. Determinants of the size of the species pool from which a local population is drawn must, therefore, include biogeographic and evolutionary processes occurring over broad areas. Speciation, extinction, habitat shifts and spatial variability in species distribution patterns are best studied at larger spatial scales and offer a framework to interpret local variability in species richness. These bridges between local and regional viewpoints offer many new challenges to understanding factors controlling the development and evolution of helminth communities.

There is a growing empirical and theoretical framework directed at understanding variability in helminth community structure. Patterns of helminth community structure reflect a history of accumulation of parasite species and a period over which interactions may occur to mould the development of the community. Helminths of amphibians and reptiles offer excellent systems to examine contemporary patterns of helminth community structure, and the processes responsible for that structure. Of all the vertebrate classes, amphibians and reptiles are extremely amenable to both comparative and field and laboratory experimental, manipulative approaches to investigate ecological and evolutionary mechanisms governing helminth community development. For this reason, herps should be influential, and controversial, in the further development of theories concerning parasitic systems and the way we view community ecology as a whole.

ACKNOWLEDGEMENTS

This work was supported by contract DE-AC09-76SRO-819 between the U.S. Department of Energy and the University of Georgia. I thank R. Wolfe, A. Dancewicz, and J. Wallin for help in field collections; J. Mobley for library assistance in compiling the many surveys used in the analyses; J. Coleman for her advice and assistance in the graphics; and P. Dixon and J. E. Pinder, III, for statistical advice. I would also like to acknowledge the helpful suggestions provided by R. Anderson, A. O. Bush, G. W. Esch, T. Goater, C. R. Kennedy, P. Mulvey, and J. H. K. Pechmann that vastly improved this manuscript.

REFERENCES

Alford, R. A. (1986) Habitat use and positional behavior of Anuran larvae in a Northern Florida temporary pond. *Copeia*, **1986**, 408–23.

Anderson, R.C. (1984) The origins of zooparasitic nematodes. *Can. J. Zool.*, **62**, 317–28.

Avery, H. W. (1987) Roles of Diet Protein and Temperature in the Nutritional Energetics of Juvenile Slider Turtles, *Trachemys scripta*. M.S. Thesis, State University College at Buffalo, Buffalo, NY.

Avery, R. A. (1971) Helminth parasite populations in newts and their tadpoles. *Freshwater Biology*, **1**, 113–19.

Baker, M. (1984) Nematode parasitism in amphibians and reptiles. *Can. J. Zool.*, **62**, 747–57.

Behler, J. L. and King, F. W. (1979) *The Audubon Society Field Guide to North American Reptiles and Amphibians*, Alfred A. Knopf, Inc., New York.

Benes, E. S. (1985) Helminth parasitism in some central Arizona lizards. *Southwest. Nat.*, **30**, 467–73.

Bohnsack, J. A. and Talbot, F. H. (1980) Species packing by reef fishes on Australian and Caribbean reefs: an experimental approach. *Bull. Marine Sci.*, **30**, 710–23.

Brandt, B. B. (1936) Parasites of certain North Carolina Salientia. *Ecol. Monogr.*, **6**, 491–532.

Brooks, D. R. (1979a) Testing the context and extent of host–parasite coevolution. *Syst. Zool.*, **28**, 299–307.

Brooks, D. R. (1979b) Testing hypotheses of evolutionary relationships among parasites: the digeneans of crocodilians. *Am. Zool.*, **19**, 1225–38.

Brooks, D .R. (1980) Allopatric speciation and non-interactive parasite community structure. *Syst. Zool.*, **29**, 192–203.

Brooks, D. R. (1981a) Raw similarity measures of shared parasites: an empirical tool for determining host phylogenetic relationships? *Syst. Zool.*, **30**, 203–7.

Brooks, D. R. (1981b) Hennig's parasitological method: a proposed solution. *Syst. Zool.*, 30, 229–49.

Bundy, D. A. P., Vogel, P. and Harris, E. A. (1987) Helminth parasites of Jamaican anoles (Reptilia: Iguanidae): a comparison of the helminth fauna of 6 *Anolis* species. *J. Helminth.*, **61**, 77–83.

Bush, A. O. and Holmes, J. C. (1986) Intestinal helminths of lesser scaup ducks: an interactive community. *Can. J. Zool.*, **64**, 142–52.

Campbell, R. A. (1968) A comparative study of the parasites of certain Salientia from Pocahontas State Park, Virginia. *Virginia J. Sci.*, **19**, 13–20.

Chabaud, A. G. (1981) Host range and evolution of nematode parasites of vertebrates. *Parasitology*, **82**, 169–70.

Coggins, J. R. and Sajdak, R. A. (1982) A survey of helminth parasites in the salamanders and certain anurans from Wisconsin. *Proc. Helminthol. Soc. Wash.*, **49**, 99–102.

Collins, R. F. (1969) The helminths of *Natrix*, spp. and *Agkistrodon piscivorous piscivorous* (Reptilia: Ophidia) in Eastern North Carolina. *J. Elisha Mitchell Sci. Soc.*, **85**, 141–4.

Conant, R. (1975) *A Field Guide to Reptiles and Amphibians of Eastern and Central North America*, Houghton Mifflin Co., Boston.

Connell, J. H. (1980) Diversity and the coevolution of competitors, or the ghost of competition past. *Oikos*, **35**, 131–8.

Connell, J. H. (1983) On the prevalence and relative importance of interspecific competition: evidence from field experiments. *Am. Nat.*, **122**, 661–96.

Connell, J. H. (1985) The consequences of variation in initial settlement vs. post-settlement mortality in rocky intertidal communities. *J. Exp. Marine Biol. Ecol.*, **93**, 11–45.

Connor, E. F. and McCoy, E. D. (1979) The statistics and biology of the species–area relationship. *Am. Nat.*, **113**, 791–833.

Cornell, H. V. (1985) Local and regional richness of cynipine gall wasps on California oaks. *Ecology*, **66**, 1247–60.

Crawford, K. M., Spotila, J. R. and Standora, E. A. (1983) Operative environmental temperature and basking behavior of the turtle, *Pseudemys scripta*. *Ecology*, **64**, 989–99.

Crompton, D. W. T. and Nesheim, M. C. (1976) Host–parasite relationships in the alimentary tract of domestic birds. *Adv. Parasitol.*, **14**, 95–194.

Detterline, J. L., Jacob, J. S. and Wilhelm, W. E. (1984) A comparison of helminth endoparasites in the cottonmouth (*Agkistrodon piscivorous*) and three species of water snakes (*Nerodia*). *Trans. Am. Microsc. Soc.*, **103**, 137–43.

Dogiel, V. A. (1964) *General Parasitology*, Oliver and Boyd, Edinburgh and London.

Dronen, N. O., Jr (1978) Host–parasite population dynamics of *Haematoloechus coloradensis* Cort, 1915 (Digenea: Plagiorchidae). *Am. Midl. Nat.*, **99**, 330–49.

Dubinina, M. N. (1949) Ecological studies on the parasite fauna of *Testudo horsefeldi* Grau in Tadzhikistan (in Russian). *Parazit. Sbornik*, **11**, 61–97.

Duellmann, W. E. and Trueb, L. (1986) *Biology of Amphibians*, McGraw-Hill Book Company, New York.

Dunbar, J. R. and Moore, J. D. (1979) Correlations of host specificity with host habitat in helminths parasitizing the plethodontids of Washington County, Tennessee. *J. Tenn. Acad. Sci.*, **54**, 106–9.

Dunham, A. E. (1981) Populations in a fluctuating environment: the comparative population ecology of the iguanid lizards *Sceloporus merriamia* and *Urosaurus ornatus*. *Misc. Publ. Mus. Zool., Univer. Michigan*, **158**, 1–62.

Dyer, W. G., Brandon, R. A. and Price, R. L. (1980) Gastrointestinal helminths in relation to sex and age of *Desmognathus fuscus* (Green, 1818) from Illinois. *Proc. Helminthol. Soc. Wash.*, **47**, 95–9.

Ernst, C. H. and Ernst, E. M. (1980) Relationships between North American turtles of the *Chrysemys* complex as indicated by their endoparasitic helminths. *Proc. Biol. Soc. Wash.*, **93**, 339–45.

Esch, G. W. and Gibbons, J. W. (1967) Seasonal incidence of parasitism in the painted turtle, *Chrysemys picta marginata* Agassiz. *J. Parasitol.*, **53**, 818–21.

Esch, G.W., Gibbons, J. W. and Bourque, J. E. (1975) An analysis of the relationship between stress and parasitism. *Am. Midl. Nat.*, **93**, 339–53.

Esch, G. W., Gibbons, J. W. and Bourque, J. E. (1979a) The distribution and abundance of enteric helminths in *Chrysemys s. scripta* from various habitats on the Savannah River plant in South Carolina. *J. Parasitol.*, **65**, 624–32.

Esch, G. W., Gibbons, J. W. and Bourque, J. E. (1979b) Species diversity of helminth parasites in *Chrysemys s. scripta* from a variety of habitats in South Carolina. *J. Parasitol.*, **65**, 633–8.

Feder, M. E. (1983) Integrating the ecology and physiology of plethodontid salamanders. *Herpetologica*, **39**, 291–310.

Feder, M. E. and Lynch, J. F. (1982) Effects of latitude, season, elevation, and microhabitat on field body temperatures of neotropical and temperate zone salamanders. *Ecology*, **63**, 1657–64.

Fischthal, J. H. (1955a) Helminths of salamanders from Promised Land State Forest Park, Pennsylvania. *Proc. Helminthol. Soc. Wash.*, **22**, 46–9.

Fischthal, J. H. (1955b) Ecology of worm parasites in South-Central New York salamanders. *Am. Midl. Nat.*, **53**, 176–83.

Fowler, S. V. and Lawton, J. H. (1982) The effects of host–plant and local abundance on the species richness of agromyzid flies attacking British umbellifers. *Ecol. Entomol.*, **7**, 257–65.

Freeland, W. J. (1979) Primate social groups as biological islands. *Ecology*, **60**, 719–28.

Gaines, S., Brown, S. and Roughgarden, J. (1985) Spatial variation in larval concentrations as a cause of spatial variation in settlement for the barnacle, *Balanus glandula*. *Oceologia*, **67**, 267–72.

Gibbons, J. W. and Semlitsch, R. D. (1986) Activity patterns. In *Snakes: Ecology and Evolutionary Biology*, (eds, R. A. Seigel, J. T. Collins and S. S. Novak), Macmillan Co., New York, pp. 396–421.

Goater, T. M., Esch, G. W. and Bush, A. O. (1987) Helminth parasites of sympatric salamanders ecological concepts at infracommunity, component, and compound community levels. *Am. Midl. Nat.*, **118**, 289–300.

Grant, B. W. and Dunham, A. E. (1988) Thermally imposed time constraints on the activity of the desert lizard *Sceloporus merriami*. *Ecology*, **69**, 167–76.

Hadley, N. F. and Szarek, S. R. (1981) Productivity of desert ecosystems. *Biol. Sci.*, **31**, 747–53.

Hanski, I. (1982) Dynamics of regional distribution: the core and satellite species hypothesis. *Oikos*, **38**, 210–21.

Hardin, E. L. and Janovy, J., Jr (1988) Population dynamics of *Distoichometra bufonis* (Cestoda: Nematotaeniidae) in *Bufo woodhousii*. *J. Parasitol.*, **74**, 360–65.

Harwood, P. D. (1932) The helminths parasitic in the Amphibia and Reptilia of Houston, Texas, and vicinity. *Proc. U.S. Nat. Museum*, **18**, 1–77.

Holmes, J. C. (1983) Evolutionary relationships between parasitic helminths and their hosts. In *Coevolution*, (eds, D. J. Futuyma and M. Slatkin), Sinauer Associates, Sunderland, MA, pp. 161–85.

Holmes, J. C. (1987) The structure of helminth communities. *Intern. J. Parasitol.*, **17**, 203–8.

Holmes, J. C. and Price, P. W. (1980) Parasite communities: the roles of phylogeny and ecology. *Syst. Zool.*, **29**, 203–13.

Holmes, J. C. and Price, P. W. (1986) Communities of parasites. In *Community Ecology: Pattern and Process*, (eds, D. J. Anderson and J. Kikkawa), Blackwell Scientific Publications, Oxford, UK, pp. 187–213.

Huey, R. B. (1982) Temperature, physiology, and the ecology of reptiles. In *Biology of the Reptilia*, vol. 12, (eds, C. Gans and F. H. Pough), Academic Press, New York, pp. 25–91.

Huey, R.B. and Pianka, E. R. (1981) Ecological consequences of foraging mode. *Ecology*, **62**, 991–9.

Jackson, T. and Beaudoin, R. L. (1967) Comparison of the parasite fauna in two metamorphic stages of the red-spotted newt *Notophthalmus viridescens viridescens*. *Proc. Pa. Acad. Sci.*, **40**, 70–75.

Jacobson, K. C. (1987) Infracommunity Structure of Enteric Helminths in the Yellow-bellied Slider *Trachemys scripta scripta*. M.S. Thesis, Wake Forest University, Winston-Salem, NC. 46 pp.

Jarroll, E. L., Jr (1980) Population dynamics of *Bothriocephalus rarus* (Cestoda) in *Notophthalmus viridescens*. *Am. Midl. Nat.*, **103**, 360–66.

Kennedy, C. R. (1978) An analysis of the metazoan parasitocoenoses of brown trout *Salmo trutta* from British Lakes. *J. Fish Biol.*, **13**, 255–63.

Kennedy, C. R., Bush, A. O. and Aho, J. M. (1986) Patterns in helminth communities: why are birds and fish different? *Parasitology*, **93**, 205–15.

Kuris, A. M., Blaustein, A. R. and Alió, J. J. (1980) Hosts as islands. *Am. Nat.*, **116**, 570–86.

Landewe, J. E. (1963) Helminth and Arthropod Parasites of Salamanders from Southern Illinois. M.S. Thesis, Southern Illinois University, Carbondale, IL. 47 pp.

Lawton, J. H. (1982) Vacant niches and unsaturated communities: a comparison of bracken herbivores at 2 sites on 2 continents. *J. Anim. Ecol.*, **51**, 573–95.

Lawton, J. H., Cornell, H., Dritschilo, W. and Hendrix, S. D. (1981) Species as islands: comments on a paper by Kuris *et al.* *Am. Nat.*, **117**, 623–7.

Lawton, J. H. and Strong, D. R. (1981) Community patterns and competition in folivorous insects. *Amer. Nat.*, **118**, 317–38.

Likens, G. E. (1983) A priority for ecological research. *Bull. Ecol. Soc. America*, **64**, 234–43.

Limsuwan, C. and Dunn, M. C. (1978) A survey of helminth parasites from turtles in Rutherford County, Tennessee. *J. Tenn. Acad. Sci.*, **53**, 111–14.

MacArthur, R. H. and Wilson, E. O. (1967) *The Theory of Island Biogeography*,

Princeton University Press, Princeton, NJ.
Martin, D. R. (1972) Distribution of helminth parasites in turtles native to Southern Illinois. *Trans. Ill. Acad. Sci.*, **65**, 61–7.
Pearce, R. C. and Tanner, W. W. (1973) Helminths of *Sceloporus* lizards in the Great Basin and Upper Colorado Plateau of Utah. *Great Basin Nat.*, **33**, 1–18.
Pianka, E. R. (1986) *Ecology and Natural History of Desert Lizards*, Princeton University Press, Princeton, NJ.
Platt, T. R. (1977) A survey of the helminth fauna of two turtle species from Northwestern Ohio. *Ohio J. Sci.*, **77**, 97–8.
Price, P. W. (1980) *Evolutionary Biology of Parasites*, Princeton University Press, Princeton, NJ.
Price, P. W. (1984a) Communities of specialists: vacant niches in ecological and evolutionary time. In *Ecological Communities: Conceptual Issues and the Evidence*, (eds, D. R. Strong, D. S. Simberloff, L. Abele and A. B. Thistle), Princeton University Press, Princeton, NJ, pp. 510–23.
Price, P. W. (1984b) Alternative paradigms in community ecology. In *A New Ecology: Novel Approaches to Interactive Systems*, (eds, P. W. Price, C. N. Slobodchikoff and W. S. Gaud), John Wiley and Sons, New York, pp. 353–83.
Price, P. W. and Clancy, K. M. (1983) Patterns in number of helminth parasite species in freshwater fishes. *J. Parasitol.*, **69**, 449–54.
Price, R. L. and Buttner, J. K. (1982) Gastrointestinal helminths of the central newt, *Notophthalmus viridescens louisianensis* Wolterstorff, from Southern Illinois. *Proc. Helminthol. Soc. Wash.*, **49**, 285–8.
Rankin, J. S. (1937) An ecological study of parasites of some North Carolina salamanders. *Ecol. Monogr.*, **7**, 169–269.
Rankin, J. S. (1945) An ecological study of the helminth parasites of amphibians and reptiles of Western Massachusetts and vicinity. *J. Parasitol.*, **31**, 142–50.
Rau, M. E. and Gordon, D. M. (1978) Overwintering of helminths in the Garter snake (*Thamnophis sirtalis sirtalis*). *Can. J. Zool.*, **56**, 1765–7.
Rau, M. E. and Gordon, D. M. (1980) Host specificity among the helminth parasites of four species of snakes. *Can. J. Zool.*, **58**, 929–30.
Rausch, R. (1947) Observations on some helminths parasitic in Ohio turtles. *Am. Midl. Nat.*, **38**, 434–42.
Ricklefs, R. E. (1987) Community diversity: relative roles of local and regional processes. *Science*, **235**, 167–71.
Rohde, K. (1979) A critical evaluation of intrinsic and extrinsic factors responsible for niche restriction in parasites. *Am. Nat.*, **114**, 648–71.
Root, R. B. (1973) Organization of a plant–arthropod association in simple and diverse habitats: the fauna of collards (*Brassica oleracea*). *Ecol. Monogr.*, **43**, 95–124.
Rosenzweig, M. and Abramsky, Z. (1986) Centrifugal community organization. *Oikos*, **46**, 339–48.
Salt, G. W. (1982) *Ecology and Evolutionary Biology: A Round Table on Research*, University of Chicago Press, Chicago.
Schad, G. A. (1963) Niche diversification in a parasitic species flock. *Nature*, **198**, 404–6.

Schiebe, J. S. (1987) Climate, competition, and the structure of temperate zone lizard communities. *Ecology*, **68**, 1424–36.

Schoener, T. W. (1974) Resource partitioning in ecological communities. *Science*, **185**, 27–39.

Schoener, T. W. (1983) Field experiments on interspecific competition. *Am. Nat.*, **122**, 240–85.

Seigel, R. (1986) The Foraging Ecology and Resource Partitioning Patterns of Two Species of Garter Snakes. PhD Dissertation, University of Kansas, Lawrence, U.S.A.

Sokal, R. R. and Rohlf, F. J. (1981) *Biometry*, W. H. Freeman and Company, San Francisco.

Stock, T. M. and Holmes, J. C. (1987) Host specificity and exchange of intestinal helminths among four species of grebes (Podicipedidae). *Can. J. Zool.*, **65**, 669–76.

Stock, T. M. and Holmes, J. C. (1988) Functional relationships and microhabitat distributions of enteric helminths of grebes (Podicipedidae): the evidence for interactive communities. *J. Parasitol.*, **74**, 214–27.

Strong, D. R., Lawton, J. H. and Southwood, T. R. E. (1984) *Insects on Plants: Community Patterns and Mechanisms*, Blackwell Scientific Publishers, Oxford.

Telford, S. R., Jr (1970) A comparative study of endoparasitism among some Southern California lizard populations. *Am. Midl. Nat.*, **83**, 516–54.

Terborgh, J. W. and Faaborg, J. (1980) Saturation of bird communities in the West Indies. *Am. Nat.*, **116**, 178–95.

Toft, C. A. (1981) Feeding ecology of Panamanian Litter Anurans: patterns in diet and foraging mode. *J. Herpetol.*, **15**, 139–44.

Toft, C. A. (1985) Resource partitioning in amphibians and reptiles. *Copeia*, **1985**, 1–21.

Vogel, P. and Bundy, D. A. P. (1987) Helminth parasites of Jamaican anoles (Reptilia: Iguanidae): variation in prevalence and intensity with host age and sex in a population of *Anolis lineatopus*. *Parasitology*, **94**, 399–404.

Wilbur, H. M. (1987) Regulation of structure in complex systems: experimental temporary pond communities. *Ecology*, **68**, 1437–52.

8

Helminth communities in avian hosts: determinants of pattern

Albert O. Bush

8.1 INTRODUCTION

Historical perspectives

In 1983, Rausch published an excellent review on the biology of avian helminths. In that review he included a substantial summary on the study of avian helminths to which readers are referred for a more detailed historical account. My intent in this section is to highlight, briefly, important events leading towards the study of helminth communities in avian hosts.

The study of avian helminths is an old one. Perhaps the first record of helminthiasis in avian hosts was that in the 13th century when Pepagomenos observed worms under the nictitating membrane of birds used in the sport of falconry (Rausch, 1983 citing Huber, 1906). But it was not until five centuries later, in the late 1700s, that significant studies on helminths in avian hosts began to appear. These studies dealt largely with classification and taxonomy of helminths in domesticated hosts. This trend in systematics continued through the mid 19th century but with two important additions – data on free-ranging animals were provided as were data on some helminth life cycles. Early in the 20th century the literature on helminth parasites of birds flourished. Though taxonomy was still the focus, authors began to provide expanded coverage on such topics as life cycles, transmission dynamics and biogeographic patterns of helminth parasites in relation to the birds which they infect. Excellent examples would include Cram's (1927) monograph on nematodes, her (1931) detailed descriptions of developmental stages of avian nematodes or her (Cram *et al.*, 1931) chapter on parasites in bobwhite quail (*Colinus virginianus*). The value of such works can be attested to by the fact that they are still cited in the recent ecological literature (e.g. Moore *et al.*, 1986, 1988).

The decades following the early 20th century saw the emergence of faunistic studies where publications of surveys on helminths in a single (or several) avian species from the same (or different) geographical location(s) became common. Initially qualitative, in the late 1950s and early 1960s there was an increasing trend towards quantitative data collection (for a particularly thorough review, see Dogiel, 1964 and references therein).

One other feature of historical note is that population studies on helminths in avian hosts are uncommon. Unlike other vertebrate host groups (or even invertebrate intermediate hosts) where studies on the population ecology of a single or several helminth species were common, ecological studies of helminths in birds have been primarily directed at the community level. There are some rare exceptions (e.g. Korpaczewska, 1963; Avery, 1969; Thompson, 1985).

A synoptic historical view of helminth communities in avian hosts would thus acknowledge a vast quantity of historical data relating to individual helminth species, an apparent wealth of faunistic data available and a concentration on community level analyses.

Current perspectives

My intent in the following sections will be to examine the fundamental questions which have been asked of data on helminth communities in avian hosts. I do not intend an exhaustive (or even thorough) review of the literature; two recent and excellent reviews are already available (Rausch, 1983; Moore, 1987). Rather, my intent will be to select specific studies which are readily accessible to me, which demonstrate a divergence of opinion, and with which I am most familiar.

In spite of the apparent wealth of data available, there are two major problems associated with trying to use the extant literature. First, although the importance of analyses directed at the infracommunity are obvious, many authors continue to treat all hosts from a specific collection as a community and thus compare (e.g. with similarity indices) different host species or the same host species from different areas. These studies are difficult to interpret since the 'community' is nothing more than an arbitrary, collapsed data set obscuring patterns within biological communities (i.e. the individual host or infracommunity). A second problem is that many studies focus on only one helminth taxon (e.g. cestodes) in a host species and thus a true description of the community is not available.

Infracommunity patterns

Analyses directed at the infracommunity level are rare. Such studies all focus attention on the pattern of distribution of helminth species within an individual host. Initial studies at the component community level (e.g. Threlfall, 1965, 1968; Crompton and Nesheim, 1976; Czaplinski, 1974) confirmed what had been observed for helminths in other vertebrate classes – within the intestinal environment, helminth species occupied specific and predictable locations.

More recent studies have addressed the question of whether helminth communities are random aggregations of species or whether they have structure which is repetitive and thus predictable. Such studies typically examine the distribution of helminth species' infrapopulations and attempt to determine if there is evidence for, or against, interactive site selection. Hair and Holmes (1975) concluded that there was little evidence for interactive site selection among helminth species in lesser scaup ducks, *Aythya affinis*, (but see comments by Bush and Holmes, 1986b). In contrast, most other studies on helminths in avian hosts have confirmed interactive communities (Avery, 1969; Riley and Owen, 1975; Pojmanska, 1982; Bush and Holmes, 1986b; Stock and Holmes, 1987a, 1988; Goater and Bush, 1988; Edwards and Bush, 1989) and all authors have invoked inter- or intraspecific competition as a likely mechanism for structuring communities.

One feature common to the above studies is of considerable note. All purported interactions within or between helminth species involved either congeners or taxa belonging to the same guild. Furthermore, where data are available, most of the particular helminth species involved are specialists in the particular host species and many appear to use the same intermediate host.

Component level patterns

Analyses of helminth communities at this level are quite common. The almost universal questions being addressed are related to the patterns of distribution of helminths among host species. The nature of the questions vary, but four specific determinants of patterns have been addressed with reasonable regularity (geographical, age-related, gender-related and seasonal/yearly).

Many studies have attempted to identify geographic patterns in the distribution of helminth species within a particular host species. Threlfall (1968) compared the helminth fauna of herring gulls (*Larus argentatus*) collected from Newfoundland with those collected in Great Britain. He noted that the former were depauperate with respect to number of species and number of individuals and he suggested that most of the differences were due to the more restricted diet of Newfoundland birds. De Jong (1976) examined mallards (*Anas platyrhynchos*) in the Netherlands concluding that birds on the wintering grounds had fewer helminths (both species and individuals) than birds on the breeding grounds and suggested that either a lack of transmission on the wintering grounds, or the rigours of migration, might account for such patterns. Buscher's (1965, 1966) data on helminths in four species of migratory waterfowl collected in North America can be interpreted similarly. Wallace and Pence (1986) concluded that helminth communities in migratory blue-winged teal (*Anas discors*) collected in the autumn (post-breeding) and the spring (post-wintering) were largely similar, with no replacement of the breeding ground helminth fauna on the wintering grounds (with significant reductions in the number of individuals but no substantive losses of helminth species). Bush and Holmes (1986a) examined habitat-associated patterns in the helminth communities of lesser scaup ducks collected from 13 different lakes in Alberta (breeding grounds) and concluded that lake of origin did have some influence on the helminth community and that communities varied more among lakes than among ecological regions (prairie, aspen parkland, boreal forest). Forrester *et al.* (1984) compared helminth communities in bobwhite quail from 11 localities in Florida. They found broadly similar patterns in the distribution of helminth species among the different localities, with no significant correlation of geographic distance between samples with similarity values between samples.

Another major pattern which has been addressed at the component community level is the influence of host age on the helminth communities. Threlfall (1968) noted that herring gull chicks had significantly fewer helminth species and individuals than adults. He concluded that this difference was a function of time – chicks simply had not been exposed long enough to acquire the adult complement of helminths. Bakke (1972a) noted that juvenile common gulls (*Larus canus*) had fewer digeneans than either immatures or adults. He suggested that seasonal aspects, geographical factors and food habits could all account for such differences. De Jong (1976) found that prevalence and intensity of helminth infection was higher among juvenile ducks than in adult ducks. De Jong argued that the greater variety of food taken by ducklings, and a possible lack of immunity during the first few weeks of life, were responsible. Buscher (1965, 1966) and Wallace and Pence (1986) also noted higher prevalences of helminths in juvenile ducks and they too invoked a lack of immunity in juvenile birds as a plausible explanation. Forrester *et al.* (1984) found no significant differences in prevalence of helminth species between chick and adult bobwhites although chicks were significantly more heavily infected with one particular species. Moore *et al.* (1987) examined the intestinal helminths of bobwhites and concluded that age was occasionally an important feature influencing helminth prevalence and density.

The influence of gender on the structure of helminth communities has also been examined. Threlfall (1968) suggested (but did not provide evidence for) significant differences between helminth communities in male and female herring gulls, concluding that different food habits could account for the observed differences. Cornwell and Cowan (1963) found significant differences in the helminths of male and female canvasback ducks (*Aythya valisineria*) during the breeding season; during that time, males feed extensively on algae while females and their broods ingest high quantities of invertebrates. Wallace and Pence (1986) found no significant differences in prevalence or intensity of helminths between male and female blue-winged teal. Likewise, neither Forrester *et al.* (1984) nor Moore *et al.* (1987) found host gender to be an important variate in explaining differences in bobwhite quail helminth communities.

A final pattern which has been addressed with some regularity is the influence of seasonal or yearly variation. (As a cautionary note, recognize that host age and both season and year are continuous variables which are not independent of one another, nor is seasonal pattern necessarily independent of a migratory pattern.) Threlfall (1968) found that the prevalence of some helminth species in gulls showed a peak in the spring followed by declining prevalence through the rest of the collection period. Other helminth species exhibited a reverse pattern. Bakke (1972b) found the same patterns for digeneans in common gulls. Hair (1975) examined helminth communities in lesser scaup from Cooking and Hastings Lakes, Alberta on a seasonal basis over a two-year period. He found no significant seasonal or yearly patterns in helminths in adult birds. Forrester *et al.* (1984) argued that significant variation in prevalence of helminths in bobwhite quail was attributable to yearly variation and suggested density-dependent mechanisms might be involved. Moore *et al.* (1987) compared bobwhite quail from the same location, but separated by a 15 year interval. They report significant changes in the prevalence and intensity of some species and suggest an apparent local extinction of another. Although they argue that monoxenous helminths may be influenced by density-dependent mechanisms, they emphasize that abiotic factors must also be considered.

One observation appears clear at the component community level. I have intentionally selected literature which shows that there are no unambiguous patterns which would allow one to summarily dismiss the importance of any biotic or abiotic varible. However, a common thread to most of the interpretations presented above is host diet which in turn is a function of habitat. Depending on the nature of the questions asked, for any given system one must carefully interpret the data with respect to the biology of the hosts and their relationship to the environment.

Current questions

Based on the observations made above, a pertinent question, similar to that addressed by several others in this book, is – to what extent can helminth communities in avian hosts be invaded by additional individuals or species? Such questions are difficult since they engender the 'empty niche' concept (e.g. Price, 1980, 1984; Lawton, 1982, 1984) and controversy (e.g. Herbold and Moyle, 1986). My intent is to avoid that controversy (though I believe the concept to have considerable heuristic value) and argue that two individual hosts, of the same species, should be able to harbour the same number of individuals of the same helminth species, all other things being equal. This latter phrase is of course the key and, although quite simplistic and intuitive, I do not believe such an hypothesis has been explicitly stated. Nevertheless, it is but variations on that theme under which most authors have sought 'pattern'. Seldom will all things be equal and, by selecting intentional inequalities, we can begin to tease apart the determinants of pattern in helminth communities.

To examine the question of invasibility, I will examine patterns of helminth species richness and abundance in willets, *Catoptrophorus semipalmatus*, collected on their breeding grounds (freshwater) and their wintering grounds (saltwater). Because of the nature of the question, I will focus my attention on infracommunities and component level communities, with major emphasis on the latter. Detailed analyses of the infracommunities will be presented elsewhere. Though I will make occasional reference to them, I lack sufficient data to do justice to patterns associated with compound communities. I will examine the influence of age, sex and geographical location of host collection on the structure of the helminth communities. I will also examine the guild structure of the helminth communities to determine the degree to which guilds are similar in populations from freshwater versus those from saltwater. If willet helminth communities are saturated, I would predict the same, or at least similar, levels of species packing across their geographic distribution. I would also predict that helminth communities from across the willet's geographic distribution would show the same, or similar, number of guilds.

8.2 DATA SETS AND ANALYSES

I selected willets as a host species for several reasons: (a) they are abundant,

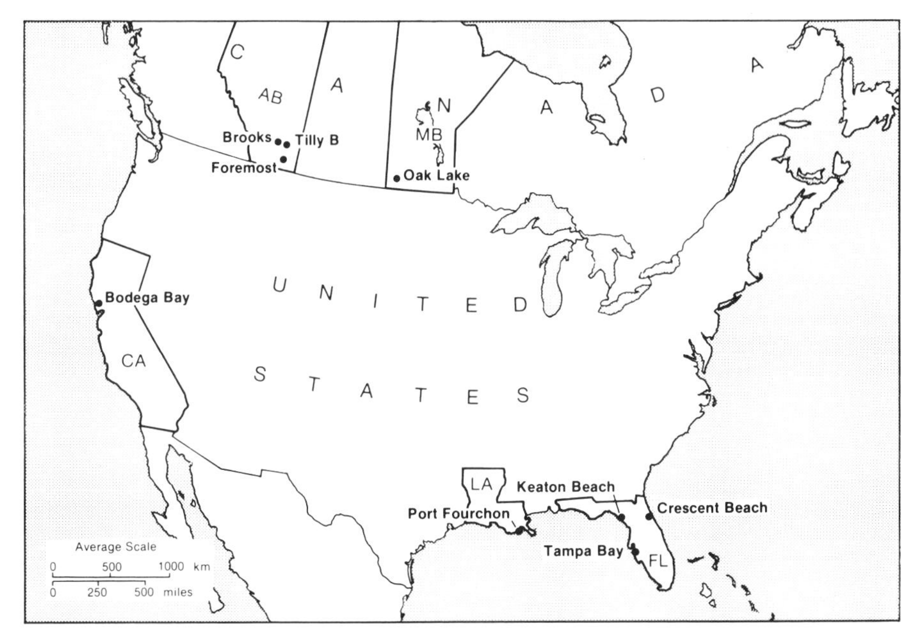

Fig. 8.1 Collection localities of willets from North America.

(b) all their helminths are acquired by ingestion (thus, prey selection is likely to be an important determinant to the presence of helminth species), (c) as intracontinental migrants, they undergo a major shift in environments and habitat types (and such shifts will influence the types of prey available), and (d) preliminary studies suggested high species richness with tight species packing.

Willets were collected on their saltwater wintering grounds and on their freshwater breeding grounds (Fig. 8.1) during the appropriate seasons from 1981 through 1988. To reduce seasonal effects, all winter collections were conducted during the last week in January through the first week of February; summer collections were conducted during the first week of June (except for Oak Lake juveniles which were collected in August). Three populations were sampled along the coast of Florida. Ten birds were collected at Crescent Beach (29°45′N, 81°15′W), ten were collected at Tampa Bay (27°40′N, 82°32′W) and five birds were collected at Keaton Beach (29°45′N, 83°30′W). Five birds were collected near Port Fourchon, Louisiana (28°7′N, 90°11′W). Nine birds were collected at Bodega Bay, California (38°16′N, 123°5′W). (Data from these birds were graciously provided by Dr. Hilda Ching.) Three populations were sampled on their breeding grounds in Alberta. Five birds were collected at Foremost

(49°30′N, 111°28′W), 19 birds were collected at Tilly B (50°34′N, 111°36′W) and five adult and five chicks (two < one week old, three < two weeks old) were collected near Brooks (50°56′N, 112°2′W). Five adult and seven juveniles (approximately three months old) were collected at Oak Lake, Manitoba (49°45′N, 99°55′W).

Treatment of hosts after collection and subsequent necropsy techniques follow procedures outlined by Bush and Holmes (1986a). Every individual helminth was enumerated with one exception – the number of *Parvatrema borealis* from one California bird was estimated. For the purposes of this chapter, I examine only the 'intestinal' helminths, that is, those helminths present from the junction of the gizzard–duodenum through the cloaca, including the ceca.

For some analyses, data from California were not available (e.g. length of intestine) and these hosts were thus excluded. For those localities where data were available for young of the year birds, I treated them as a separate group from adults since I am unaware of any method to determine at what age they might be expected to harbour a helminth fauna typical of adult hosts. I also excluded such hosts from analyses involving host size. I excluded data from California birds for guild analysis since I did not do the necropsies and thus could not assign species to a location (mucosal/lumenal) within the intestine. I also excluded chicks and juveniles from such analyses since they would have had no exposure to the marine environment; thus, the importance of any guilds, or members of such guilds, potentially associated with that environment, would be reduced.

To describe community richness, I used mean (± s.e.) data on Brillouin's diversity (using Stirling's approximation for factorials and natural logs) since I was interested in rare species, evenness (the degree to which individuals are evenly spread among the species present) and per cent of the intestine unoccupied; I used measures of central tendency for both numbers of species and numbers of individuals. To examine patterns of predictability, I used both quantitative (per cent similarity) and qualitative (Jaccard's coefficient) measures. I used the model developed by Ryti and Gilpin (1987) to determine the degree to which helminths could successfully colonize hosts (transmission potential) and the degree to which hosts could be colonized by helminths (susceptibility). For these analyses, I treated each geographical location and age group as an archipelago and each bird within an archipelago as an island.

Most raw data were aggregated (variance/mean ratio > 1); I therefore used nonparametric analyses for hypothesis testing. Spearman's rho was used for all correlations; two-sample testing was done using the Mann–Whitney test and multiple sample testing was done using the Kruskal–Wallis test. Multiple comparison *a posteriori* tests were done using the conservative procedure (unequal sample sizes) of Hollander and Wolfe (1973). The Kruskal–Wallis H (for $k > 3$) was calculated using an F approximation

(Pearson and Hartley, 1972). To determine the relationship between host size and numbers of both species and individuals, linear regression was used. I used two measures of host size: length of the intestine to the nearest mm and the cube root of individual bird mass (thus making it a linear variable and reducing variation due to daily fluctuations). Residual plots showed that untransformed data provided the best fit to the model for number of species, but that transformed ($\log_{10}[N + 1]$) data provided the best fit to the models for number of individuals.

I use frequency to refer to the number of hosts infected and I use density (or abundance) to refer to the number of individuals per host (this includes uninfected hosts). I treat each geographical location (and, when appropriate, age category) as a population. Helminth communities in each bird from the same population are treated as replicates of communities in other birds from the same population. I use guild to refer to a subset of the infracommunity – those organisms that feed in the same fashion, irrespective of their taxonomic affinity. For clarity, I treat freshwater and saltwater as discrete environments.

8.3 RESULTS

General

A total of 85 willets was examined. One less than one week old chick was uninfected, the second less than one week old chick had three individuals of two species. All other hosts were infected with three or more species. There was a mean of 7.4 (± 0.3) species (median = 7), with a range from 0–15 across the willet's entire geographical range. Density averaged 517.7 (± 87.2) (median = 248), with a range from 0–4490. In freshwater ($n = 46$), there was a mean of 7.5 (± 0.5) species (median = 7), with a range from 0–15. Average density was 362.3 (± 95.5) (median = 173), with a range from 0–3901. In saltwater ($n = 39$), there was a mean of 7.3 (± 0.5) species (median = 6), with a range from 4–14. Average density was 701 (± 149.1) (median = 380), with a range from 52–4490. There were no significant differences in the number of helminth species when all freshwater versus all saltwater hosts were compared (Mann–Whitney $U = 811$, $p > 0.4$), but saltwater hosts had significantly more individuals (Mann–Whitney $U = 1295$, $p < 0.001$).

Component communities

Details on helminth communities in individual populations of birds are provided in Tables 8.1 (freshwater) and 8.2 (saltwater). Sample sizes varied from five to 19 at the different populations.

Table 8.1 The distribution of helminths among the six populations of willets collected on their breeding grounds on the prairies of Canada. The values of *n* and *N* reflect, respectively, the frequency and total number of individual helminths for all birds collected in freshwater environments

			Alberta			
	Fresh-water		*Brooks adult*		*Brooks chick*	
	n	N	*Mean ± S.E.*	*Median*	*Mean ± S.E.*	*Median*
Trematoda						
Plagiorchis elegans (A)* (1)† (T)‡	32	839	10.0 ± 3.5	8.0	7.6 ± 3.9	4.0
Notocotylus sp. (B) (1) (T)	21	609	6.2 ± 1.9	7.0	75.0 ± 37.8	44.0
Odhneria odhneri (B) (2) (T)	21	283	5.6 ± 2.9	4.0	---§	
Echinostoma sp. (A) (1) (T)	14	2879	2.4 ± 1.5	1.0	---	
Stictodora hancocki (A) (2) (T)	10	770	71.6 ± 46.3	0.0	---	
Parorchis acanthus (C) (2) (T)	8	15	0.2 ± 0.2	0.0	---	
Maritrema patulus (A) (2) (T)	3	26	---		---	
Notocotylidae (C) (1) (T)	3	13	---		---	
Microphallus nicolli (A) (2) (T)	1	13	---		---	
Stictodora n.sp. (A) (2) (T)	1	5	---		---	
Microphallus pygmoeum (B) (2) (T)	1	43	8.6 ± 8.6	0.0	---	
Paramaritremopsis stunkardi (A) (2) (T)	1	868	173.6 ± 173.6	0.0	---	
Strigeidae (A) (1) (T)	1	3	---		---	
Trematode A (A) (3) (T)	1	26	5.2 ± 5.2	0.0	---	
Cestoda						
Anomotaenia n.sp. (A) (1) (L)	40	7686	82.0 ± 30.6	81.0	---	
Anomotaenia gallinaginis (A) (1) (L)	28	481	31.8 ± 8.9	34.0	9.0 ± 6.9	1.0
Aploparaxis sp. (A) (1) (L)	24	628	54.2 ± 28.9	25.0	6.4 ± 3.1	4.0
Kowalewskiella cingulifera (A) (2) (L)	12	168	6.2 ± 6.2	0.0	---	
Hymenolepis albertensis (A) (1) (M)	12	389	14.2 ± 11.3	5.0	4.4 ± 4.2	0.0
Hymenolepis microskrjabini (A) (1) (M)	10	63	5.0 ± 2.0	5.0	2.6 ± 2.4	0.0
Dictymetra radiaspinosa (A) (1) (L)	8	146	---		---	
Anomotaenia sp. (A) (1) (L)	8	39	0.4 ± 0.4	0.0	0.6 ± 0.4	0.0
Hymenolepis amphitricha (A) (1) (L)	7	47	---		---	
Lateriporus skrjabini (A) (1) (L)	6	25	---		---	
Hymenolepis hopkinsi (A) (1) (M)	6	43	---		3.4 ± 2.3	1.0
Ophrecocotyle insignis (A) (2) (M)	6	42	1.0 ± 1.0	0.0	---	
Kowalewskiella glareolae (A) (2) (L)	4	88	---		---	
Hymenolepididae sp. 1 (A) (3) (M)	2	2	0.2 ± 0.2	0.0	---	
Aploparaxis n.sp. (A) (3) (L)	1	8	---		---	
Microsomacanthus sp. (A) (3) (M)	1	1	---		---	
Dilepididae sp. 1 (A) (1) (M)	1	1	---		0.2 ± 0.2	0.0
Dilepididae sp. 2 (A) (3) (M)	1	1	0.2 ± 0.2	0.0	---	
Hymenolepididae sp. 2 (A) (3) (M)	1	4	---		---	
Acanthocephala						
Polymorphus marilis (A) (1) (L)	23	352	48.0 ± 17.7	57.0	5.2 ± 2.4	2.0
Parafilicollis sp. (A) (2) (L)	1	1	0.2 ± 0.2	0.0	---	
Nematoda						
Capillaria sp. A (A) (1) (N)	9	14	0.4 ± 0.2	0.0	---	
Capillaria sp. B (B) (3) (N)	15	46	4.0 ± 1.4	4.0	---	

*A = predominantly small intestinal; B = predominantly cecal; C = predominantly large intestinal/cloacal.

† = life cycle known or believed to be completed in freshwater; 2 = life cycle known or believed to be completed in salt water; 3 = life cycle unknown.

Alberta				*Manitoba*			
Tilly B		*Foremost*		*Oak Lake adult*		*Oak Lake juvenile*	
Mean ± S.E.	*Median*	*Mean ± S.E.*	*Median*	*Mean ± S.E.*	*Median*	*Mean ± S.E.*	*Median*
3.2 ± 1.2	1.0	52.6 ± 25.4	21.0	2.4 ± 2.2	0.0	59.3 ± 11.7	56.0
0.6 ± 0.4	0.0	31.6 ± 15.2	16.0	1.0 ± 1.0	0.0	4.0 ± 2.3	1.0
7.6 ± 2.8	3.0	12.4 ± 10.9	1.0	9.6 ± 8.4	2.0	---	
0.6 ± 0.3	0.0	559.0 ± 559.0	0.0	---		8.7 ± 4.6	3.0
9.3 ± 5.6	0.0	47.0 ± 46.5	0.0	---		---	
0.5 ± 0.2	0.0	0.2 ± 0.2	0.0	0.6 ± 0.6	0.0	---	
1.1 ± 1.0	0.0	---		1.0 ± 1.0	0.0	---	
---		---		---		1.9 ± 1.3	0.0
0.7 ± 0.7	0.0	---		---		---	
---		1.0 ± 1.0	0.0	---		---	
---		---		---		---	
---		---		---		---	
---		---		---		0.4 ± 0.4	0.0
---		---		---		---	
168.1 ± 87.8	41.0	687.6 ± 170.8	603.0	70.6 ± 44.3	35.0	41.6 ± 26.7	18.0
5.8 ± 1.9	1.0	0.2 ± 0.2	0.0	5.8 ± 3.6	0.0	19.4 ± 6.9	19.0
15.5 ± 11.3	1.0	5.0 ± 2.5	3.0	---		0.7 ± 0.6	0.0
4.2 ± 2.1	0.0	7.4 ± 2.7	8.0	4.0 ± 2.9	0.0	---	
0.7 ± 0.4	0.0	---		56.6 ± 31.9	18.0	---	
1.3 ± 0.7	0.0	---		---		---	
0.9 ± 0.9	0.0	9.6 ± 9.6	0.0	0.6 ± 0.6	0.0	11.1 ± 6.4	1.0
0.6 ± 0.6	0.0	---		---		3.1 ± 2.5	0.0
1.7 ± 1.3	0.0	2.4 ± 2.4	0.0	---		0.4 ± 0.3	0.0
0.6 ± 0.4	0.0	---		2.6 ± 1.7	1.0	---	
0.3 ± 0.3	0.0	3.8 ± 3.8	0.0	---		0.1 ± 0.1	0.0
1.1 ± 1.0	0.0	1.2 ± 1.0	0.0	2.0 ± 2.0	0.0	---	
1.0 ± 0.8	0.0	14.0 ± 14.0	0.0	---		---	
---		---		0.2 ± 0.2	0.0	---	
0.4 ± 0.4	0.0	---		---		---	
0.1 ± 0.1	0.0	---		---		---	
---		---		---		---	
---		---		---		---	
---		---		0.8 ± 0.8	0.0	---	
1.9 ± 1.0	0.0	---		9.6 ± 5.4	6.0	0.3 ± 0.2	0.0
---		---		---		---	
0.4 ± 0.2	0.0	0.6 ± 0.4	0.0	---		0.1 ± 0.1	0.0
0.6 ± 0.2	0.0	1.8 ± 0.7	3.0	1.2 ± 0.6	1.0	---	

‡ = Guild (T = trematode; L = lumenal absorber; M = mucosal absorber; N = nematode).
§- - - = helminth not found in that population.

Table 8.2 The distribution of helminths among the five populations of willets collected on their wintering grounds along the coast of North America. The values of *n* and *N* reflect, respectively, the frequency and total number of individual helminths for all birds collected in saltwater respectively

	Saltwater		*Crescent Beach*	
	n	N	*Mean* ± *S.E.*	*Median*
Trematoda				
Odhneria odhneri (B)* (2)† (T)‡	30	3996	208.2 ± 73.5	117.0
Maritrema patulus (A) (2) (T)	16	1627	19.7 ± 13.1	0.0
Parvatrema sp. (A) (2) (T)	14	4593	5.0 ± 4.5	0.0
Microphallus nicolli (A) (2) (T)	11	2561	157.6 ± 144.2	3.5
Parorchis acanthus (C) (2) (T)	11	18	0.5 ± 0.3	0.0
Stictodora hancocki (A) (2) (T)	10	4724	16.3 ± 13.8	0.0
Ascorhytis charadriformis (A) (2) (T)	9	1931	---	
Maritrema laricola (A) (2) (T)	9	3098	---	
Stictodora n.sp. (A) (2) (T)	8	1098	2.8 ± 1.6	0.0
Levinseniella gymnopocha (B) (2) (T)	8	344	---	
Himasthla quissentensis (A) (2) (T)	6	24	2.3 ± 1.3	0.5
Gynaecotyla adunca (A) (2) (T)	6	42	1.8 ± 1.1	0.0
Diacetabulum riggini (A) (2) (T)	6	388	---	
Levinseniella hunteri (B) (2) (T)	5	161	---	
Numeniotrema kinsellai (A) (2) (T)	5	150	---	
Microphallus turgidus (B) (2) (T)	5	108	5.5 ± 5.4	0.0
Stephanoprora n.sp. (A) (2) (T)	4	174	---	
Probolocoryphe lanceolata (A) (2) (T)	3	64	6.4 ± 4.7	0.0
Maritrema prosthometra (A) (2) (T)	2	20	---	
Microphallus n.sp. (A) (2) (T)	2	4	---	
Parvatrema borealis (A) (2) (T)	1	100	---	
Asocotyle ampullacea (A) (2) (T)	1	13	1.3 ± 1.3	0.0
Stephanoprora denticulata (A) (2) (T)	1	4	0.4 ± 0.4	0.0
Cloacitrema michiganense (C) (2) (T)	1	1	---	
Stictodora cursitans (A) (2) (T)	1	1	---	
Notocotylus sp. (B) (1) (T)	1	1	0.1 ± 0.1	0.0
Ascocotyle sp. (A) (2) (T)	1	1	0.1 ± 0.1	0.0
Trematode B (A) (3) (T)	1	61	---	
Cestoda				
Anomotaenia n.sp. (A) (1) (M)	23	1204	51.6 ± 30.0	27.5
Ophreocotyle insignis (A) (2) (L)	14	331	5.5 ± 3.9	0.5
Hymenolepididae sp. 3 (A) (3) (?)	8	55	---	
Kowalewskiella cingulifera (A) (2) (L)	6	10	0.5 ± 0.3	0.0
Hymenolepis amphitricha (A) (1) (L)	2	11	---	
Aploparaxis sp. (A) (1) (L)	1	2	---	
Dilepididae sp. 1 (A) (1) (M)	1	3	0.3 ± 0.3	0.0
Dilepididae sp. 2 (A) (3) (M)	1	8	---	
Acanthocephala				
Parafilicollis sp. (A) (2) (L)	9	89	7.5 ± 3.3	3.0
Prosthorhynchus sp. (A) (2) (L)	4	21	2.1 ± 0.9	0.0
Nematoda				
Capillaria sp. A (A) (1) (N)	5	20	0.7 ± 0.6	0.0
Capillaria sp. B (B) (3) (N)	29	200	7.2 ± 2.9	1.0
Spiruridae sp. 1 (A) (3) (N)	1	1	---	

*A = predominantly small intestinal; B = predominantly cecal; C = predominantly large intestinal/cloacal.

† = life cycle known or believed to be completed in freshwater; 2 = life cycle known or believed to be completed in salt water; 3 = life cycle unknown.

Florida				*Louisiana*		*California*	
Tampa Bay		*Keaton Beach*		*Port Fourchon*		*Bodega Bay*	
Mean ± S.E.	*Median*	*Mean ± S.E.*	*Median*	*Mean ± S.E.*	*Median*	*Mean ± S.E.*	*Median*
150.6 ± 32.9	158.5	50.8 ± 20.8	38.0	29.2 ± 9.6	24.0	0.9 ± 0.9	0.0
56.5 ± 27.8	33.0	173.0 ± 72.6	79.0	– – –§		– – –	
19.1 ± 12.9	2.0	2.4 ± 1.6	0.0	868.0 ± 854.3	1.0	– – –	
0.1 ± 0.1	0.0	196.8 ± 82.4	238.0	– – –		– – –	
0.3 ± 0.2	0.0	0.8 ± 0.5	0.0	1.0 ± 0.8	0.0	0.1 ± 0.1	0.0
0.2 ± 0.1	0.0	911.8 ± 364.8	546.0	– – –		– – –	
– – –		– – –		– – –		214.6 ± 93.6	133.0
– – –		– – –		– – –		344.2 ± 136.3	226.0
– – –		214.0 ± 179.6	32.0	– – –		– – –	
– – –		– – –		– – –		38.2 ± 17.8	26.0
– – –		0.2 ± 0.2	0.0	– – –		– – –	
– – –		2.2 ± 2.0	0.0	2.6 ± 2.6	0.0	– – –	
0.8 ± 0.8	0.0	76.0 ± 23.1	65.0	– – –		– – –	
– – –		9.6 ± 7.2	0.0	22.6 ± 18.4	8.0	– – –	
– – –		30.0 ± 13.0	33.0	– – –		– – –	
– – –		10.6 ± 5.7	5.0	– – –		– – –	
– – –		– – –		34.8 ± 27.1	6.0	– – –	
– – –		– – –		– – –		– – –	
– – –		– – –		4.0 ± 3.5	0.0	– – –	
– – –		– – –		– – –		0.4 ± 0.3	0.0
– – –		– – –		– – –		11.1 ± 11.1	0.0
– – –		– – –		– – –		– – –	
– – –		– – –		– – –		– – –	
– – –		– – –		– – –		0.1 ± 0.1	0.0
– – –		0.4 ± 0.4	0.0	– – –		– – –	
– – –		– – –		– – –		– – –	
– – –		– – –		– – –		– – –	
– – –		12.2 ± 12.2	0.0	– – –		– – –	
41.4 ± 9.8	40.5	41.8 ± 41.8	0.0	13.0 ± 8.7	4.0	– – –	
0.1 ± 0.1	0.0	24.4 ± 4.1	23.0	30.6 ± 14.5	32.0	– – –	
– – –		– – –		– – –		6.1 ± 3.3	2.0
0.5 ± 0.3	0.0	– – –		– – –		– – –	
– – –		– – –		2.2 ± 1.7	0.0	– – –	
0.2 ± 0.2	0.0	– – –		– – –		– – –	
– – –		– – –		– – –		– – –	
– – –		1.6 ± 1.6	0.0	– – –		– – –	
1.2 ± 1.2	0.0	– – –		0.2 ± 0.2	0.0	0.1 ± 0.1	0.0
– – –		– – –		– – –		– – –	
– – –		2.6 ± 1.5	2.0	– – –		– – –	
1.5 ± 0.7	0.5	15.8 ± 6.1	16.0	3.0 ± 1.0	2.0	2.1 ± 0.8	1.0
– – –		– – –		0.2 ± 0.2	0.0	– – –	

‡ = Guild (T = trematode; L = lumenal absorber; M = mucosal absorber; N = nematode).
§– – – = helminth not found in that population.

There were significant differences associated with species richness (Kruskal–Wallis $H = 34.5$, $p < 0.001$), density (Kruskal–Wallis $H = 43.3$, $p < 0.001$) and per cent of intestine unoccupied (Kruskal–Wallis $H = 23.2$, $p = 0.006$) when comparisons were made between all populations; differences in diversity were close (Kruskal–Wallis $H = 17.34$, $p = 0.067$), but differences in evenness were not (Kruskal–Wallis $H = 11.16$, $p > 0.4$) (Figs 8.2, 8.3 and 8.4). Adult willets from Brooks and willets from Keaton Beach had significantly more helminth species than any other populations, but the difference in helminth species richness between willets from those two populations was not significant. There were significantly more individuals in the adult willets from Brooks and willets from Foremost, Keaton Beach and Bodega Bay than in any of the other populations, but these four populations did not differ significantly from one another. Populations from Brooks (adults) and Keaton Beach had significantly less unoccupied space in the intestine than any other population except Foremost; they did not differ from each other. Adult willets from Brooks and willets from Keaton Beach had significantly higher diversity than all other populations; they did not differ significantly from each other.

There were no significant differences between sexes with respect to the number of species (Mann–Whitney $U = 218$, $p > 0.3$), number of individuals (Mann–Whitney $U = 207$, $p > 0.5$), diversity (Mann–Whitney $U = 184$, $p = 1.0$) evenness (Mann–Whitney $U = 175$, $p > 0.8$) or the per cent of the intestine unoccupied (Mann–Whitney $U = 176$, $p > 0.8$).

There were significantly more individuals in adult birds than in chicks and juveniles (Mann–Whitney $U = 656$, $p = 0.006$). Between the same two categories, there were no significant differences in number of species (Mann–Whitney $U = 497$, $p > 0.5$), diversity (Mann–Whitney $U = 452$, $p > 0.9$), evenness (Mann–Whitney $U = 399$, $p > 0.6$) or the per cent of the intestine unoccupied (Mann–Whitney $U = 341$, $p > 0.5$).

Length of intestine and host mass varied between individual adult birds. However, there was no relationship between the length of the intestine and either the number of species ($Y = 4.0 + 0.05X$, $p > 0.3$) or the number of individuals ($Y = 2.7 - 0.003X$, $p > 0.7$). Neither was there a significant relationship between the number of species ($Y = 27.0 - 2.9X$, $p > 0.1$) nor the number of individuals ($Y = 6.3 - 0.6X$, $p > 0.08$) and the cube root of the host's mass. Therefore, helminth species richness and density are not a function of host size (i.e. the largest birds, or those with the longest intestines, do not necessarily have richer communities).

With one exception, qualitative similarities among birds from the same population were higher than similarities between populations (Fig. 8.5), suggesting that transmission of specific helminth species is higher within most populations than between populations. The pattern is somewhat less clear with quantitative similarity, where seven populations had higher

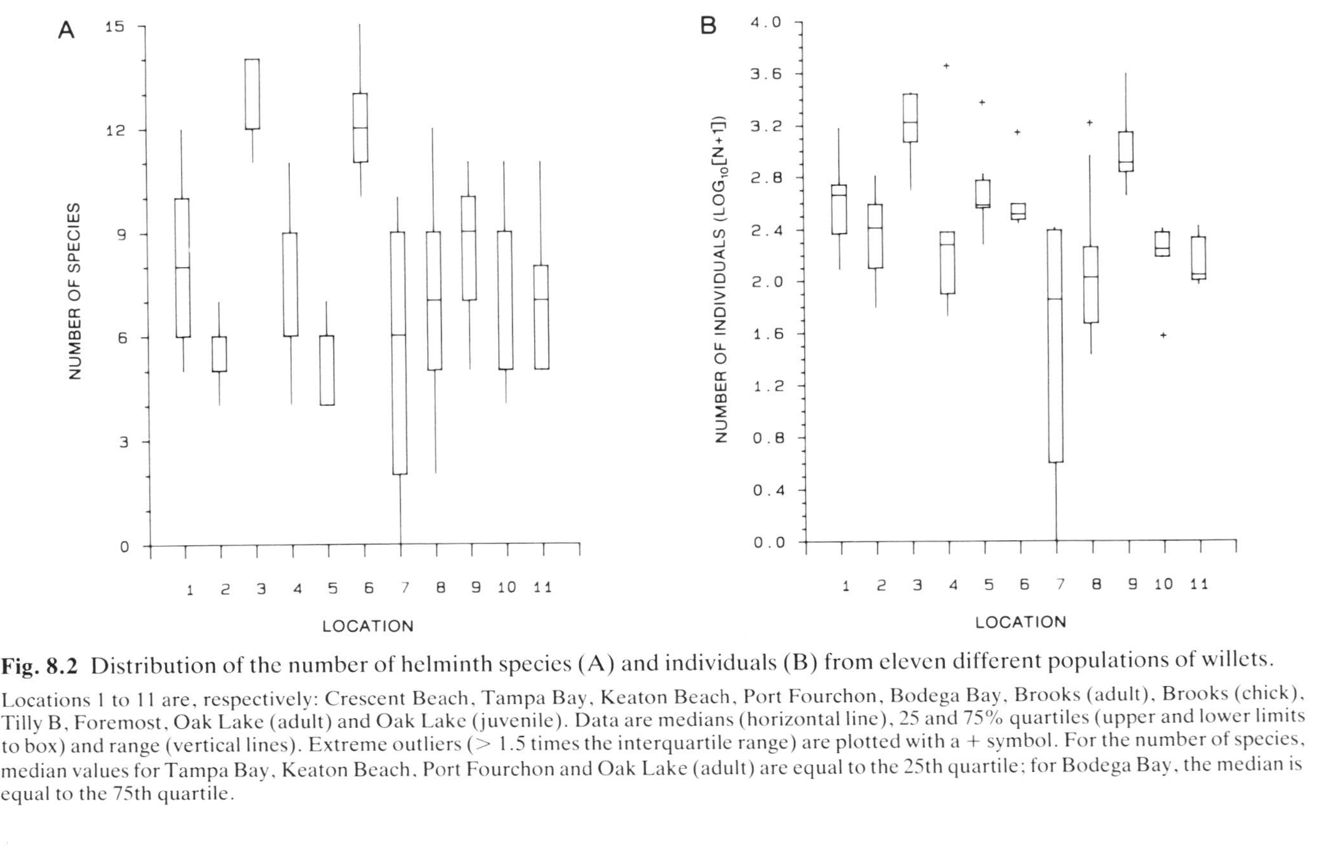

Fig. 8.2 Distribution of the number of helminth species (A) and individuals (B) from eleven different populations of willets.

Locations 1 to 11 are, respectively: Crescent Beach, Tampa Bay, Keaton Beach, Port Fourchon, Bodega Bay, Brooks (adult), Brooks (chick), Tilly B, Foremost, Oak Lake (adult) and Oak Lake (juvenile). Data are medians (horizontal line), 25 and 75% quartiles (upper and lower limits to box) and range (vertical lines). Extreme outliers (> 1.5 times the interquartile range) are plotted with a + symbol. For the number of species, median values for Tampa Bay, Keaton Beach, Port Fourchon and Oak Lake (adult) are equal to the 25th quartile; for Bodega Bay, the median is equal to the 75th quartile.

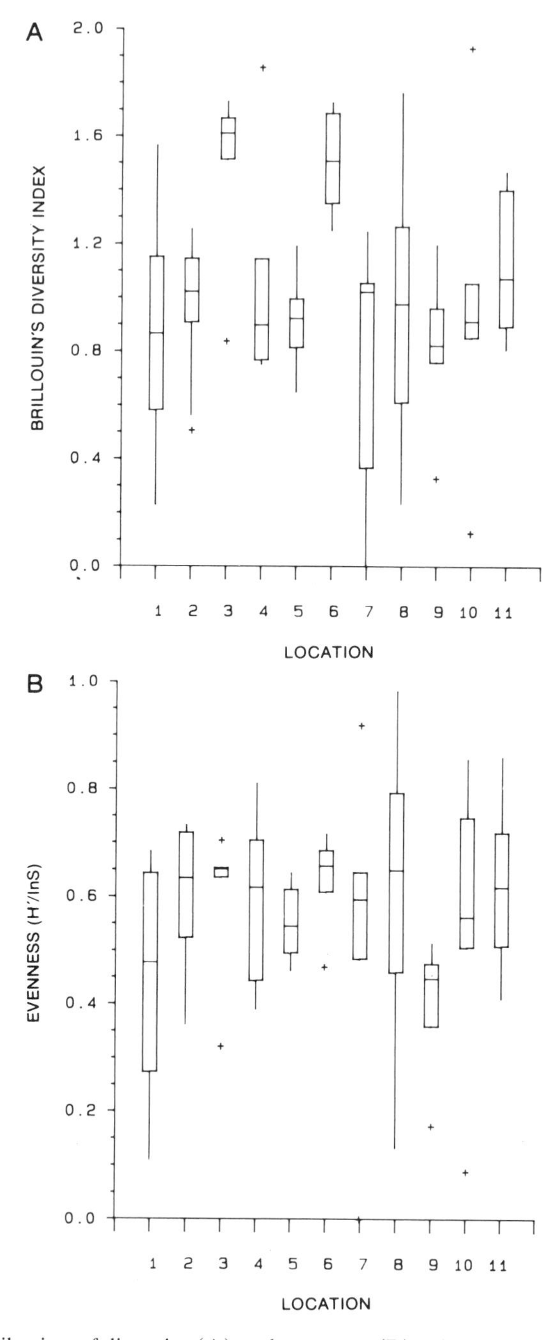

Fig. 8.3 Distribution of diversity (A) and evenness (B) values from eleven different populations of willets. See Fig. 8.2 for location codes and interpretation of data. The median value for evenness in the population from Keaton Beach is nearly equal to the 75th quartile.

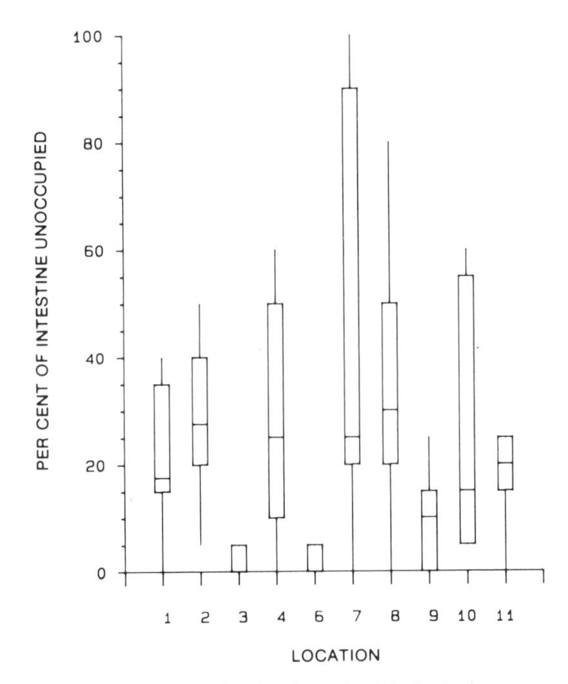

Fig. 8.4 Distribution of empty space in the intestinal helminth communities from eleven different populations of willets. See Fig. 8.2 for location codes and interpretation of data. The median values for Keaton Beach and Brooks (adult) are equal to zero.

within population similarities than between population similarities. This again suggests higher availability within most populations than between populations; it also emphasizes that helminth individuals are neither evenly spread within, nor between, populations. Helminth communities in chicks from Brooks showed very low similarities. Recall, however, that the two chicks (less than one week old) were very depauperate and all similarities with the uninfected chick are obviously zero. If those two individuals are excluded from consideration, qualitative (72.2% ± 9.1) and quantitative (81.1% ± 2.2) similarities are extremely high.

Table 8.3 presents the results of the Ryti and Gilpin colonization model applied to the eleven populations. There was no absolute pattern in any of the five parameters that would clearly separate freshwater and saltwater hosts. However, four of the five saltwater populations (Bodega Bay, Keaton Beach, Tampa Bay and Port Fourchon) rank in the top five values for explained variance and for the rate of helminth species' decrease across hosts. Chicks from Brooks, which are the best estimate of early colonization, show the highest value for orderedness, the second highest value for

PER CENT SIMILARITY

	CRESCENT BEACH	TAMPA BAY	KEATON BEACH	PORT FOURCHON	BODEGA BAY	FOREMOST	OAK LAKE (ADULT)	BROOKS (ADULT)	OAK LAKE (JUVENILE)	TILLY B	BROOKS (CHICK)
CRESCENT BEACH	0.30 +/- 0.09; 0.31 +/- 0.05	0.42 +/- 0.06	0.22 +/- 0.06	0.15 +/- 0.04	0.07 +/- 0.03	0.19 +/- 0.05	0.16 +/- 0.06	0.12 +/- 0.04	0.16 +/- 0.04	0.00	0.00
TAMPA BAY	0.27 +/- 0.05	0.62 +/- 0.05; 0.53 +/- 0.05	0.30 +/- 0.06	0.21 +/- 0.04	0.10 +/- 0.02	0.24 +/- 0.04	0.19 +/- 0.03	0.15 +/- 0.03	0.14 +/- 0.02	0.00	0.00
KEATON BEACH	0.25 +/- 0.02	0.31 +/- 0.02	0.58 +/- 0.05; 0.62 +/- 0.05	0.13 +/- 0.05	0.05 +/- 0.03	0.15 +/- 0.05	0.13 +/- 0.05	0.09 +/- 0.03	0.06 +/- 0.01	0.00	0.00
PORT FOURCHON	0.17 +/- 0.02	0.23 +/- 0.02	0.19 +/- 0.02	0.19 +/- 0.07; 0.43 +/- 0.06	0.19 +/- 0.04	0.32 +/- 0.06	0.33 +/- 0.06	0.27 +/- 0.04	0.04 +/- 0.01	0.07 +/- 0.02	0.00
BODEGA BAY	0.05 +/- 0.01	0.08 +/- 0.00	0.06 +/- 0.01	0.19 +/- 0.02	0.69 +/- 0.05; 0.65 +/- 0.05	0.22 +/- 0.05	0.28 +/- 0.06	0.20 +/- 0.03	0.01 +/- 0.01	0.09 +/- 0.02	0.00
FOREMOST	0.16 +/- 0.02	0.22 +/- 0.03	0.17 +/- 0.02	0.26 +/- 0.03	0.24 +/- 0.03	0.53 +/- 0.12; 0.41 +/- 0.04	0.44 +/- 0.07	0.32 +/- 0.05	0.11 +/- 0.05	0.07 +/- 0.02	0.00
OAK LAKE (ADULT)	0.19 +/- 0.02	0.21 +/- 0.02	0.18 +/- 0.03	0.23 +/- 0.03	0.27 +/- 0.02	0.29 +/- 0.03	0.31 +/- 0.09; 0.30 +/- 0.08	0.27 +/- 0.05	0.04 +/- 0.01	0.05 +/- 0.01	0.00
BROOKS (ADULT)	0.16 +/- 0.02	0.17 +/- 0.02	0.16 +/- 0.01	0.30 +/- 0.03	0.30 +/- 0.02	0.33 +/- 0.04	0.35 +/- 0.03	0.32 +/- 0.09; 0.51 +/- 0.04	0.14 +/- 0.05	0.16 +/- 0.02	0.00
OAK LAKE (JUVENILE)	0.26 +/- 0.02	0.24 +/- 0.02	0.20 +/- 0.02	0.10 +/- 0.02	0.01 +/- 0.01	0.10 +/- 0.02	0.14 +/- 0.02	0.12 +/- 0.02	0.50 +/- 0.06; 0.45 +/- 0.05	0.00	0.00
TILLY B	0.00	0.00	0.00	0.13 +/- 0.03	0.19 +/- 0.03	0.15 +/- 0.03	0.15 +/- 0.03	0.28 +/- 0.04	0.00	0.39 +/- 0.06; 0.26 +/- 0.03	0.00
BROOKS (CHICK)	0.07 +/- 0.02	0.05 +/- 0.02	0.09 +/- 0.02	0.05 +/- 0.01	0.00	0.02 +/- 0.01	0.04 +/- 0.01	0.04 +/- 0.01	0.05 +/- 0.01	0.00	0.28 +/- 0.17; 0.29 +/- 0.14

JACCARD COEFFICIENT

Fig. 8.5 Similarity values for helminth communities between willet populations. Values represent mean (± s.e.) for all possible pairwise comparisons within or between populations. Values above the diagonal represent quantitative similarities. Dotted lines separate freshwater and saltwater populations such that the upper left triangle (above the diagonal) represents quantitative comparisons within and between saltwater populations; the lower right triangle represents quantitative comparisons within and between freshwater populations and the upper rectangle represents quantitative comparisons between freshwater and saltwater comparisons. Similar qualitative comparisons are presented below the diagonal.

Table 8.3 Helminth species occurrence patterns on eleven archipelagos (populations of birds). Populations are ordered by the explained variance

Host population	*Fitted parameters*			*Explained variance (%)*	*B/C*	*Proportion of ls*
	a	b	c			
Brooks (chick)	−327.05	70.55	30.23	100.00	1.297	0.53
Bodega Bay	−143.49	17.56	4.97	64.98	2.888	0.50
Keaton Beach	−25.19	2.78	1.51	54.67	0.462	0.63
Tampa Bay	−7.01	0.60	0.76	52.14	0.605	0.42
Port Fourchon	−8.98	1.65	0.65	50.88	0.976	0.55
Brooks (adult)	−5.57	1.04	0.25	45.90	0.895	0.53
Oak Lake (juvenile)	−6.24	0.81	0.49	43.74	0.819	0.50
Foremost	−5.86	1.13	0.38	41.83	0.829	0.47
Oak Lake (adult)	−6.45	1.34	0.44	38.49	0.951	0.43
Crescent Beach	−4.72	0.53	0.26	29.81	0.887	0.36
Tilly B	−4.76	0.31	0.28	29.68	0.812	0.27

dispersion of colonizing success (B/C ratio), the highest values for the three fitted parameters (a, b, c) and they are tied for the third highest value for the overall helminth species load (proportion of 1s) within populations. This latter value is significantly reduced by the chick that was less than one week old and was uninfected. If that individual is removed, then chicks from Brooks also have the highest overall helminth species load (67%).

There are four obvious feeding guilds within the communities. Trematodes are mucosal and feed by engulfing gut tissue, also by absorbing nutrients directly across their body surface. I will refer to them as the trematode guild. Cestodes and acanthocephalans feed strictly by absorbing materials across their body surfaces. Among these strict absorbers are large species which attach to the mucosa but whose bodies (absorptive surfaces) are lumenal in position. I will refer to these as lumenal absorbers. In contrast are those absorbers which are small and whose entire bodies are intimately associated with the mucosa. I will refer to these as mucosal absorbers. The fourth guild includes the nematodes which are mucosal and feed strictly by engulfing gut tissue or contents. I will refer to them as the nematode guild. There were significantly more species of the trematode guild in populations from saltwater environments compared with those from freshwater environments (Mann–Whitney $U = 772$, $p < 0.001$). In contrast, species which are lumenal absorbers were significantly more frequent in populations from freshwater compared with populations from saltwater (Mann–Whitney $U = 79$, $p < 0.001$). Neither mucosal absorbers nor the nematode guild showed any significant differences in frequencies between populations from the two environments (Mann–Whitney $U = 469$, $p > 0.5$ and $U = 613$, $p > 0.1$, respectively).

Infracommunities

Although data presented in Fig. 8.2A show a relatively consistent pattern of species richness (except as noted previously for adults from Brooks and Keaton Beach), data from infracommunities are less clear-cut. Fig. 8.6 is a comparative, diagrammatic representation of the distribution of helminth species along the intestines of two birds, selected specifically to show a species-rich and a species-poor infracommunity. The communities are also from the two most closely-matched hosts available in my data. Both were adult males from saltwater, with intestinal lengths of 68.4 cm in the bird from Keaton Beach and 69.5 cm in the bird from Crescent Beach. For this comparison, numbers of individual helminths within each of the species are ignored. Clearly, the middle and posterior regions of the intestine of the bird from Crescent Beach appear underutilized.

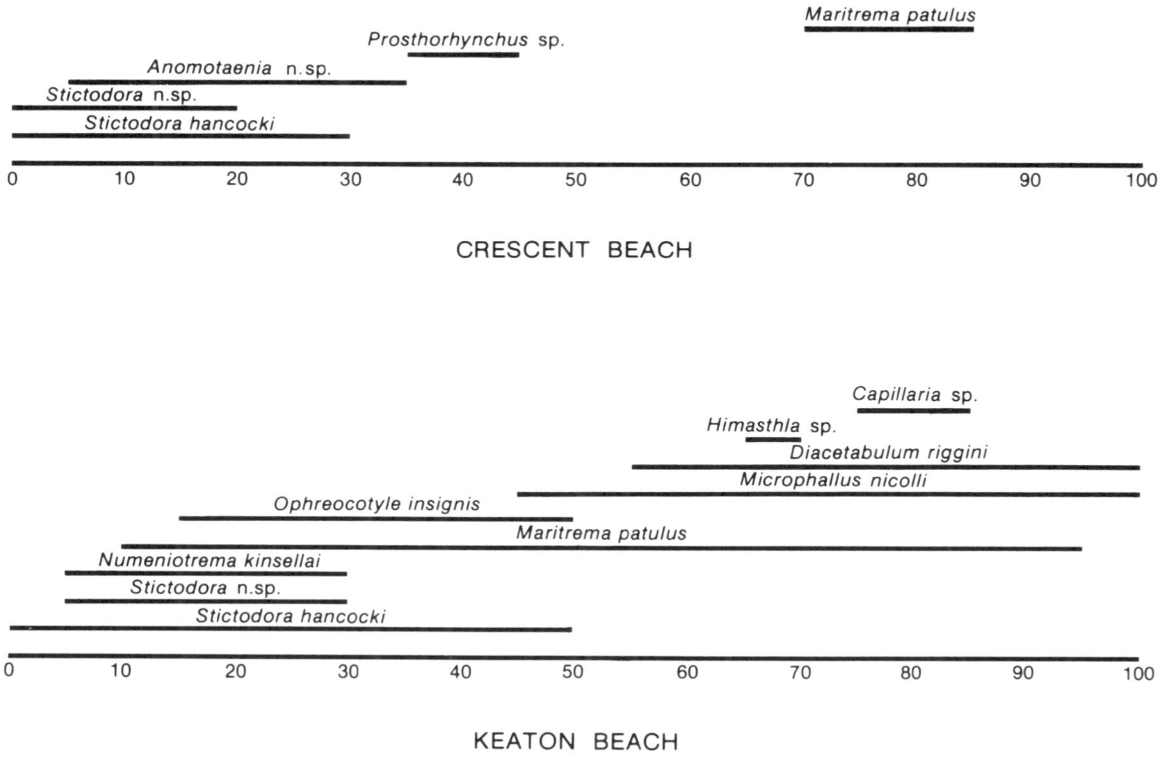

Fig. 8.6 Absolute distribution of helminths in two infracommunities from saltwater environments. Solid lines indicate the extent of the linear distribution for each species. Number of individuals within species are ignored.

It is obvious from the similarity values presented in Fig. 8.5, that hosts exposed to the contrasting environments of freshwater and saltwater nevertheless share some helminth species. The quantitative similarities for the Port Fourchon population show this best, where the within population similarity is less than the similarity of that population versus three freshwater populations. Data presented in Table 8.1 show that many birds collected on their breeding grounds in freshwater harboured helminths that are transmitted in saltwater environments. Many of these helminths are microphallid (particularly *Odhneria odhneri*) or heterophyid (particularly *Stictodora* spp.) trematodes, but the cestode *Ophreocotyle insignis* was also relatively frequent. The pattern is quite different for helminth communities in birds collected on their wintering grounds in saltwater. Here, only one helminth with a freshwater transmission pattern (*Anomotaenia* n. sp.) makes a substantive contribution to the community.

Both *Ophreocotyle* and *Anomotaenia* occupy the anterior 50% of the small intestine. Both are approximately the same length (± 9 cm) when gravid. Both appear to destrobilate and thus cease egg production when they are in a host which is in an inappropriate environment for transmission (i.e *Ophreocotyle* in hosts collected in freshwater were represented by scoleces only; *Anomotaenia* in hosts collected in saltwater were represented by either scoleces or occasionally by immature strobila less than 2 cm long). There were no significant differences in the number of *Ophreocotyle* in infected birds from freshwater versus those from saltwater (Mann–Whitney $U = 62$, $p > 0.1$) or in the number of *Anomotaenia* in infected birds from freshwater versus those from saltwater (Mann–Whitney $U = 364$, $p > 0.2$). The position of the medians of the distribution of the populations of *Ophreocotyle* in communities from freshwater versus communities from saltwater were not significantly different (Mann–Whitney $U = 46$, $p > 0.7$). The positions of the medians of the distribution of the populations of *Anomotaenia* in communities from freshwater versus communities from saltwater were not significantly different (Mann–Whitney $U = 374$, $p > 0.3$).

8.4 DISCUSSION

Component communities

General patterns

Helminth communities in willets are highly variable and species rich. They exhibit a level of community complexity which appears to be characteristic of other avian hosts – at least those associated with aquatic environments (Kennedy *et al.*, 1986).

Several of the determinants of pattern which I specifically addressed are unimportant in this system. Gender of the host has virtually no impact on the structure of the helminth communities. This is not particularly surprising; I am unaware of any data suggesting that male and female willets behave in different fashions on either their breeding or wintering grounds.

At first glance, age might appear to be an important determinant of community pattern. Very young birds are depauperate and not representative of more mature communities. By the time willets are between one and two weeks old however, their helminth communities approach the complexity evident in older birds. Within three months, their helminth communities (barring those helminths with saltwater life cycles) are comparable to communities in mature birds. Thus age, potentially an important variable, soon becomes irrelevant.

Host specificity

Few authors have examined the specificity of helminths to their hosts in an attempt to dissect patterns within communities. The determination of host specialists is often subjective; it can be improved with increased knowledge about the compound community. Helminths which can be regarded as specialists in the communities found in willets suggest an interesting pattern. In willet communities from freshwater, three species of cestodes (*Anomotaenia* n. sp., *Anomotaenia* sp. and *Aploparaxis* sp.) appear to be specialists. In addition, willets also harbour some helminths which are specialists in other host species, usually related shorebirds (e.g. *Dictymetra radiaspinosa*, which is a specialist in upland sandpipers, *Bartramia longicauda*). In contrast, willet communities from saltwater contain no host specialists (other than *Anomotaenia* n. sp. and *Aploparaxis* sp., which they acquired in freshwater environments), nor do they appear to harbour helminths which are specialists in other host species. The life cycle of a potential fourth willet specialist, *Capillaria* sp. B, is unknown.

These data suggest, as noted by Holmes (Chapter 5) and Kennedy (Chapter 6), a distinct uniqueness of helminth transmission patterns with respect to freshwater versus saltwater environments. Helminths, particularly trematodes, with marine life cycles appear to opt for spreading the risk among a large number of host species. Many of the microphallid trematodes which become gravid in willets in saltwater environments also become gravid in other scolopacids and charadriids, and also in rails, herons, egrets, ibises and even in mammals such as the rice rat (*Oryzomys palustris*) (personal observation). Even the marine-transmitted cestodes *Ophreocotyle insignis, Kowalewskiella cingulifera* and *K. glareolae* are reported from, and become gravid in, large numbers of avian hosts. Helminths, particularly cestodes, with freshwater life cycles appear to opt for specialization in a single, or a narrow group of host species. Those which are specialists in related shorebirds (e.g. *Dictymetra radiaspinosa*) occasionally become gravid in willets; those which are specialists in more distantly related hosts (e.g. *Lateriporus skrjabini*, a specialist in lesser scaup ducks) do not become gravid in willets. Based on these data, it appears that host specialists are more likely to be found in freshwater environments while host generalists are more likely to be found in saltwater environments. What ecological or evolutionary selection pressures might lead to these different adaptive modes in freshwater and saltwater are, as yet, unclear.

Regional patterns of species richness

As the data above suggest, the one determinant of pattern which appears to be quite meaningful at the component community level is the environment from which the hosts were collected (Tables 8.1 and 8.2). At the regional level, there are obvious and significant differences in the composition of the helminth communities between hosts collected in freshwater environments from those collected in saltwater environments. While populations from both environments have comparable number of species, those from saltwater environments have considerably more individuals. Not surprisingly, most species found in populations from saltwater environments have life cycles which require saltwater inhabiting intermediate hosts; in contrast, however, many species found in populations from freshwater environments also have life cycles requiring saltwater inhabiting intermediate hosts (these are almost exclusively trematodes – about 40% of the total infections in freshwater hosts by trematodes [with known life cycles] are species which have saltwater life cycles). Such observations, and patterns shown in Figs 8.2–8.4, suggest an essentially constant level of species packing irrespective of major environmental differences. The greater density in communities from saltwater environments is a function of the domination of trematodes in that system (I will expand on this below when I discuss guilds).

Observations on differences in species richness on a regional scale are not particularly novel. Dogiel (1964), using a large number of studies conducted throughout the Soviet Union, went so far as to classify parasites of birds as being ubiquitous, southern forms (transmitted on wintering grounds), northern forms (transmitted on breeding grounds) or as migration species. Recent authors have followed his lead. Hood and Welch (1980) examined parasites of red-winged blackbirds (*Agelaius phoeniceus*) from Manitoba (breeding) and Arkansas (wintering). They reported almost 100% prevalence of helminths in hosts from the breeding grounds versus 57% prevalence on the wintering grounds. They noted a significant change in diet (from plant material on the wintering grounds to animal material on the breeding grounds) and concluded that such differences were the reason for the increased prevalence on the breeding grounds. Tallman *et al.* (1985) studied the trematode fauna of two intercontinental migrants, the solitary sandpiper (*Tringa solitaria*) and the pectoral sandpiper (*Calidris melanotos*) on their wintering (Ecuador), migratory (Louisiana and South Dakota) and breeding grounds (Ontario), concluding that none of the collecting locations was a major source of trematode infection.

Data on species richness in willets do not support conclusions reached in these studies, but then neither of the above studies examined hosts collected from such disparate environments. Willets in freshwater environments feed primarily on aquatic insects and amphipods; those in saltwater environments feed primarily on mollusks and crustaceans (personal observation). This change in diet does not result in a concomitant change in species richness. Rather, the pattern is one of constant species richness but an alternation in guilds.

Regional patterns of guild composition

Alternation of guilds in helminth communities of willets between freshwater and saltwater environments is striking (Tables 8.1 and 8.2). Communities in hosts from freshwater environments are composed largely of three guilds: the lumenal absorbers (the large cestodes and acanthocephalans), the mucosal absorbers (small cestodes) and the mucosal trematode guild. The former two guilds are almost entirely composed of members with freshwater life cycles while the latter contains a large number of residual species persisting from the saltwater environment. The lumenal dwelling large absorbers dominate the helminth communities in hosts from freshwater environments. Communities in hosts from saltwater environments are also composed largely of the same three guilds. In marked contrast to the pattern noted above, it is the mucosal inhabiting trematode guild that dominates the communities. This guild is composed entirely of species with saltwater life cycles. In short, the pattern is one of the constant presence of two guilds (mucosal absorbers and the nematode guild) regardless of the environment. The two other guilds are present and which of these will dominate depends on the environment. Communities in hosts from freshwater environments are saturated with absorbers but are relatively underutilized by the trematode guild. In contrast, communities in hosts from saltwater environments are saturated by the trematode guild but are underutilized by lumenal absorbers.

I am not familiar with any studies which examine guild structure of helminth communities in avian hosts on a broad, regional basis. However, reanalysis of the data for lesser scaup in Bush (1980) shows no detectable differences of guild alternation in populations of hosts collected from different ecological regions. At the component community level, Bush and Holmes (1986b) suggested that two guilds (mucosal absorbers and lumenal absorbers) were saturated but that the trematode guild was not. Font (personal communication), who is examining lesser scaup on their wintering grounds in a saltwater environment, has found the communities to be comparatively depauperate, but with a marked increase in the trematode guild coupled with a substantial decline of both mucosal and lumenal absorbers.

Data on guilds in willets, in contrast to data on species richness, lend partial support for the existence of unsaturated communities. If we consider populations from saltwater environments, there is a clear lack of large lumenal absorbers. Those which are present (Tables 8.1 and 8.2) are neither frequent nor abundant. (But see below for comments on biomass.) In contrast, in populations from freshwater environments, though there are reasonable numbers of the trematode guild, there is clearly room for more (based on observations from birds collected in saltwater environments). These observations raise two questions – are the communities unsaturated and why are there alternations of dominance by different guilds in different environments? My interpretation of the data is that the communities are saturated and that guild alternation is a function of the environment. My argument, to both questions, is simply lack of availability (i.e. of colonizers, in ecological time, to further saturate communities and of guilds to cross environmental barriers). There are some lumenal absorbers in the saltwater populations as noted above, but all are generalists in a wide variety of hosts (this is not true for the same guild in freshwater populations). Furthermore, based on unpublished studies of other hosts from the same environment, I am not aware of any other lumenal absorbers that are even available. Such a guild appears to be the provenance of freshwater environments. Likewise, there do not appear to be any additional members of the trematode guild available in the freshwater environments (based on unpublished observations of 11 other charadriiform birds collected in the same localities at the same time). Where data are available, all species found in this study (including those in freshwater populations which are known to have a saltwater life cycle) are, as previously noted, host generalists. My interpretation is that the apparently unsaturated guilds are not biologically meaningful, they simply reflect lack of availability in ecological time. Dogiel (1964) made a similar argument for helminths in humans. He suggested that humans act as hosts for about 150 species of helminths, yet any individual rarely harbours more than two to three species concurrently. He further noted (1964), 'It would be fruitless to search for parasites like guinea worm, *Ancylostoma duodenale, Leishmania*, etc., characteristic of the southern latitudes, among the permanent residents of Leningrad'. His point is well-taken; before one can argue that communities are unsaturated, one must demonstrate that there are potential invaders.

Patterns in local populations

For most populations, there was a constant level of species richness regardless of whether the populations were from freshwater or saltwater environments, though two populations, adult birds from Brooks and birds from Keaton Beach, had significantly more species (Fig. 8.2A). A third population, Foremost, had more species than the remaining populations, but the differences were not statistically significant. This trend was paralleled for diversity and the per cent of the intestine unoccupied. The pattern was also the same for density of individuals (but with the addition of the population from Bodega Bay). My interpretation is that willets typically harbour about six to seven species of intestinal helminths. More can be accommodated, as is shown by the populations from Brooks and Keaton Beach and, to some extent, Foremost. The interesting question then becomes, why do those three populations have more individuals and species (resulting in less empty space), than the other eight populations? I believe the answer lies in the nature of the local habitat. Keaton Beach is a salt-marsh with a very limited mud-flat that can be used for feeding by willets. Observations over several weeks showed that the willets actively fed by circling back and forth in that limited area, making three to four passes per hour. As I did not collect the willets from Bodega Bay, I cannot comment on them, but all other willets from saltwater habitats fed in a unidirectional pattern and, once they had foraged past a specific point, they did not return to that point again during the day. The pattern of foraging observed for the Keaton Beach population was consistent with the populations from Brooks and Foremost. Both of the latter populations foraged along the margins of very small (less than one hectare) sloughs. In contrast, all other freshwater populations were associated with lakes greater than 500 hectares and they foraged in a linear fashion along the margin of those lakes, seldom passing the same point twice during a day of observation. In short, I am suggesting that the limited amount of suitable foraging habitat for the populations at Keaton Beach, Brooks and Foremost lead to higher densities of helminths. As the hosts are foraging, they are also seeding that limited habitat with helminth eggs, there is a more captive pool of potential intermediate hosts and thus a greater potential exposure for all birds in that population. Other host species, using the same limited area, also seed the habitat with eggs from their helminth communities. Provided that such helminths can establish in willets, repeated sampling will increase the probability that willets will acquire these helminths, thus increasing mean helminth species richness in those host populations. Those populations where birds feed in a linear fashion sample a much larger pool of potential intermediate hosts which themselves are less likely to encounter the reproductive products of helminths. At first glance, data on the chicks from Brooks would not support such an hypothesis; however, if we consider only the three chicks that were between one and two weeks old, they had a mean of more than eight species (and they could never have any helminths with marine life cycles since they could not have been exposed to appropriate intermediate hosts).

Similarities, explained variance and the rate of helminth species' decrease across hosts, were typically higher in populations of willets from saltwater environments. These three observations further emphasize the importance of the large suites of helminths (mostly microphallid trematodes) which are host generalists in the saltwater system. Those same trematodes, plus the cestodes *Anomotaenia* n. sp. and *Ophreocotyle insignis* and the nematodes *Capillaria* spp. A and B, provide considerable similarity within and, interestingly, between populations (regardless of environment). Some similarity values for helminth communities in willets between different populations were even higher than similarity values within populations (Fig. 8.5). A particularly good example is the willet population from Port Fourchon where within population quantitative similaritiy (19%) was less than the comparative similarity of that population and three populations from freshwater environments (range of 27–33% quantitative similarity). This apparent anomaly is due to the similar densities of host generalists, coupled with the very uneven distribution of individual helminths in the Port Fourchon population. Three of the five hosts from that population contained only 6.5% of the total number of helminths from that population. Such uneven distributions in density (all hosts were basically infected with the same species) might be explained in several ways, but I think two are most likely. First, the hosts with the greater densities may be selecting infected intermediate hosts with a greater frequency than hosts with low densities. Alternatively, the distribution of infective stages in the intermediate hosts may be aggregated such that feeding on one infected intermediate host might result in the establishment of a single helminth; feeding on a different infected intermediate host might result in the establishment of several hundred helminths. Such observations point to what I believe to be a crucial focal point for understanding communities of helminths in their final hosts – to what extent is the community structure in the final host determined, or fixed, by events that occur in the intermediate hosts? There are no clear answers yet, but Chapters 3 and 4 attest to the interest, and exciting progress, in such areas.

Infracommunities

My intent for this chapter was to dwell on structure at the component community level since historical trends in publishing emphasize that this is where most of the interest lies. I have, however, presented evidence for two observations at the infracommunity level which I believe are germane to considerations at the component level.

I have argued that, considered in ecological time, helminth communities in willets are saturated with respect to the component level. A similar argument is less tenable at the level of the individual host. In Fig. 8.6, I presented evidence from two individuals, matched as closely as possible in physical characteristics, suggesting the apparent underutilization of resources. But simple number of species may be misleading. The community from Keaton Beach had more individuals; it also had many members of the trematode guild, one lumenal absorber and lacked any small mucosal absorbers. The community from Crescent Beach had few members of the trematode guild but had many individual mucosal absorbers (*Anomotaenia* n. sp. destrobilates in saltwater) and a moderate number of lumenal absorbers. Is the community from Crescent Beach unsaturated? Perhaps, but it could be that the presence of the mucosal absorbers and the lumenal absorbers in that community is equivalent to the higher level of species packing apparent in the community from Keaton Beach. With respect to biomass, the community from Crescent Beach is certainly more saturated. We do not yet have the data or techniques to address the importance of biomass versus density but Stock and Holmes (1987a) have recently made significant inroads. More definitive answers to such questions will require different techniques (preferably experimental studies on these, thus far, untractable systems); others will require considerably more data of an appropriate nature.

When the distribution of helminth populations within avian infracommunities has been examined, most studies provided evidence for interactions between some species, usually members of the same guild. Evidence for interaction was usually based on niche shifts, of one or more species, in response to the presence of another species. The distributions and state of maturity of the two cestodes, *Anomotaenia* n. sp. and *Ophreocotyle insignis*, in willet infracommunities present an interesting case for speculation in which an alternative mechanism of cohabitation might be involved. I have suggested that the distribution of the two species of cestode are the same within the communities, but that their active growth phases appear to be complementary. What this may represent is an evolved 'mutual accommodation'. If, for whatever reason, both species require the resources of the anterior gut, but such resources are limiting, they become potential competitors. However, for *Ophreocotyle* to produce eggs in freshwater environments or for *Anomotaenia* to do the same in saltwater environments, is a waste of biotic potential. Perhaps the two species are temporally partitioning the resources and thus avoiding competition. This would require that scoleces be energetically 'cheap' compared with strobilae. Read (1959) and Roberts (1961) present evidence suggesting that, although scoleces have a high per gram rate of metabolism, they have a low rate of actual nutrient use because of their small size. If such speculation is correct, mutual accommodation would allow both species to reside in their appropriate region of the intestine concurrently, but the presence of the scoleces of the one species would have little impact on the fitness of the other species. To provide support for such speculation, one would have to show first, that the patterns are not simply the result of chance complementarity and second, that at least some of the *Anomotaenia* n. sp. producing eggs in freshwater environments during the current year were the same individuals that produced eggs in that environment during the previous year. A reciprocal pattern would need to be demonstrated for *Ophreocotyle*.

8.5 CONCLUSIONS

I have attempted to show that there are many data available on helminths in avian hosts and that many interesting determinants of patterns have been explored. Though many of the surveys are not based on infracommunities and are thus not appropriate for some types of community level analyses, they nevertheless provide us with information on what helminth species are found in which hosts. And this, like the life cycle work conducted on many helminths through historical and contemporary times, will be an asset for those who attempt to dissect patterns and mechanisms in the future.

I specifically addressed several possible determinants of pattern and my conclusions are not too dissimilar from others addressing the same patterns in avian helminth communities. The age of the host appears to have little bearing on the structure of the community, at least through time. But why should it? None of the helminths which I have discussed show any evidence for stimulating the immune system of the host (unlike some nematodes of mammals where there is clearly an immunogenic response and a definitive pattern associated with age). Likewise, again unlike mammalian systems (where there is a placental relationship and nursing) chicks will not be born with a complement of helminths and will not receive antibodies from a nursing female.

Also, like many previous authors, I am unable to find any evidence that would suggest any gender-related helminths. To demonstrate such patterns,

I believe one must find an avian species in which males and females feed on different food items or use different microhabitats within a specific habitat (e.g. as did Cornwell and Cowan, 1963). Given such a host, gender-related helminths might become important as possible determinants of community structure.

With respect to the development of avian helminth communities, my interpretation is that you have what you eat. Almost all helminths in birds are acquired through ingestion. Therefore, the overall environment with its included habitats will set the stage for the survival, and thus potential transmission, of helminths with direct life cycles and for the survival, and thus potential transmission, of intermediate stages. It is thus environments with their inclusive habitats that I suggest are the major determinants of pattern in avian helminth communities.

My suggestion that environment and habitat are important for understanding the presence or absence of helminth species is certainly not novel; Dogiel (1964) expressed the same thoughts over a quarter of a century ago. What perhaps is novel, is the degree of importance which I attach to different environments, and to their mosaic of habitats, in determining the patterns observed in avian helminth communities. My conclusions have been encouraged by recent papers on larval marine invertebrate settlement rates (Connell, 1985; Gaines and Roughgarden, 1985) which have led to the concept of 'supply-side' ecology (Lewin, 1986) and 'screens' or 'filters' (Holmes, 1986; Chapter 5) although I consider screens to be substantially less inclusive than envisaged by Holmes.

My interpretation of the determinants of pattern in helminth communities of avian hosts is as follows. I believe that environments with their varied habitats, determine the 'supply' of helminths that are available for colonization. One only has to scan Rausch's (1983) section on helminths of passerine birds to quickly realize that such hosts are remarkably depauperate of helminths. But there are some extraordinary exceptions e.g. consider the species richness (primarily generalist trematodes) found in Macgillivray's seaside sparrows (*Ammospiza maritima*) (Hunter and Quay, 1953). A young willet, born on the prairies of Canada, will not have the microphallid trematodes so characteristic of adult willets; the 'supply' just isn't there. Once the 'supply' has been determined at a gross level, the screens or filters proposed by Holmes come into play. Such screens or filters will determine what subset of those helminths available will actually colonize any given host species. For example, I have argued that willets can acquire host specialists from other host species. Willets collected at Tilly B were almost constantly in the company of red-necked grebes (*Podiceps grisegena*) and avocets (*Recurvirostra americana*). I have never found any tatriid cestodes, so characteristic of grebes (Stock and Holmes, 1987b, 1988) in willets. The tatriids are there; I have even found them in the avocets, but the screen(s) preventing their colonization of willets appear(s) insurmountable. Finally, with the component community determined by supply and screens, the actual structure of the infracommunity will be determined by either biotic interaction or stochastic processes.

My initial question asked whether willet communities were saturated or invasible. My conclusion is that they are saturated; however, given the appropriate habitat and compound community, they can be invaded. My arguments largely reflect invasibility during ecological time. It would be easy (and perhaps correct) to suggest that the eight host populations which have unoccupied regions of the intestine could be colonized in evolutionary time. But, just as Connell (1980) has warned us about the 'ghost of competition past', we should be concerned about untestable predictions based on the 'promise of evolution in the future'. (See Boucot's (1983) argument that communities remain static through long intervals of evolutionary time.)

As I interpret patterns in willet infra and component communities, the important questions thus become: Can additional species colonize in ecological time? Will they colonize in evolutionary time? If they can colonize, under what circumstances, and from whom will they come?

The analysis of structure in helminth communities is really just beginning; much of what I have suggested in this chapter (and what others have also suggested in their chapters) is preliminary. When we learn more about these fascinating systems, we will be able to draw more robust conclusions. I think we have much to learn from our colleagues who study free-living communities. We are becoming overwhelmed with the importance of numbers, as once were the ecologists studying free-living systems; but they are progressing to more meaningful analyses based on such concepts as biomass or guild structure. We too have begun to consider biomass (e.g. Stock and Holmes, 1987a) and guild structure (Bush and Holmes, 1986b) and such analyses will provide a more comprehensive perspective from which we will generate the paradigms of the future. Such exciting areas as community structure in invertebrate intermediate hosts (Chapters 3 and 4), the importance of host social groups (Moore *et al.*, 1988), the identification of screens (Holmes, 1986; Chapter 5), the importance of host/helminth phylogeny (Brooks, 1979; Mitter and Brooks, 1983) and the evolution of helminth communities (Price, 1986) will provide exciting research and debate in the ensuing decades.

ACKNOWLEDGEMENTS

Many students, too numerous to name, have assisted me with the acquisition of data on helminth communities in shorebirds. I heartily thank them all. Many colleagues, also too numerous to name, have challenged my ideas and freely shared their own ideas, helping me to try to interpret the patterns (or lack thereof) which I see in helminth communities. To them too, my sincerest thanks. I am most grateful to Hilda Ching for providing data on willets from California and to Donald Forrester and Bill Font for providing logistic support to collect willets in Florida and Louisiana, respectively. I thank Dr Randall Ryti for modifying his colonization model to make it appropriate for my data. I thank John Aho, Jerry Esch, John Holmes, Clive Kennedy and Peter Price for commenting on an earlier draft of this chapter. All hosts were collected under appropriate provincial/state and Canada/U.S.A. permits. Initial support for my work on willets came, in part, from N.S.E.R.C. Canada Grant No. A3050 to Dr William Threlfall, Memorial University of Newfoundland while I was in receipt of an N.S.E.R.C. Canada PDF. Since then, my work on helminth communites in birds has been supported by N.S.E.R.C. Canada Grant No. A8090.

REFERENCES

Avery, R. A. (1969) The ecology of tapeworm parasites in wildfowl. *Wildfowl*, **20**, 59–69.

Bakke, T. A. (1972a) Studies of the helminth fauna of Norway XXIII: The common gull, *Larus canus* L., as final host for Digenea (Platyhelminthes). II. The relationship between infection and sex, age and weight of the common gull. *Nor. J. Zool.*, **20**, 189–204.

Bakke, T. A. (1972b) Studies of the helminth fauna of Norway XXII: the common gull, *Larus canus* L., as final host for Digenea (Platyhnelminthes). I. The ecology of the common gull and the infection in relation to season and the gulls habitat, together with the distribution of the parasites in the intestines. *Nor. J. Zool.*, **20**, 165–88.

Boucot, A. J. (1983) Area–dependent–richness hypotheses and rates of parasite/pest evolution. *Am. Nat.*, **121**, 294–300.

Brooks, D. R. (1979) Testing hypotheses of evolutionary relationships among parasites: the digeneans of crocodilians. *Am. Zool.*, **19**, 1225–38.

Buscher, H. N. (1965) Dynamics of the intestinal helminth fauna in three species of ducks. *J. Wildl. M.*, **29**, 772–81.

Buscher, H. N. (1966) Intestinal helminths of the blue-winged teal, *Anas discors* L., at Delta, Manitoba. *Can. J. Zool.*, **44**, 113–16.

Bush, A. O. (1980) *Faunal Similarity and Infracommunity Structure in the Helminths of Lesser Scaup*, PhD thesis, University of Alberta, Edmonton.

Bush, A. O. and Holmes, J. C. (1986a) Intestinal helminths of lesser scaup ducks: patterns of association. *Can. J. Zool.*, **64**, 132–41.

Bush, A. O. and Holmes, J. C. (1986b) Intestinal helminths of lesser scaup ducks: an interactive community. *Can. J. Zool.*, **64**, 142–52.

Connell, J. H. (1980) Diversity and the coevolution of competitors, or the ghost of competition past. *Oikos*, **35**, 131–8.

Connell, J. H. (1985) The consequences of variation in initial settlement vs. post-settlement mortality in rocky intertidal communities. *J. Exp. Mar. Biol.*, **93**, 11–45.

Cornwell, G. W. and Cowan, A. B. (1963) Helminth populations of the canvasback (*Aythya valisineria*) and host–parasite–environmental interrelations. *Trans. North Am. Wildl. Nat. Resour. Conf.*, **28**, 173–98.

Cram, E. B. (1927) Bird parasites of the nematode suborders Strongylata, Ascaridata, and Spirurata. *Bull. U.S. Natl. Mus.*, No. 140.

Cram, E. B. (1931) Developmental stages of some nematodes of the Spiruroidea parasitic in poultry and game birds. *U.S. Dept. Agric. Tech. Bull.*, No. 227.

Cram, E. B., Jones, M. F. and Allen, E. A. (1931) Intestinal parasites and parasitic diseases of the bobwhite. In *The bobwhite quail: its habits, preservation and increase*, (ed. H. L. Stoddard), Charles Scribner's Sons, New York, pp. 229–313.

Crompton, D. W. T. and Nesheim, M. C. (1976) Host–parasite relationships in the alimentary tract of domestic birds. *Adv. in Parasitol.*, **14**, 95–194.

Czaplinski, B. (1974) On the necessity and methodology of studies on the topical specificity of tapeworms. *Wiad. Parazyt.*, **20**, 699–701.

De Jong, N. (1976) Helminths from the mallard (*Anas platyrhynchos*) in the Netherlands. *Neth. J. Zool.*, **26**, 306–18.

Dogiel, V. A. (1964) *General Parasitology*, (English trans. by Z. Kabata) Oliver and Boyd Ltd, Edinburgh.

Edwards, D. D. and Bush, A. O. (1989) Helminth communities in avocets: importance of the compound community. *J. Parasitol.*, **75**, 225–38.

Forrester, D. J., Conti, J. A., Bush, A. O. *et al.* (1984) Ecology of helminth parasitism of bobwhites in Florida. *Proc. Helminthol, Soc. Wash.*, **51**, 255–60.

Gaines, S. and Roughgarden, J. (1985) Larval settlement rate. *Proc. Natl. Acad. Sci. U.S.A.*, **82**, 3707–11.

Goater, C. P. and Bush, A. O. (1988) Intestinal helminth communities in long-billed curlews: the importance of congeneric host-specialists. *Holarctic Ecol.*, **11**, 140–5.

Hair, J. D. (1975) *The Structure of the Intestinal Helminth Communities of Lesser Scaup (Aythya affinis)*, PhD Thesis, University of Alberta, Edmonton.

Hair, J. D. and Holmes, J. C. (1975) The usefulness of measures of diversity, niche width and niche overlap in the analysis of helminth communities in waterfowl. *Acta Parasitol. Polon.*, **23**, 253–69.

Herbold, B. and Moyle, P. B. (1986) Introduced species and vacant niches. *Am. Nat.*, **128**, 751–60.

Hollander, M. and Wolfe, D. A. (1973) *Nonparametric Statistical Methods*, John Wiley and Sons, New York.

Holmes, J. C. (1986) The structure of helminth communities. In *Parasitology – Quo Vadit?*, (ed. M. J. Howell), Proceedings 6th International Congress of Parasitology, Australian Academy of Science, Canberra, pp. 203–8.

Hood, D. E. and Welch, H. E. (1980) A seasonal study of the parasites of the red-winged blackbird (*Agelaius phoeniceus* L.) in Manitoba and Arkansas. *Can. J. Zool.*, **58**, 528–37.

Huber, J. Ch. (1906) Demetrios Pepagomenos über die würmer in den augen der jagdfalken. *Zool. Annal.*, **2**, 71–3.

Hunter, W. S. and Quay, T. L. (1953) An ecological study of the helminth fauna of Macgillivray's seaside sparrow, *Ammospiza maritima macgillivraii* (Audubon). *Am. Midl. Nat.*, **50**, 407–13.

Kennedy, C. R., Bush, A. O. and Aho, J. M. (1986) Patterns in helminth communities: why are birds and fish different? *Parasitology*, **93**, 205–15.

Korpaczewska, W. (1963) Formation of a population structure and cestode complexes in water birds. *Acta Parasitol. Polon.*, **11**, 337–44.

Lawton, J. H. (1982) Vacant niches and unsaturated communities: a comparison of bracken herbivores at sites on two continents. *J. Anim. Ecol.*, **51**, 573–95.

Lawton, J. H. (1984) Non-competitive populations, non-convergent communities, and vacant niches: the herbivores of bracken. In *Ecological Communities: conceptual issues and the evidence*, (eds, D. R. Strong, D. Simberloff, L. G. Abele and A. B. Thistle), Princeton University Press, New Jersey, pp. 67–100.

Lewin, R. (1986) Supply-side ecology. *Science*, **234**, 25–7.

Mitter, C. and Brooks, D. R. (1983) Phylogenetic aspects of coevolution. In *Coevolution* (eds, D. J. Futuyma and M. Slatkin), Sinauer Associates, Inc., Massachusetts, pp. 65–98.

Moore, J. (1987) Some roles of parasitic helminths in trophic interactions – a view from North America. *Rev. Chilena de Hist. Nat.*, **60**, 159–79.

Moore, J., Freehling, M. and Simberloff, D. (1986) Gastrointestinal helminths of the northern bobwhite in Florida: 1968 and 1983. *J. Wildl. Dis.*, **22**, 497–501.

Moore, J., Freehling, M., Horton, D. and Simberloff, D. (1987) Host age and sex in relation to intestinal helminths of bobwhite quail. *J. Parasitol.*, **73**, 230–33.

Moore, J., Simberloff, D, and Freehling, M. (1988) Relationships between bobwhite quail social-group size and intestinal helminth parasitism. *Am. Nat.*, **131**, 22–32.

Pearson, E. S. and Hartley, H. O. (1972) *Biometrika Tables for Statisticians*, vol. II, Cambridge University Press, Cambridge.

Pojmanska, T. (1982) The co-occurrence of three species of *Diorchis* Clerc, 1903 (Cestoda: Hymenolepididae) in the European coot, *Fulica atra* L. *Parasitology*, **84**, 419–29.

Price, P. W. (1980) *Evolutionary Biology of Parasites*, Princeton University Press, New Jersey.

Price, P. W. (1984) Communities of specialists: vacant niches in ecological and evolutionary time. In *Ecological Communities: Conceptual Issues and the Evidence*, (eds, D. R. Strong, D. Simberloff, L. G. Abele and A. B. Thistle), Princeton University Press, New Jersey, pp. 510–24.

Price, P. W. (1986) Evolution in parasite communities. In *Parasitology – Quo Vadit?* (ed. M. J. Howell), Proceedings 6th International Congress of Parasitology, Canberra, Australian Academy of Science, pp. 209–14.

Rausch, R. L. (1983) Biology of avian parasites. In *Avian Biology*, (eds, D. S. Farner, J. R. King and K. C. Parkes), vol. VII, Academic Press, New York, pp. 367–442.

Read, C. P. (1959) The role of carbohydrates in the biology of cestodes. VIII. Some conclusions and hypotheses. *Exp. Parasitol.*, **8**, 365–82.

Riley, J. and Owen, R. W. (1975) Competition between two closely related *Tetrabothrius* cestodes of the fulmar (*Fulmarus glacialis* L.) *Z. Parasitenk.*, **46**, 221–8.

Roberts, L. S. (1961) The influence of population density on patterns and physiology of growth in *Hymenolepis diminuta* (Cestoda: Cyclophyllidea) in the definitive host. *Exp. Parasitol.*, **11**, 332–71.

Ryti, R. T. and Gilpin, M. E. (1987) The comparative analysis of species occurrence patterns on archipelagos. *Oecologia*, **73**, 282–7.

Stock, T. M. and Holmes, J. C. (1987a) *Dioecocestus asper* (Cestoda: Dioecocestidae): an interference competitor in an enteric helminth community. *J. Parasitol.*, **73**, 1116–23.

Stock, T. M. and Holmes, J. C. (1987b) Host specificity and exchange of intestinal helminths among four species of grebes (podicipedidae). *Can. J. Zool.*, **65**, 669–76.

Stock, T. M. and Holmes, J. C. (1988) Functional relationships and microhabitat distribution of enteric helminths of grebes (Podicipedidae): the evidence for interactive communities. *J. Parasitol.*, **74**, 214–27.

Tallman, E. J., Corkum, K. C. and Tallman, D. A. (1985) The trematode fauna of two intercontinental migrants: *Tringa solitaria* and *Calidris melanotos* (Aves: Charadriiformes). *Am. Midl. Nat.*, **113**, 374–83.

Thompson, A. B. (1985) Transmission dynamics of *Profilicollis botulus* (Acanthocephala) from crabs (*Carcinus maenas*) to eider ducks (*Somateria mollissima*) on the Ythan Estuary, N.E. Scotland. *J. Anim. Ecol.*, **54**, 605–16.

Threlfall, W. (1965) Studies on the Helminth Parasites of Herring Gulls (*Larus argentatus pontopp*) on the Newborough Warren Nature Reserve, Anglesey. PhD thesis, University of Wales.

Threlfall, W. (1968) Studies on the helminth parasites of the American herring gull (*Larus argentatus* Pont.) in Newfoundland. *Can. J. Zool.*, **46**, 1119–26.

Wallace, B. M. and Pence, D. B. (1986) Population dynamics of the helminth community from migrating blue-winged teal: loss of helminths without replacement on the wintering grounds. *Can. J. Zool.*, **64**, 1765–73.

9

Helminth community of mammalian hosts: concepts at the infracommunity, component and compound community levels

Danny B. Pence

9.1 INTRODUCTION

The literature on biotic and abiotic factors affecting transmission, survival and reproduction of helminths in domestic animals and humans, and on helminth surveys of wild mammals, are legion. However, of the examples cited in the review on communities of parasites by Holmes and Price (1986), few are drawn from studies on the helminths of mammals. This is in contrast to some other vertebrate groups discussed in this book, especially fish and birds. In this chapter, I will use information from the literature, examples taken from my own research and data provided by colleagues, to present views on the community ecology of helminths in mammals. I will attempt to present this in the context of our present understanding of certain concepts of parasite ecology at the infracommunity, component and compound community levels.

9.2 STUDY SITES AND ANALYSIS OF DATA

I will specifically focus on helminths of black bear (*Ursus americana*), coyotes (*Canis latrans*), and white-tailed deer (*Odocoileus virginianus*) because: (a) all are large free-ranging terrestrial species endemic to much of North America, (b) they are ecotone species that are actively invading new habitats at the periphery of their ranges, and (c) they are common species with locally high population densities. Gastrointestinal helminths of black bear are examined from four localities in the southeastern United States (Pence *et al.*, 1983). Information on gastrointestinal helminth species of coyotes come from studies at six localities across North America: the Gulf Coastal Prairies of Texas and Louisiana (Custer and Pence, 1981), Rolling Plains of Texas (Pence and Meinzer, 1979), Brushlands of southern Texas (Pence and Windberg, 1984), western Tennessee (Van Den Bussche *et al.*, 1987), forested regions of Alberta (Holmes and Podesta, 1968) and Riding Mountain National Park of Manitoba (Samuel *et al.*, 1978). Data on abomasal helminths of white-tailed deer from the coastal plain of Georgia and mesic areas of Arkansas and Louisiana along the Mississippi River floodplain are from records of the Southeastern Cooperative Wildlife Disease Study (University of Georgia, Athens, GA); deer from central and western Texas are from Waid *et al.* (1985) and Stubblefield *et al.* (1987), respectively. Additional data on gastrointestinal helminths of surface-foraging, burrow-dwelling rodents and lagomorphs from a short grass prairie in the Southern High Plains of Texas are also presented for comparison (Rodenberg and Pence, 1978). Specific details on localities, collection procedures, and other aspects of these data are presented in the respective references.

Several attributes are used to describe community structure in these helminth communities. I use frequency (n) to denote the occurrence of a helminth species within, or across, a host population. Density (N) refers to the total number of individuals of a particular helminth species. Measures of central tendency were calculated for both numbers of species and individuals; values included both infected and uninfected hosts. To describe community diversity, I calculated Brillouin's index (based on natural logs) and evenness (Pielou, 1975). Predictability of community composition of black bears and coyotes was determined using qualitative (Jaccard's coefficient) and quantitative (per cent similarity) measures of similarity. Statistical analyses were performed using SYSTAT (Wilkinson, 1988) and SAS statistical programs (SAS, 1985). Kruskal–Wallis one-way ANOVA and/or one- or two-way ANOVA and factorial ANOVA and MANOVA on ranked data were used to examine the effects of host age, sex, season and locality on the number of species and relative density of helminths at the component community level. Chi-square tests (with Yates correction) were used for goodness of fit of observed numbers of helminth species in the host infracommunities to a Poisson distribution. The null hypothesis of no significant differences was accepted at $p > 0.05$.

9.3 THE INFRACOMMUNITY

It is at the infracommunity level that any interactions influencing the distribution and abundance of helminths must take place. Holmes (1973) reviewed the evidence for interspecific interactions and site segregation in the helminths with some examples from mammals (i.e. Kisielewska, 1970a, b). Holmes and Price (1986) developed the assumptions necessary to predict one of two kinds of parasite infracommunities, interactive or isolationist. These two extremes are indicative of either a high or low probability, respectively, of colonization of the host. Using differences in community structure among freshwater fish and aquatic birds, Kennedy *et al.* (1986) identified criteria that might result in establishment of an isolationist versus an interactive community. These included: (a) simple digestive tract, (b) low vagility, (c) simple or specialized diet, (d) nonselective feeding on prey species which are not intermediate hosts for parasites, and (e) exposure to few parasites with direct life cycles. These distinctions were further developed and are more fully explained in Goater *et al.* (1987). Based on general community theories, Lotz and Font (1985) reasoned that interactions will be a significant force determining helminth community structure only when species are near carrying capacity and resources are limiting. Under these conditions, one would expect to find several infrapopulations with densities approaching the highest value observed. Thus, the distribution pattern of each species among hosts could provide another useful measure for predicting isolationist or interactive community structure.

Isolationist versus interactive communities

I have applied the above criteria to examine patterns of helminth community richness and abundance in three species of mammals: (a) abomasal nematodes of two populations of white-tailed deer from mesic localities in Arkansas and Louisiana (Table 9.1), (b) gastrointestinal helminths from the black bear in the southeastern United States (Table 9.3) and (c) gastrointestinal helminths from coyotes in the Brushlands of southern Texas (Table 9.4). Collectively, these studies should offer a valuable comparison to determine factors influencing where mammal helminth communities position themselves along the isolationist–interactive continuum.

Following the assumptions of Kennedy *et al.* (1986), the prediction is that the abomasal nematode community of white-tailed deer should tend toward a high probability for colonization (Table 9.5). This is a highly invasive, ecotone species rapidly expanding its range in North America (high vagility). As a terrestrial ruminant, white-tailed deer possess a complex

Table 9.1 Nematodes from the abomasum of 126 adult white-tailed deer from five localities in the southern United States

	Coastal		*Mesic*				*Xeric*			
	Georgia		*Louisiana*		*Arkansas*		*Central Texas*		*Western Texas*	
Nematode species	$\bar{X} \pm S.E.$	n	$\bar{X} \pm S.E.$	n	$\bar{X} \pm S.E.$	n	$\bar{X} \pm S.E.$	n	$\bar{X} \pm S.E.$	n
Haemonchus contortus	388 ± 171	4	60 ± 0	10	–	–	105 ± 26	62	8 ± 4	36
Spiculopteragia odocolei	494 ± 69	5	–	–	106 ± 35	36	13 ± 7	11	–	–
Spiculopteragia pursglovi	–	–	1337 ± 294	100	1014 ± 146	100	–	–	–	–
Ostertagia dikmansi	–	–	163 ± 39	50	79 ± 20	55	–	–	–	–
Ostertagia mossi	93 ± 26	5	437 ± 114	80	177 ± 15	91	–	–	–	–
Ostertagia ostertai	–	–	–	–	–	–	1 ± 0	5	–	–
Cooperia sp.	–	–	–	–	–	–	2 ± 0	1	–	–
Trichostrongylus askivali	179 ± 119	3	618 ± 1	10	157 ± 29	45	–	–	–	–
Trichostrongylus axei	–	–	–	–	55 ± 0	9	–	–	–	–
Trichostrongylus dosteri	703 ± 349	4	–	–	–	–	–	–	–	
Total number of species	5		5		6		4		1	
Density/host	1857 ± 734		2615 ± 448		1588 ± 245		121 ± 33		8 ± 4	
Number of hosts	5		10		11		86		14	

digestive tract but forage on a variety of grasses, woody shrubs and forbs. The large number of helminth species found in the abomasum are direct life cycle trichostrongylid nematodes acquired by ingestion of infective larva. Accordingly, this host could acquire a helminth community dominated by deterministic (Bush, 1980) or core species. However, the number of helminth species/host followed a Poisson distribution ($\chi^2 = 1.79$ and 2.67, $N = 10$ and 11, $p > 0.05$, for both populations of deer), indicating a random pattern of colonization by the respective nematode species.

High densities of five and six species of abomasal nematodes, respectively, were observed in the two deer populations. In terms of niche utilization, the number of helminth species found per population was relatively high compared to the total of nine species reported from the abomasum of white-tailed deer in the southeastern United States. This probably indicates a high utilization of fundamental niches available in these two deer populations. Additionally, it should be noted from Table 9.2 that less than 50% of the helminth infracommunities from Louisiana and Arkansas had greater than or equal to 60% realized utilization of fundamental niches (≥ 3 of 5 and ≥ 4 of 6 helminth species present in $> 50\%$ of the infracommunites in Louisiana and Arkansas, respectively). Because of limitations in these data sets, I have neither calculated mean species diversity nor have I determined if there is exploitation of temporary opportunities by the abomasal nematodes of deer. Likewise, I have no information regarding the distribution of these nematodes along their resource gradient or of their individual response to other guild members.

Table 9.2 Percentages of maximum number of helminth species in a sample of 10 and 11 white-tailed deer from mesic areas in Louisiana and Arkansas, respectively

	Deer number										
Abomasal nematode	1	2	3	4	5	6	7	8	9	10	11
Louisiana											
Haemonchus contortus	100	0	0	0	0	0	0	0	0	0	–
Spiculopteragia pursglovi	18	19	25	94	28	20	69	41	100	35	–
Osteragia dikmansi	0	63	73	0	27	0	0	27	100	0	–
Osteragia mossi	0	35	36	20	23	0	5	55	100	35	–
Trichostrongylus askavali	0	0	0	0	0	0	0	0	100	0	–
Arkansas											
Spiculopteragia odocoilei	21	43	100	0	43	0	0	0	0	0	0
Spiculopteragia pursglovi	30	46	100	28	30	51	83	67	51	70	24
Ostertagia dikmansi	27	0	25	25	0	100	0	56	0	0	57
Ostertagia mossi	23	15	22	7	15	100	39	0	25	30	33
Trichostrongylus askavali	11	0	0	0	32	100	27	0	0	0	23
Trichostrongylus axei	0	0	0	0	0	0	100	0	0	0	0

Table 9.3 Gastrointestinal helminths of 104 black bear from four localities in the southeastern United States

			Northern mountains		*Southern*				*Northern coastal plains*	
					Mountains		*Coastal plains*			
Helminth taxa	n	N	X̄ ± *S.E.*	*Med**	X̄ ± *S.E.*	*Med*	X̄ ± *S.E.*	*Med*	X̄ ± *S.E.*	*Med*
Digenea										
Pharyngostomoides procyonis	2	7	–	–	–	–	–	–	0.1 ± 0.1	0.0
Acanthocephala										
Macracanthorhynchus ingenes	49	386	0.1 ± 0.1	0.0	0.9 ± 0.8	0.0	2.4 ± 1.1	1.0	6.2 ± 0.1	3.0
Nematoda										
Strongyloides sp.	66	16,492	0.3 ± 0.1	0.0	36.5 ± 30.0	0.0	53.8 ± 32.5	1.0	279.8 ± 105.0	18.5
Capillaria putorii	26	154	0.6 ± 0.2	0.0	1.6 ± 0.05	0.0	0.1 ± 0.1	0.0	2.1 ± 1.7	0.0
Gongylonema pulchrum	41	128	1.7 ± 0.5	1.0	2.5 ± 1.5	0.0	0.7 ± 0.4	0.0	0.8 ± 0.3	0.0
Physaloptera rara	14	84	0.1 ± 0.1	0.0	–	–	2.9 ± 2.9	0.0	0.8 ± 0.4	0.0
Cyathospirura sp.	3	8	0.4 ± 0.4	0.0	–	–	–	–	–	–
Ancylostoma caninum	3	3	–	–	0.1 ± 0.1	0.0	–	–	0.1 ± 0.1	0.0
Placoconus lotoris	25	98	0.6 ± 0.5	0.0	1.2 ± 0.5	0.0	1.8 ± 1.2	0.0	0.9 ± 0.2	0.0
Molineus barbatus	36	449	0.1 ± 0.1	0.0	1.2 ± 0.5	0.0	0.8 ± 0.3	0.0	7.7 ± 4.6	0.0
Lagochilascaris sp.	2	17	–	–	–	–	–	–	2.3 ± 2.2	0.0
Baylisascaris transfuga	30	314	8.0 ± 3.0	4.0	4.7 ± 1.4	2.0	5.9 ± 5.2	0.0	–	–
Toxascaris leonina	2	5	–	–	–	–	–	–	0.1 ± 0.1	0.0
Total number of species			2.6 ± 0.3		2.8 ± 1.6		2.4 ± 0.3		3.1 ± 0.2	
Total number of individuals ± 114.8			11.8 ± 3.1		48.7 ± 35.3		68.1 ± 33.6			300.9
Brillouin's diversity index			0.4 ± 0.1		0.5 ± 0.8		0.3 ± 0.1		0.4 ± 0.1	
Evenness index			0.5 ± 0.4		0.5 ± 0.1		0.5 ± 0.1		0.4 ± 0.1	
Host sample size			19		17		14		54	

* Median number of individuals per species.

However, species abundance patterns were highly aggregated within most individual deer, with different species reaching their largest infrapopulation size in different deer (Table 9.2). Only for three individuals (deer nine in Louisiana and deer three and six in Arkansas) did the largest infrapopulations of two or more nematode species reside with the same individual (Table 9.2). These are the helminth infracommunities in which interactions could potentially occur. It could be argued that lower densities or absence of species from different helminth infracommunities may result from competitive interactions (Goater and Bush, 1988). However, if this was occurring, why were the maximum densities of several species concentrated in one or two infracommunities? Alternatively, from Table 9.2 it is apparent that the density of one nematode species may be high, but the densities of the rest of the helminth species in that infracommunity represent only a fraction of the highest value observed for that respective species. Thus, based only on the distribution of helminth numbers within the deer populations, white-tailed deer would appear to exhibit characteristics tending toward an isolationist community structure.

The black bear has a simple (for a mammal), but muscular digestive tract and is a nonselective omnivore feeding on many prey species; dietary items result from herbivory, insectivory, scavanging and carnivory. Despite its omnivorous diet and high vagility, the helminth community is dominated by direct life cycle nematodes. The frequency distribution of all helminth species did not differ significantly from a Poisson distribution ($\chi^2 = 10.97$, $N = 104$, $p > 0.05$), indicating a random pattern of colonization. These factors, plus the absence of a recurrent group of helminth species (Pence *et al.*, 1983), indicate a comparatively low probability for colonization by the respective helminth species in the community (Table 9.5).

Across the four southeastern localities where bears were examined, the gastrointestinal helminth community only had eight commonly occurring species (> 10% prevalence of infection) (Table 9.3). Of these, six species are usually monoxenous in life history while only two species utilize arthropod intermediate hosts (Table 9.3). Pence *et al.* (1983) described the details of this helminth community and concluded that it resembled a raccoon helminth fauna, with only one species (*Baylisascaris transfuga*) specific to the Ursidae. Although a recurring group of helminth species was not observed, rank correlations indicated a postive relationship between the relative density of most helminth species pairs. The relationship did not hold for *B. transfuga* where density varied negatively with four other helminth species. These relationships were believed to result from differences across localities.

Based on low values for frequencies, mean densities, mean diversity and mean number of species (Table 9.3), the helminth fauna of the black bear is depauperate compared to that of other carnivores in the southeastern

Table 9.4 Gastrointestinal helminths of 752 coyotes from six localities in North America

			Texas			
			Gulf coastal prairies		*Rolling plains*	
Helminth taxa	n	N	X ± *S.E.*	*Med*	X ± *S.E.*	*Med*
Digenea						
Alaria arisaemoides	28	2166	–	–	–	–
Alaria marcianae	169	31 098	–	–	3.0 ± 1.6	0.0
Eucestoda						
Diphyllobothrium sp.	3	8	–	–	–	–
Taenia pisiformis	438	7082	29.7 ± 3.4	21.0	2.7 ± 0.5	0.0
Taenia multiceps	45	1321	0.1 ± 0.1	0.0	0.8 ± 0.7	0.0
Taenia macrocystis	27	168	1.5 ± 0.8	0.0	–	–
Taenia taeniaeformis	1	2	0.1 ± 0.1	0.0	–	–
Taenia twitchelli	1	1	–	–	–	–
Echniococcus grandulosus	5	5873	–	–	–	–
Echinococcus multilocularis	7	63 827	–	–	–	–
Mesocestoides lineatus	180	3851	0.1 ± 0.1	0.0	13.7 ± 3.5	1.0
Rhabdometra odiosa	1	1	–	–	–	–
Acanthocephala						
Oncicola canis	126	5498	–	–	0.8 ± 0.2	0.0
Pachysentis canicola	35	148	–	–	1.0 ± 0.3	0.0
Nematoda						
Capillaria sp.	1	1	–	–	–	–
Trichuris vulpis	151	3949	0.3 ± 0.2	0.0	–	–
Trichuris sp.	1	4	–	–	–	–
Physaloptera rara	283	3747	0.1 ± 0.1	0.0	16.8 ± 6.3	2.0
Pterygodermaites affinis	54	551	–	–	3.8 ± 0.9	0.0
Protospirurua numidica	58	697	–	–	–	–
Physocephalus sp.	2	4	–	–	–	–
Cyrnea sp.	1	1	–	–	–	–
Synhimanthus sp.	1	2	–	–	–	–
Ancylostoma caninum	526	21 051	23.4 ± 2.9	14.0	19.8 ± 3.7	7.0
Uncinaria stenocephala	21	446	–	–	–	–
Toxascaris leonima	263	4842	0.1 ± 0.1	0.0	26.3 ± 3.1	11.0
Toxocara canis	1	1	–	–	–	–
Dermatoxys veligera	2	4	–	–	<0.1 ± 0.1	0.0
Syphacia sp.	1	8	–	–	<0.1 ± 0.1	0.0
Total number of species			2.0 ± 0.1		4.3 ± 0.1	
Total number of individuals			55.3 ± 4.3		89.0 ± 9.9	
Brillouin's diversity index			0.42 ± 0.04		0.80 ± 0.03	
Evenness index			0.59 ± 0.04		0.64 ± 0.02	
Host sample size			78		150	

Texas		*Western Tennessee*		*Canada*			
Southern brushlands				*Alberta*		*Manitoba*	
X ± *S.E.*	*Med*	X ± *S.E.*	*Med*	X ± *S.E.*	*Med*	X ± S.E.	Med
–	–	–	–	14.8 ± 9.1	0.0	43.8 ± 15.5	0.0
171.4 ± 45.8	8.5	–	–	3.1 ± 1.4	0.0	4.9 ± 1.7	0.0
–	–	–	–	0.2 ± 0.1	0.0	–	–
7.2 ± 1.1	2.0	9.1 ± 5.2	2.0	3.6 ± 1.5	0.0	14.4 ± 5.2	3.5
6.7 ± 3.2	0.0	–	–	–	–	–	–
0.3 ± 0.1	0.0	–	–	<0.1 ± 0.1	0.0	–	–
–	–	–	–	–	–	–	–
–	–	–	–	<0.1 ± 0.1	0.0	–	–
–	–	–	–	0.5 ± 0.5	0.0	171.9 ± 134.4	0.0
–	–	–	–	–	–	1877.3 ± 1547.6	0.0
10.1 ± 1.8	2.0	–	–	–	–	–	–
<0.1 ± 0.1	0.0	–	–	–	–	–	–
30.4 ± 6.5	0.0	–	–	–	–	–	–
–	–	–	–	–	–	–	–
–	–	–	–	–	–	–	–
<0.1 ± 0.01	0.0	14.7 ± 2.4	1.0	–	–	–	–
<0.1 ± 0.1	0.1	–	–	–	–	–	–
2.5 ± 0.6	0.0	2.9 ± 0.4	0.0	–	–	<0.1 ± 0.1	0.0
–	–	–	–	–	–	–	–
3.9 ± 0.8	0.0	–	–	–	–	–	–
<0.1 ± 0.1	0.0	–	–	–	–	–	–
0.1 ± 0.1	0.0	–	–	–	–	–	–
0.1 ± 0.1	0.0	–	–	–	–	–	–
83.0 ± 7.9	40.0	5.9 ± 1.0	1.0	0.1 ± 0.1	0.0	<0.1 ± 0.1	0.0
–	–	–	–	3.2 ± 1.3	0.0	8.8 ± 4.6	0.0
2.4 ± 0.7	0.0	0.4 ± 0.1	0.0	3.7 ± 1.3	0.0	6.4 ± 1.7	1.0
–	–	–	–	<0.1 ± 0.1	0.0	–	–
–	–	–	–	–	–	–	–
–	–	–	–	–	–	–	–
4.7 ± 0.1		2.2 ± 0.1		1.7 ± 0.1		2.9 ± 0.2	
317.9 ± 69.1		33.0 ± 9.1		28.9 ± 15.6		2127.5 ± 1711.0	
0.81 ± 0.03		0.41 ± 0.02		0.27 ± 0.05		0.38 ± 0.05	
0.59 ± 0.02		0.49 ± 0.02		0.38 ± 0.06		0.41 ± 0.06	
177		267		46		34	

United States. While limited, the available information indicates helminth communities of black bear exhibit characteristics which are isolationist in nature (Table 9.5).

Coyotes are medium-sized carnivores, well suited to ecotone habitats; they are highly vagile and are rapidly expanding into areas at the periphery of their range. Being omnivores, their diet is very diverse; they feed nonselectively and most of the prey species can serve as intermediate hosts for helminths. Focusing only on the helminth community from coyotes in the Brushlands of Texas (Pence and Windberg, 1984), 16 helminth species were recovered, with only three having direct life cycles (Table 9.4). Although the pattern of helminth species distribution follows a Poisson ($\chi^2 = 107.68$,

Table 9.5 Criteria for predicting an isolationist versus an interactive community as applied to helminth communities from white-tailed deer, black bear and coyotes (see text for details)

Criteria	*Isolationist community*	*White-tailed deer**	*Black bear*	*Coyote*	*Interactive community*
Probability of colonizing the host	Low				High
Digestive tract	Simple	→	←	←	Complex
Vagility	Low	→	→	→	High
Diet	Simple	→	→	→	Complex
Feeding Selectivity	Non-selective	?→	←?	?→	Selective
Helminth life cycles	Direct	←	←	→	Indirect
Pattern of colonization	Random	←	←	←	Non-random
Niche utilization	Unsaturated non-equilibrium community				Saturated equilibrium community
Mean species diversity	Low	?	←	→	High
Number of high density species	Low	→	←	→	High
Exploitation of temporary opportunities	No	?	?	→	Yes
Fundamental versus realized niches	Low ratio	→	←	→	High ratio
Distribution along resource gradient	Individually distributed	?	?	?	Evenly distributed
Response to other guild members	Insensitive	?	?	?	Responsive
General tendency	Isolationist	?→	←	→	Interaction

* Abomasal helminths only.

$N = 177$, $p > 0.05$), there is a recurrent group of six common intestinal helminth species (D. B. Pence, unpublished data). All of the above predict a high probability for colonization by the helminth species in this coyote population.

Compared to either white-tailed deer or black bear, the helminth community of coyotes appears rich in species, and densities of the nine common helminth species are high, as is community diversity (Table 9.4). There is also some evidence at the component community level that at least certain helminth species tend to exploit temporary opportunities (i.e. much greater densities of certain common species occur in younger animals). Finally, many of the fundamental niches along the gastrointestinal tract seem to have been occupied. Representative species of four different guilds are present: absorber (three cestodes and one acanthocephalan in the small intestine), grazer (one digenean in the duodenum and two nematodes in the stomach and small intestine), blood-feeding (one hookworm in the small intestine) and a lumen-dwelling ascarid in the small and large intestines.

Unfortunately, the two other major criteria for predicting an isolationist versus interactive community as proposed by Holmes and Price (1986) have not been examined in the helminth community of this coyote population. These include: (a) the linear distribution of helminth species along the gastrointestinal tract, and (b) the response of the respective species in a particular guild to other guild members. Studies on the helminth communities of coyotes from other localities, such as those by Hirsch and Gier (1974), Pence and Meinzer (1979), and Custer and Pence (1981), indicate species interactions are important, but these analyses tested data that were summed across infracommunities. Based on the present information, I believe that the helminth community of this coyote population exhibit characteristics indicative of an interactive community (Table 9.5).

Conclusions

I wish to emphasize that most of the above predictions have not been rigorously tested, if tested at all. Indeed, these kinds of studies on mammal helminth communities are rare. Hobbs (1980) presented evidence for active site segregation in helminths of pikas (*Ochotona princeps*) supporting an interactive community structure. In contrast, Lotz and Font (1985) concluded that contemporary interactions were of minor importance in structuring infracommunities of enteric helminths of the big brown bat (*Eptesticus fuscus*). Thus, the evidence is equivocal. Herein, I present additional information on helminth infracommunities from three species of mammals in which patterns follow many of the predictions for an isolationist community in a large herbivore and carnivore and for an interactive community in a medium-sized carnivore and a large herbivore. Clearly, considerably more data are required before we can understand the patterns and processes important in shaping mammalian helminth infracommunities.

9.4 THE COMPONENT COMMUNITY LEVEL

There is a plethora of survey papers on the component helminth communities of mammals. At best, most of these provide some information on the frequency of occurrence and densities of helminth species for given host populations or sometimes across localities.

An inherent concept at the component community level is the a priori expectation that the helminth infracommunity of one host individual should be a replicate of every other host individual in that species. In reality, this is rarely observed and it is at this community level that many studies, including some on the helminths of mammals, have examined variability in species richness across a multitude of biotic, temporal and spatial variables. How these features can influence the transmission dynamics and distribution patterns of helminths is of critical importance if we are to understand helminth community dynamics.

Within a given locality, many studies have addressed the effects of host factors such as age, sex or physiological condition and temporal factors (seasons) on helminth frequency (i.e. Kisielewska, 1970a; Haukisalmi *et al.*, 1987) and/or on relative densities (i.e. Pence and Meinzer, 1979; Franson *et al.*, 1978; Stone and Pence, 1978; Custer and Pence, 1981; Pence and Dowler, 1979; Corn *et al.*, 1985; Waid *et al.*, 1985; Pence *et al.*, 1983; Pence and Windberg, 1984; Forrester *et al.*, 1987; Kinsella and Pence, 1987; Pence *et al.*, 1988; Waid and Pence, 1988) in a variety of mammals. Both as main and interactive effects, changes in host age over seasons seem to be the most often cited factor affecting frequency and density of helminth species. In contrast, few studies have examined the temporal long-term changes in composition focusing on persistence and stability of the helminth community (i.e. Kisielewska, 1970a; Keith *et al.*, 1985; Haukisalmi *et al.*, 1987; Spratt, 1987). In general, helminth frequencies and densities, but not species' occurrences, wax and wane over time with changing environmental conditions and fluctuations in host population density.

The importance of host-related and temporal factors in influencing the helminth community in mammals is illustrated by the coyote data set from the Brushlands of southern Texas (Table 9.4). There were significant age-related and seasonal effects on the mean number of helminth species, as well as on the ranked density values of total individuals (Tables 9.6 and 9.7). This resulted from a greater number of species and individuals in juvenile coyotes collected in the fall. Of the individual species, four of nine

Table 9.6 Means (± S.E.) for number of total species, number of total individuals, and number of individuals of nine common species of helminths in different host age and sex subpopulations across seasons for the 177 sample data set for coyotes from the Brushlands of southern Texas

	Autumn				*Spring*			
	Juvenile		*Adult*		*Juvenile*		*Adult*	
Helminths	*Female*	*Male*	*Female*	*Male*	*Female*	*Male*	*Female*	*Male*
Total species	6.1 ± 0.2	6.2 ± 0.4	5.3 ± 0.3	4.8 ± 0.34	4.2 ± 0.3	4.1 ± 0.3	3.8 ± 0.3	3.7 ± 0.3
Total individuals	1091.0 ± 251.5	849.0 ± 141.0	121.6 ± 18.2	99.9 ± 29.0	191.5 ± 48.1	344.6 ± 94.2	97.5 ± 37.4	127.2 ± 20.6
Alaria marcianae	684.2 ± 253.1	416.4 ± 124.7	42.7 ± 9.6	10.3 ± 5.0	104.2 ± 39.7	216.7 ± 91.4	60.7 ± 34.9	47.6 ± 17.5
Mesocestoides lineatus	17.2 ± 6.6	15.8 ± 9.9	10.9 ± 4.7	9.4 ± 4.6	13.4 ± 5.5	15.2 ± 7.4	1.1 ± 0.3	2.7 ± 1.8
Taenia pistiformis	2.5 ± 0.9	4.3 ± 1.7	6.0 ± 2.6	11.8 ± 4.4	5.5 ± 2.0	3.8 ± 1.4	12.8 ± 4.9	7.5 ± 2.1
Taenia multiceps	0.8 ± 0.6	58.6 ± 37.6	2.4 ± 1.5	0.4 ± 0.3	0.4 ± 0.2	4.5 ± 3.8	0.2 ± 0.1	6.8 ± 3.3
Oncicola canis	144.4 ± 23.7	83.5 ± 47.9	14.3 ± 5.9	40.2 ± 24.8	1.0 ± 0.4	2.1 ± 1.5	0.1 ± 0.1	0.8 ± 0.5
Toxascaris leonina	3.7 ± 2.4	2.7 ± 2.1	5.7 ± 3.9	3.7 ± 1.7	1.0 ± 0.5	0.7 ± 0.6	0.9 ± 0.6	0.9 ± 0.5
Ancylostoma caninum	225.0 ± 27.9	243.1 ± 30.6	32.4 ± 9.1	19.1 ± 5.6	62.8 ± 11.4	97.2 ± 22.5	20.0 ± 3.9	59.6 ± 11.0
Physaloptera rara	6.5 ± 2.5	9.5 ± 4.5	3.8 ± 2.4	2.9 ± 1.1	0.3 ± 0.2	0.2 ± 0.2	0.3 ± 0.2	0.1 ± 0.1
Protospirura numidica	8.5 ± 3.0	14.1 ± 7.1	3.4 ± 1.5	3.4 ± 1.4	2.0 ± 1.1	3.9 ± 2.6	0.7 ± 0.4	0.9 ± 0.3

Table 9.7 Values of the F statistic generated by MANOVA and factorial ANOVA for main and interactive effects of host age and sex across season factors from the 177 sample data set of ranked densities for numbers of total individuals (all species) and for each of nine common species of helminths from coyotes in the Brushlands of southern Texas

Helminth	*Age*	*Sex*	*Age–sex*	*Season*	*Age–season*	*Sex–season*	*Age–sex–season*
MANOVA	13.80*	2.17	1.40	22.78*	7.31*	1.39	1.04
Factorial ANOVA							
Alaria marcianae	35.14*	0.62	1.63	6.19*	7.21*	2.05	2.53
Mesocestoides lineatus	4.67*	7.65*	0.87	1.89	0.54	3.80	0.56
Taenia pistiformis	1.06	0.09	0.05	3.05	0.06	0.61	0.23
Taenia multiceps	0.03	3.42	0.23	0.56	1.77	1.22	3.11
Oncicola canis	41.24*	0.00	5.19*	220.43*	20.51*	0.00	1.26
Toxascaris leonina	1.84	0.16	0.02	7.42*	1.27	0.26	0.01
Ancylostoma caninum	120.15*	1.80	0.13	7.84*	32.59*	8.14*	2.36
Physaloptera rara	9.94*	0.01	0.06	123.39*	7.32*	1.76	0.09
Protospirura numidica	7.77*	0.52	1.76	16.96*	1.09	0.24	0.18

* Significant at $p \leq 0.05$.

gastrointestinal helminths in the community (*Alaria marcianae, Oncicola canis, Ancylostoma caninum and Physaloptera rara*) had significantly higher densities in the autumn-collected juvenile coyotes. Also, *A. caninum* had significantly higher densities in juvenile male than in female coyotes during the autumn. All the remaining helminth species, except *Taenia pisiformis* and *Taenia multiceps*, had significantly higher densities across either, or both, the main effects of age and/or seasons (Table 9.7).

Many of the documented changes in helminth densities across host and short-term temporal factors are discussed by Pence and Windberg (1984) who analysed these data by a slightly different and less sensitive method. These differences can be attributed mostly to changing immune status and other factors (e.g. transmammary transmission of helminths such as *A. marciane* (Shoop and Corkum, 1983)) related to age over time (seasons) in a highly age-structured host population. Moreover, these analyses emphasize that significant changes can occur in number of species as well as in densities of the individual species in the helminth community. In contrast to migratory birds, species within the helminth community of a given host population are not lost during certain seasons or from specific host subpopulations.

Investigators should be aware of the above differences when comparing helminth communities in host populations from different localities. To accurately depict community structure, collections should include all the important demographic and temporal variables, or be qualified accordingly. The former conditions are rarely met, and while the latter provides important information on differences in species composition, there is some loss of detail in community structure (density data). Such is the case in the data sets on white-tailed deer (adults collected in different seasons), black bear (mostly autumn collections) and coyotes (autumn and winter collections in Tennessee, Manitoba and Alberta).

When examined at different localities, major shifts in helminth community structure have been noted for many host species. Investigations usually attempt to quantitatively document changes in species richness, dominance and faunal similarity between helminth communities in the same mammalian host from different localities (i.e. Tenora, 1967; Holmes and Podesta, 1968; Mollhagen, 1978; Pence and Meinzer, 1979; Custer and Pence, 1981; Pence *et al.*, 1983). However, these studies failed to determine community richness and similarity at the infracommunity level, but were derived using summed distributions across helminth infracommunities.

Fedynich *et al.* (1986) found only one commonly occurring helminth species in beaver (*Castor canadensis*) from eastern Texas, compared with several frequently occurring high-density helminth species from beaver in the northern latitudes. It was proposed that the number of species and their densities in the helminth community follows Brown's (1984) hypothesis that a species abundance pattern decreases from its epicenter of origin toward the periphery of its range. Although transmission, total reproductive effort, and survival of a helminth species are related to host population density, they may be affected by other factors, especially environmental extremes at the periphery of, or within, a host's range. This is well illustrated in the compositional changes in abomasal nematodes of white-tailed deer between five locations in the southern United States (Table 9.1). Most pronounced is a dramatic decline in the number of species and their densities from mesic habitats in the east to increasingly xeric habitats in the west. The loss, without replacement of both species and densities, can be explained best by the inability of most species of trichostrongylid nematode larvae to survive in more xeric environments. Only *Spiculopteragia odocoilei*, a specialist nematode of white-tailed deer, was recovered at a low frequency and density from the central Texas deer population. The generalist species, *Haemonchus contortus*, was found in white-tailed and sympatric mule deer (*Odocoileus hemionus*) from western Texas. Population densities of deer were approximately the same at each location, except in western Texas where the population density is lower. All these populations are within the native range of white-tailed deer in North America.

The data set on the helminth community of black bear from four contiguous localities in the southeastern United States allows examination of changes in species frequency of occurrence and density across different habitats within a more localized area (Table 9.3). Information on differences in frequency and density of helminth infection across different age

Table 9.8 Trellis diagram of means (± S.E.) for per cent similarity and Jaccards coefficients of overlap from the 104 sample data set of 13 gastrointestinal helminth species of black bear from four localities in the southeastern United States

	Jaccard's coefficients of overlap			
	Northern mountains	*Southern mountains*	*Northern coastal plains*	*Southern coastal plains*
Northern mountains	40.7 ± 2.0 (44.0 ± 2.3)	32.5 ± 1.1	13.7 ± 0.7	10.3 ± 0.3
Southern mountains	33.5 ± 1.2	28.5 ± 2.2 (28.1 ± 2.5)	16.5 ± 0.8	16.1 ± 0.4
Northern coastal plains	11.1 ± 0.9	16.9 ± 1.2	24.5 ± 2.2 (22.5 ± 3.3)	31.4 ± 0.7
Southern coastal plains	3.3 ± 0.1	16.0 ± 0.5	34.1 ± 0.8	37.5 ± 0.6 (47.9 ± 1.0)
		Per cent similarity		

classes and host gender have been previously presented (Pence *et al.*, 1983). There was not a significant difference in the mean number of individuals in the helminth community across the main and interactive effects of age and sex (ranked data, two-way ANOVA, $F = 1.23$, $p = 0.31$). There were, however, significant between-location differences in the mean number of individuals (Kruskal–Wallis one-way ANOVA, $H = 20.53$, $p < 0.005$). Values for the mean number of helminth species, mean diversity, and evenness were equivalent across the four localities (Table 9.2). The significant difference in total density of all helminth species across the four localities resulted from the greater densities of *Strongyloides* sp. in the Southern Coastal Plains.

Results for qualitative and quantitative similarity of the helminth communities of black bear across the four localities, based on all pairwise comparisons (Table 9.8), indicate a pattern similar to that described by Pence *et al.* (1983) using summed distributions across infracommunities (cluster analysis). Values for both per cent similarity and Jaccard's coefficients were highest between the southern coastal plain and northern mountain bear populations. This indicates that helminth communities between contiguous localities may vary considerably (mountains versus coastal plains). The greatest similarity between, and overlap of species in, helminth communities occurred between localities with the most similar habitat conditions. Dynamics of individual species in this helminth community are discussed by Pence *et al.* (1983).

Figure 9.1 Box plots of number of species (a), Brillouin's index (b) and evenness index (c) from the 752 sample dataset of 29 species of gastrointestinal helminths in coyotes from six localities in North America. The median is marked with a "+" near the center of the box, the upper and lower hinges are the left and right margins of the box and the end of the whiskers denote the adjacent (usually outermost) values

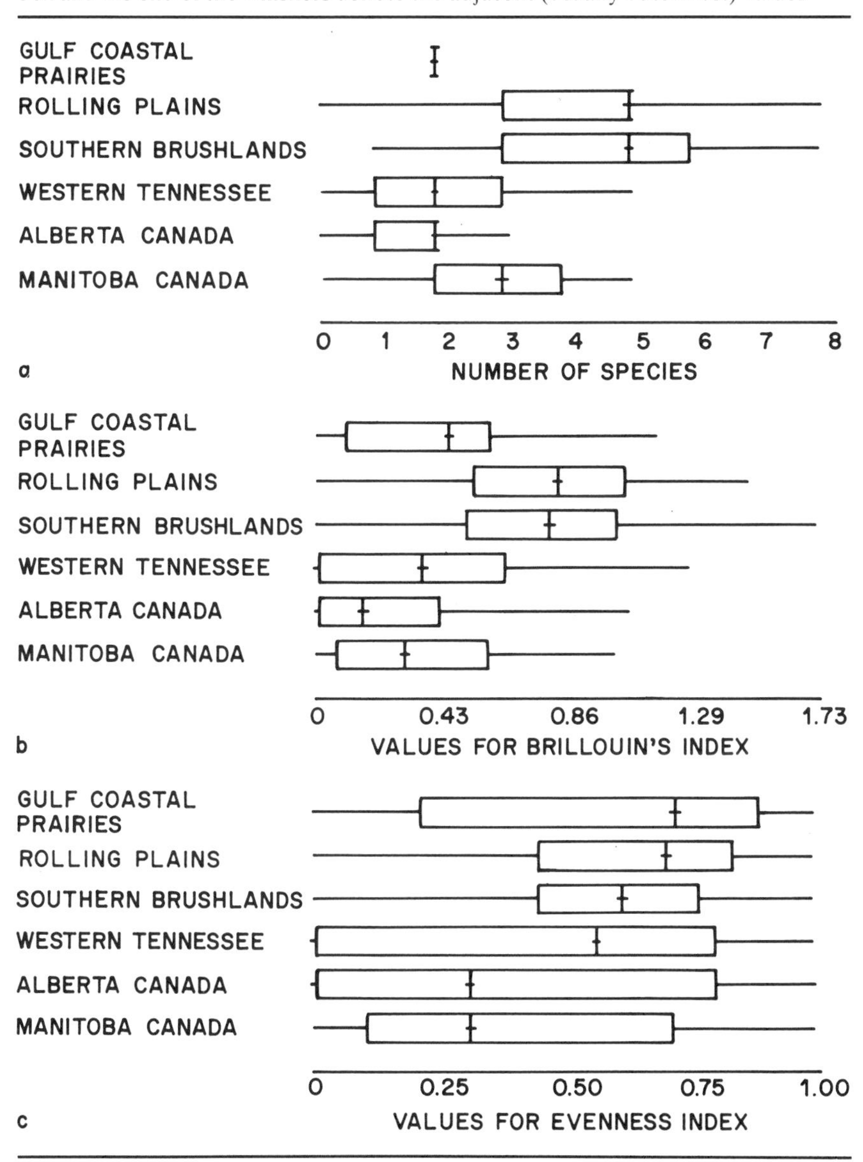

In contrast, spatial variability in helminth communities in coyotes from six isolated localities across North America was pronounced (Table 9.4). There were significant differences in the mean number of species ($H = 310.78$, $p < 0.001$), mean density of all individuals ($H = 236.88$, $p < 0.001$), Brillouin's diversity index ($H = 195.00$, $p < 0.001$) and evenness index ($H = 31.16$, $p < 0.001$). From Fig. 9.1a, it is apparent that the helminth communities from the Rolling Plains and Brushlands of Texas were similar in the median number of species in the helminth communities; those from Alberta, Tennessee and the Gulf Coastal Prairies exhibited a similar pattern. Spatial variation in diversity followed a comparable trend (Figure 9.1b), except that populations in Alberta and Manitoba had similar median diversity values to those in Gulf Coastal Prairies and Tennessee. Based on the evenness index values (Fig. 9.1c), there were similar levels of overdispersion between the two xeric areas of Texas (Rolling Plains and Brushlands); over dispersion was greatest within the two localities of the boreal forest of Alberta and Manitoba.

Comparison of faunal similarity patterns across six localities (Table 9.9) indicated trends of similarity between certain localities; similar to the black bear helminth communities, higher similarity and overlap values were observed between areas with more similar habitat types. Both mean per cent similarity and species overlap link helminth communities of: (a) Tennessee (eastern hardwood deciduous forest) and the Gulf Coastal Prairies (tall grass prairies and remnant deciduous hardwood-cypress bottomland forest), (b) the Gulf Coastal Prairies with Brushlands of southern Texas (originally tall grass prairie), (c) Brushlands with the Rolling Plains (mesquite grassland-juniper breaks), and (d) Manitoba with Alberta (boreal forest).

The coyote population in Tennessee has only recently invaded the territory of the red wolf (*Canis rufus*), following its extinction. As such, this provides an excellent opportunity for examining the establishment of a helminth community within a new locality. The helminth community in Tennessee coyotes is depauperate; there are only five helminth species compared to between nine and 16 species in other localities. All occur in at least two of the Texas locations and all but one has been reported from Canada. Four of the species are among the most frequently occurring and, in the other populations, are the numerical dominants. However, the species with the highest frequency and greatest densities in Tennessee, *Trichuris vulpis*, never occurs or has low frequencies and densities in the other five localities. This is a nematode of dogs from mesic areas in the southeastern United States; I regard it as a secondarily acquired species in this coyote population. It is the dominant species in this helminth community and has successfully exploited a niche usually vacant in coyotes from other localities.

Only three helminth species, *Taenia pisiformis*, *Ancylostoma caninum*

Table 9.9 Trellis diagram of means (± S.E.) for Jaccard's coefficients and per cent similarity from the 752 sample data set of 29 species of gastrointestinal helminths in coyotes from six localities in North America

	Jaccard's coefficients of overlap					
	Western Tennessee	*Gulf Coastal Prairies*	*Southern Texas Brushlands*	*Texas Rolling Plains*	*Manitoba Canada*	*Alberta Canada*
Western Tennessee	33.1 ± 0.2 (26.7 ± 0.2)	33.8 ± 0.1	20.3 ± 0.1	20.1 ± 0.1	12.5 ± 0.1	7.8 ± 0.1
Gulf Coastal Prairies	27.8 ± 0.1	71.3 ± 0.6 (56.3 ± 0.5)	30.7 ± 0.1	23.2 ± 0.1	19.8 ± 0.2	13.5 ± 0.2
Southern Texas Brushlands	15.9 ± 0.1	28.5 ± 0.1	42.3 ± 0.2 (40.8 ± 0.2)	30.2 ± 0.1	14.6 ± 0.1	9.8 ± 0.2
Texas Rolling Plains	14.6 ± 0.1	19.1 ± 0.1	21.9 ± 0.1	44.1 ± 0.2 (35.6 ± 0.2)	16.1 ± 0.1	9.6 ± 0.1
Manitoba Canada	8.2 ± 0.1	16.2 ± 0.3	7.7 ± 0.1	8.6 ± 0.1	32.1 ± 0.1 (16.9 ± 1.2)	21.5 ± 0.4
Alberta Canada	6.8 ± 0.1	13.6 ± 0.3	8.6 ± 0.3	9.3 ± 0.1	15.5 ± 0.5	16.3 ± 0.8 (15.5 ± 0.9)
	Per cent similarity					

and *Toxascaris leonina*, occurred across all six of the geographically diverse habitats represented in this study. However, *A. caninum* is largely replaced by the northern hookworm, *Uncinaria stenocephala*, in Alberta and Manitoba. Thus, *T. pisiformis* and *T. leonina* remain as the potential core species of Hanski (1982) in the component community of helminths of the coyote from different localities across its range. From specific localities, there may be a number of representative core species, but these are not found consistently across all the localities I examined.

From the above examples, it appears that the major factor generating diversity at the component community level in helminth communities of mammals is inextricably linked to the diversity of the habitats utilized by all the populations of a particular host species. This partly is a function of the ability of particular helminth species to survive and reproduce across all areas occupied within the range of the host. It can be argued that invading host populations at the fringe of the host species' range (i.e. white-tailed deer from xeric habitats in Texas, coyotes in Tennessee) have not had sufficient time to acquire an equivalent component level community of helminth species compared to those occupying areas of native host range. However, data from the component communities of black bear from the southeastern United States and coyotes across their native range in western North America tend to refute this (i.e. the small number of species common to all localities and loss, without replacement by equivalent guild members, of species that are locally common in specific localities; see Table 9.4).

Diversity of food habits is also an important factor responsible for generating species richness in helminth communities across different localities but assumes differing levels of importance depending on the host species (i.e. Holmes and Podesta, 1968). Predator–prey transmission of helminths is absent from the abomasal nematode community of white-tailed deer and mostly absent in the black bear helminth community. It is, however, well illustrated in the coyote helminth communities across all localities I examined. There is a shift from several species in the helminth community of coyotes that utilize terrestrial arthropod intermediate hosts in the Brushlands and Rolling Plains, toward fewer such species in the Gulf Coastal Prairies, Tennessee and Manitoba, to none in Alberta. Likewise, there appears to be a northward trend of increasing occurrence of those helminth species that utilize vertebrate intermediate hosts.

The hypotheses

Holmes and Price (1986) developed several hypotheses to facilitate explanation of helminth community structure and dynamics at the component community level. These were discussed by Goater *et al.* (1987), with the addition of the host capture hypothesis. In support of the latter is the largely raccoon helminth community in black bear as well as several examples of possible host capture by species in the component helminth communities of coyotes (i.e. *T. vulpis* and *Pachysentis canicola*). In these hosts, it is tempting to speculate that omnivory may facilitate host capture by certain helminth species (i.e. *Macracanthorhynchus ingenes* in the bear and *P. canicola* in the coyote). Alternatively, these examples could be a result of simple exchange between phylogenetically similar hosts.

There is little evidence to support the cospeciation hypothesis (Holmes and Price, 1986) in the coyote and black bear helminth communities since all the species in the former are generalists shared by other species of the Canidae and there is only one specialist, *B. transfuga*, in the bear. It could be speculated that the several species of small and medium-sized abomasal trichostrongylids of white-tailed deer may have coevolved, but these species are lost without replacement in xeric habitats and there is no equivalent assemblage of species in the closely related mule deer in western North America.

The two corollaries of MacArthur and Wilson's (1967) island biogeography theory, island size and island distance, may exert a strong influence on parasite community structure (Holmes and Price, 1986). It is possible to envision the utility of these concepts if one considers a population from specific locations as an archipelago, with each host individual representing a separate island. Thus, slight differences in the characteristics between individual islands (hosts) can potentially have a dramatic impact on the number of helminth species and the density of those species; for mammalian systems, the well-developed immune response probably strongly contributes to between host differences. In the literature, though, there are few tests of the applicability of these ideas. My data on a small sample ($N = 26$) of juvenile coyotes taken in the autumn of 1983 from the Brushlands in southern Texas are useful in his regard (see Table 9.4 for complete sample results from 1979–80 collections). The same seven common enteric species were present, suggesting some degree of compositional persistence. Included in the sample were 13 male and female juvenile coyotes; no significant differences were noted in either the mean number of parasite species (5.9 ± 0.9 and 5.5 ± 0.1 species in males and females, respectively; $t = 0.84$, $p = 0.42$) or in the total density of all species collectively (393 ± 23 and 340 ± 15 in males and females, respectively; ranked data, $t = 0.52$, $p \pm 0.61$). Comparing the number of helminth species (3 to 7, $\bar{x} = 5.8 \pm 0.1$) with the length of the intestine (1480 to 2110, 1783 ± 8 mm) indicated no significant correlation ($r = 0.01$, $p > 0.05$); also, there was no correlation with body mass (body mass^3; 1.64 ± 0.01 kg; $r = 0.09$, $p > 0.05$). Likewise, the density of all helminth species collectively (83 to 1084, 370 ± 10; data $\log_{10}$ transformed prior to analysis) showed no significant correlation with intestinal length ($r = 0.31$, $p > 0.05$) or with body mass ($r = -0.17$, $p > 0.05$). These results are not supportive of the island size hypothesis in the coyote helminth community. Equivalent data are not available for the deer or bear helminth communities.

The island distance hypothesis, and its corollary, the time hypothesis (Price, 1980), are more difficult to assess than the island size hypothesis, but they can be applied to the helminth communities that I studied. Application of any of these hypotheses to parasite communities was criticized by Kuris *et al.* (1980) because, unlike islands, hosts carry their parasites with them; however, the time hypothesis is applicable to the helminth communities of invading host species at the periphery of their range if the above is taken into consideration. To this hypothesis I would add that not only is sufficient time required to acquire a helminth community through speciation and host capture, or both, but that exchange and host capture is conditional on the availability of suitable helminth species at the compound community level. Consider, for example, the helminth fauna of white-tailed deer in xeric habitats in western Texas. This species is sympatric and undergoing hybridization with mule deer (Stubblefield *et al.*, 1987), but the abomasal helminth communities of both host species are depauperate. It is doubtful that a helminth community similar to that of the white-tailed deer in the southeastern United States will ever develop under the present environmental constraints. Alternatively, the helminth communities of the black bear in the southeastern United States and the coyote in North America seem to present examples of: (a) acquisition of species over short periods of time (i.e. *T. vulpis* in coyotes from western Tennessee; the raccoon helminth community in bears), (b) loss of species from the helminth community (i.e. *A. marcianae* in coyotes from Tennessee), and (c) probably transport of species in the helminth community into new localities (*A. caninum, T. pisiformis*, etc. in coyotes from Tennessee). However, even if two host species are phylogenetically similar, this does not imply that all helminth species will be shared with each respective host species. For example, the common dog cestode, *Dipylidium caninum* does not occur in coyote helminth communities. Probably, this is because the intermediate hosts, domestic fleas of the genus *Ctenocephalides*, rarely occur on wild canids. *Pulex simulans* is the common flea of wild canids. Such ecological barriers may prevent establishment by some helminth species.

Conclusions

At the component community level, the structure and dynamics of helminth communities in mammal populations in specific localities are most affected by changes in host age over seasons. This usually involves frequencies and densities, but not actual occurrence of species. When host populations across different localities are examined, changes in composition and densities are often observed; habitat diversity seems to be the major factor in generating species richness in the helminth community across all localities of a host's range. Aspects of all the hypotheses developed to predict component community level structure and dynamics seem to interface with discussed examples. I believe that host capture is inextricably linked to supplying some of the potential species components of the helminth communities in many mammal hosts; this largely depends on the proximity of the host species to the helminth communities of phylogenetically related host species. Host capture, simple exchange and the helminths carried with an evolving or colonizing host species provide the raw material for its developing helminth community.

9.5 THE COMPOUND COMMUNITY

Our limited information on helminth communities of mammals at the compound community level results from studies on sympatric host species (Tenora, 1967; Holmes and Podesta, 1968; Pence and Eason, 1980; Gray *et al.*, 1978; Davidson and Crow, 1983; Davidson *et al.*, 1985, 1987; Prestwood *et al.*, 1975, 1976; Stubblefield *et al.*, 1987) and from limited studies on circulation of helminths among species of small mammals (Rodenberg and Pence, 1978; Spratt, 1987).

Some inferences may be drawn from these studies. All have indicated exchange of helminth species between hosts, although this is limited to a small number of species in the compound helminth community. Pence and Eason (1980) found six of 28 helminth species shared between coyotes and bobcats (*Lynx rufus*) in the Rolling Plains of Texas (see Table 9.4 for coyote helminth data and Stone and Pence (1978) for bobcat helminth data). Estimates from this locality indicate that the coyote density is about 20 times that of the bobcat, based on equal chance trapping success (W. P. Meinzer, personal observation). In accordance with the exchange hypothesis of Holmes and Price (1986), five of the six shared species have significantly greater frequencies and four of the six have significantly greater densities in coyotes than bobcats (Pence and Eason, 1980).

Unequal exchange of helminth species at the compound community level and its correlation with host densities in mammals is evident in circulation of helminth species among surface-forgaging, burrow-dwelling rodents and lagomorphs from a short grass prairie in the Southern High Plains of Texas (Rodenberg and Pence, 1978). The most common of six host species, the cotton rat (*Sigmodon hispidus*) had the highest frequencies and densities of the three helminth species shared with the ground squirrel (*Citellus tridecemlineatus*). Although 55 and 40% of the ground squirrels and cotton rats, respectively, were infected with the stomach nematode, *Mastiphorus muris*; reproductively mature individuals of the nematode occurred in 38, 6 and 4% of the cotton rats, ground squirrels and house mice (*Mus musculus*), respectively. Equivalent infection values for the larvae were 13, 55 and 16%, respectively. Although the transmission potential for *M. muris* between cotton rats and ground squirrels is high, the chances of larvae surviving and reaching reproductive maturity is low in ground squirrels and house mice.

A second feature of this small compound helminth community was that, of the six helminth species in the component helminth community of ground squirrels, only *Moniliformis clarki* was a specialist in that host. The remainder of the species in the helminth assemblage of ground squirrels had higher frequencies and densities in either cotton rats or larvae from rabbits (see Rodenberg and Pence (1978) for details). The ground squirrel appears to be particularly suited as a reservoir (host capture?) for a number of helminth species from other mammals sharing its habitat.

These examples reinforce the hypothesis that the rate of exchange of helminth species at the compound community level is a density-dependent function of host numbers. The less abundant hosts are infected with more of the helminth species that are adapted to the more abundant hosts than vice versa. Both these cited examples (Rodenberg and Pence, 1978; Pence and Eason, 1980) support Butterworth's (1982) contention that one host's core species (e.g. coyote and cotton rat) may become another's satellite species (e.g. bobcat and ground squirrel, respectively). I found little evidence in my studies, or in the literature on helminth communities of mammals, to support the constant fauna hypothesis or the island biogeography hypothesis as applied to helminth communities of mammals at the compound community level (Holmes and Price, 1986).

9.6 SUMMARY

Compared to some vertebrate groups, little is known about the community ecology of helminths from mammalian hosts. Herein, I have presented some of my own data and have reviewed the literature in an attempt to relate certain aspects of helminth community ecology in this important group of hosts to the level of understanding of the subject in other vertebrates.

Depending on the host species, colonization patterns and other limited data indicate that some mammals may have an interactive community of helminths, although not to the extent of that described in some avian hosts. In other mammalian hosts, there is indirect evidence for an isolationist community of noninteractive helminth species. Presently, there are few critical studies on infracommunity structure of helminth communities of mammals that might corroborate these suggestions.

The most important factors affecting component community structure and pattern within localized host populations appears to be the net action of changing host age structure over seasons, with frequencies and densities changing, but not the actual occurrence of the helminth species in the community. Both occurrence of helminth species and densities of the respective species in the helminth community vary across different localities within the host's range. Habitat variability is regarded as the most important single factor with respect to species richness in the helminth communities of mammals.

There is limited evidence in support of aspects of each of the several hypotheses that have been developed to facilitate an understanding of diversity in helminth species of mammals at the component community level. In the data sets I examined, the strongest arguments can be presented for simple exchange of helminths between phylogenetically similar hosts, transport of helminth species with evolving and/or colonizing hosts and the host capture hypothesis.

Circulation of helminth species among different host species at the compound community level is important. I present some evidence in support of the exchange hypothesis.

In general, it appears that helminth communities in mammals follow many of the assumptions and predictions proposed for other vertebrate groups. However, certain aspects of the helminth communities in mammals appear to be unique; these differences may be tendered by the advanced immunological response in this vertebrate group.

REFERENCES

Brown, J. H. (1984) On the relationship between abundance and distribution of species. *Am.Nat.*, **124**, 255–79.

Bush, A. O. (1980) Faunal Similarity and Infracommunity Structure in the Helminths of Lesser Scaup. PhD thesis, University of Alberta, Edmonton, Alberta.

Butterworth, E. W. (1982) A Study of the Structure and Organization of Intestinal Communities in Ten Species of Waterfowl. Phd thesis, University of Alberta, Edmonton, Alberta.

Corn, J. L. Pence, D. G. and Warren, R. J. (1985) Factors affecting the helminth community structure of adult collard peccaries in Southern Texas. *J. Wildl. Dis.*, **21**, 254–63.

Custer, J. W. and Pence, D. B. (1981) Ecological analyses of the helminth populations of wild canids from the Gulf coastal prairies of Texas and Louisiana. *J. Parasitol.*, **67**, 289–307.

Davidson, W. R. and Crow, C. B. (1983) Parasites, diseases, and health status of sympatric populations of sika deer and white-tailed deer in Maryland and Virginia. *J. Wildl. Dis.*, **19**, 345–8.

Davidson, W. R., Crum, J. M., Blue, J. L. *et al.* (1985) Parasites, diseases, and health status of sympatric populations of fallow deer and white-tailed deer in Kentucky. *J. Wildl. Dis.*, **21**, 153–9.

Davidson, W. R., Blue, J. L., Flynn, L. B. *et al.* (1987) Parasites, diseases and health status of sympatric populations of sambar deer and white-tailed deer in Florida. *J. Wildl. Dis.*, **23**, 267–72.

Fedynich, A. M., Pence, D. B. and Urubek, R. L. (1986) Helminth fauna of beaver from central Texas. *J. Wildl. Dis.*, **22**, 579–82.

Forrester, D. J., Pence, D. B., Bush, A. O. *et al.* (1987) Ecological analysis of the helminths of round-tailed muskrats (*Neofiber alleni* True) in southern Florida. *Can. J. Zool.*, **65**, 2976–9.

Franson, J. W., Jorgenson, R. D., Boggess, E. K. and Greve, J. H. (1978) Gastrointestinal parasitism of Iowa coyotes in relation to age. *J. Parasitol.*, **64**, 303–5.

Goater, C. P. and Bush, A. O. (1988) Intestinal helminth communities in long-billed curlews: the importance of congeneric host specialists. *Holarctic Ecol.*, **11**, 140–5.

Goater, T. M., Esch, G. W. and Bush, A. O. (1987) Helminth parasites of sympatric salamanders: ecological concepts at infracommunity, component and compound community levels. *Am. Midl. Nat.*, **118**, 289–300.

Gray, G. G., Pence, D. B. and Simpson, C. D. (1978) Helminths of sympatric barbary sheep and mule deer in the Texas panhandle. *Proc. Helminthol. Soc. Wash.*, **45**, 139–41.

Hanski, I. (1982) Dynamics of regional distribution: the core and satellite species hypothesis. *Oikos*, **38**, 210–21.

Haukisalmi, V., Henttonen, H. and Tenora, F. (1987) Parasitism by helminths in the grey-sided vole (*Clethrionomys rufocanus*) in northern Finland: influence of density, habitat and sex of host. *J. Wildl. Dis.*, **23**, 233–41.

Hirsch, R. P. and Gier, H. T. (1974) Multi-species infections of intestinal helminths in Kansas coyotes. *J. Parasitol.*, **60**, 650–53.

Hobbs, R. P. (1980) Interspecific interactions among gastrointestinal helminths in pikas of North America. *Am. Midl. Nat.*, **103**, 15–25.

Holmes, J. C. (1973) Site selection for parasitic helminths: interspecific interactions, site segregation and their importance to the development of helminth communities. *Can. J. Zool.*, **51**, 333–47.

Holmes, J. C. and Podesta, R. (1968) The helminths of wolves and coyotes from the forested regions of Alberta. *Can. J. Zool.*, **46**, 1193–204.

Holmes, J. C. and Price, P. W. (1986) Communities of parasites. In *Community Ecology: Pattern and Processes*) (eds, D. J. Anderson and J. Kikkawa), Blackwell Scientific Publications, Oxford, pp. 187–213.

Keith, L. B., Cary, J. R., Yuill, T. M. and Keith, I. M. (1985) Prevalence of helminths in a cyclic snowshoe hare population. *J. Wildl. Dis.*, **21**, 233–53.

Kennedy, C. R., Bush, A. O. And Aho, J. M. (1986) Patterns in helminth communities: why are birds and fish different? *Parasitology*, **93**, 205–15.

Kinsella, J. M. and Pence, D. B. (1987) Description of *Capillaria forresteri* sp. n. (Nematoda; Trichuridae) from the rice rat (*Oryzomys palustrtis*) in Florida, with notes on its ecology and seasonal variation. *Can. J. Zool.*, **65**, 1294–7.

Kisielewska, K. (1970a) Ecological organization of helminth groupings in *Clethrionomys glareolus* (Schreb.) (Rodentia). I. Structure and seasonal dynamics of helminth groupings in a host population in the Bialowieza National Park. *Acta Parasitol. Pol.*, **18**, 121–47.

Kisielewska, K. (1970b) Ecological organization of helminth groupings in *Clethrionomys glareolus* (Schreb.) (Rodentia). V. Some questions concerning helminth groupings in the host individuals. *Acta Parasitol. Pol.*, **18**, 197–208.

Kuris, A. M., Blaustein, A. R. and Alio, J. J. (1980) Hosts as islands. *Am. Nat.*, **116**, 570–86.

Lotz, J. M. and Font, W. F. (1985) Structure of enteric helminth communities in two populations of *Eptesicus fuscus* (Chiroptera). *Can. J. Zool.*, **63**, 2969–78.

MacArthur, R. H. and Wilson, E. O. (1967) *The Theory of Island Biogeography*, Princeton University Press, Princeton.

Pence, D. B. and Dowler, R. C. (1979) Helminth parasitism in the badger, *Taxidea taxus* (Schreber, 1778), from the Western Great Plains. *Proc. Helminthol. Soc. Wash.*, **46**, 254–53.

Pence, D. B. and Eason, S. (1980) Comparison of the helminth faunas of two sympatric top carnivores from the Rolling Plains of Texas. *J. Parasitol.*, **66**, 115–20.

Pence, D. B. and Meinzer, W. P. (1979) Helminth parasitism in the coyote, *Canis latrans*, from the Rolling Plains of Texas. *Intern. J. Parasitol.*, **9**, 339–44.

Pence, D. B., Crum, J. M. and Conti, J. A. (1983) Ecological analyses of helminth populations in the black bear, *Ursus americanus*, from North America. *J. Parasitol.*, **69**, 933–50.

Pence, D. B., Warren, R. J. and Ford, C. R. (1988) Visceral helminth communities of an insular population of feral swine. *J. Wildl. Dis.*, **24**, 105–12.

Pence, D. B. and Windberg, L. A. (1984) Population dynamics across selected habitat variables of the helminth community in coyotes, *Canis latrans*, from south Texas. *J. Parasitol.*, **70**, 7356–746.

Pielou, E. C. (1975) *Ecological Diversity*, Wiley-Interscience, New York and London.

Prestwood, A. K., Kellogg, F. E., Pursglove, S. R. and Hayes, F. A. (1975) Helminth parasites among intermingling insular populations of white-tailed deer, feral cattle, and feral swine. *J. Am. Vet. Med. Assoc.*, **166**, 787–9.

Prestwood, A. K., Pursglove, S. R. and Hayes, F. A. (1976) Parasitism among white-tailed deer and domestic sheep on common range. *J. Wildl. Dis.*, **12**, 380–5.

Price, P. W. (1980) *Evolutionary Biology of Parasites*, Princeton University Press, Princeton.

Rodenberg, G. W. and Pence, D. B. (1978) Circulation of helminth species in a rodent population from the High Plains of Texas. *Occas. Papers Mus. Texas Tech Univer.*, **56**, 1–10.

Samuel, W. M., Ramalingam, S. and Carbyn, L. N. (1978) Helminths in coyotes (*Canis latrans* Say), wolves (*Canis lupus*) and red foxes (*Vulpes vulpes*) of southwestern Manitoba. *Can. J. Zool.*, **56**, 2614–17.

SAS (1985) *SAS User's Guide*, Version 5 Edition, SAS Institute Inc., Cary, NC.

Shoop, W. L. and Corkum, K. C. (1983) Transmammary infection of paratenic and definitive hosts with *Alaria marcianae* (Trematode) Mesocercariae. *J. Parasitol.*, **69**, 731–5.

Spratt, D. M. (1987) Helminth communities in small mammals in southeastern New South Wales. *Intern. J. Parasitol.*, **17**, 197–202.

Stone, J. E. and Pence, D. B. (1978) Ecology of helminth parasitism in the bobcat from West Texas. *J. Parasitol.*, **64**, 295–302.

Stubblefield, S. S., Pence, D. B. and Warren, R. J. (1987) Visceral helminth communities of sympatric mule and white-tailed deer from the Davis Mountains of Texas. *J. Wildl. Dis.*, **23**, 113–20.

Tenora, F. (1967) Ecological study on helminths of small rodents of the Rahacska Dolina Valley. *Acta. Sci. Natur. Brno*, **1**, 161–207.

Van Den Bussche, R. A., Kennedy, M. L. and Wilhelm, W. E. (1987) Helminth parasites of the coyote (*Canis latrans*) in Tennessee. *J. Parasitol.*, **73**, 327–32.

Waid, D. D. and Pence, D. B. (1988) Helminths of mountain lions (*Felis concolor*) from southwestern Texas, with a redescription of *Cylicospirura subequalis* (Molin 1860) Vevers 1922. *Can. J. Zool.*, **66**, 2110–17.

Waid, D. D., Pence, D. B. and Warren, R. J. (1985) Effects of season and physical condition on the gastrointestinal helminth community of white-tailed deer from the Texas Edward's Plateau. *J. Wildl. Dis.*, **21**, 264–73.

Wilkinson, L. (1988) *SYSTAT: The System for Statistics*, SYSTAT, Inc., Evanston.

10

Models for multi-species parasite–host communities

A. P. Dobson

10.1 INTRODUCTION

The last ten years have seen an enormous increase in the understanding of the role that parasites and pathogens play in the regulation of host abundance. Primarily, this progression has been stimulated by the theoretical framework developed to examine the population dynamics of parasite–host relationships by Anderson and May (1978, 1979, 1982) and May and Anderson (1978, 1979). These models have been extended to examine the dynamics and control of a wide range of parasites from schistosomiasis (Crombie and Anderson, 1985) and other parasitic helminths of man (Anderson 1982; Anderson and May 1985), through to the parasitic helminths of domestic livestock (Smith 1984; Grenfell *et al.*, 1987). Although considerable empirical data has accumulated from studies of free-living host–parasite communities, there have been few attempts to interpret this mass of data in terms of the models developed for simple one host, one parasite 'communities'. In part, this enigmatic divergence between empirical data and theoretical understanding may stem from hierarchical differences in the levels of complexity between the majority of empirical data and that of the models (Table 10.1). This chapter attempts to partly redress this balance by showing how the Anderson and May models can be extended to consider the dynamics of more complex parasite–host communities.

This chapter presents preliminary results from the extension of some models originally developed for simple one host, one parasite communities. The models were initially developed to determine those attributes of the

Table 10.1 Matrix of different possible combinations of parasites and hosts addressed in ecological studies from either an empirical or population dynamic perspective

Number of host species	*Number of parasite species* 1	2	Many
1	Field and lab data Range of models	Field and lab data Models	Field data Core and satellite 'null' model
2	Some field and lab data Some models	Some field data |Ø|	Some field data |Ø|
Many	Field data |Ø|	Field data |Ø|	Field data |Ø|

Ø corresponds to the empty set; no models for this type of community.

interaction between parasites and their hosts that are important in determining observed patterns of population dynamics. They are described and derived in detail elsewhere (Anderson and May, 1978; May and Anderson, 1978; Dobson 1985, 1988a, and in prep.). Rather than repeat these derivations, attention will be given to the graphical illustration of different extensions of the models, while presenting only the more transparent results of any algebraic analysis. By extending these models to examine communities that contain more than one species of parasite, it should be possible to determine the features of parasite life cycles that allow different numbers of species to coexist. The models discussed are primarily for macroparasites (the parasitic helminths and arthropods) that live in hosts that are relatively long-lived, and may be considered to have dynamics that operate on a much longer time scale than those of the parasites. It is thus assumed that host population size is relatively constant. The more important consequences of relaxing this assumption are mentioned where pertinent, whereas a full analysis of the properties of fully dynamic models for multi-species macro- and micro-parasite communities are described elsewhere (Dobson, in prep.).

Anderson and May (1978) revisited

The basic Anderson and May (1978) model for parasitic helminths utilizes three equations to describe the dynamics of the definitive host population, the adult parasites and free-living parasite infective stages.

$$\frac{dH}{dt} = (a-b)H - \alpha P \qquad (1)$$

$$\frac{dP}{dt} = \beta HW - (b+u)P - \alpha P\left[\frac{P}{H} - \frac{P^2}{H^2}\left[\frac{k+1}{k}\right]\right] \qquad (2)$$

$$\frac{dW}{dt} = \lambda P - \gamma W - HW. \qquad (3)$$

The parameters of the model are defined in Table 10.2; Fig. 10.1 illustrates the life cycle diagrammatically. If we assume that the dynamics of the free-living stages occur on sufficiently fast a time scale to always be at equilibrium with respect to the other stages of the life cycle ($dW/dt \gg dP/dt$), then we can set $dW/dt = 0$ and collapse the model down to two equations (May and Anderson, 1978). If we further assume that the hosts are long lived relative to the parasites ($u + \alpha > b$), then we can further collapse the model

Table 10.2 Definition of the parameters used in the models for this chapter (essentially these follow the notation adopted in the basic Anderson and May (1978) models). The life cycle of a hypothetical monoxenic parasite is illustrated in Fig. 10.1

Parameter	*Description*
a	Instantaneous birth rate of hosts (/host/unit of time)
b	Instantaneous death rate of host due to natural causes (/host/unit of time)
u	Instantaneous mortality rate of the adult parasite in the definitive host due to natural causes (/host/unit of time)
α	Instantaneous density dependent death rate of the parasites in their definitive host due to parasite induced host mortalities or competition for space and resources (/host/unit of time)
λ	Instantaneous birth rate of parasite transmission stages where birth results in the production of eggs which pass out of the host (/parasite/unit of time)
γ	Instantaneous mortality rate of the free-living eggs (/egg/unit of time)
β	Instantaneous rate of ingestion of the free-living eggs by the definitive host (/host/unit of time)
H_0	Transmission efficiency constant; the ratio of γ/β which varies inversely with the proportion of eggs that are successfully ingested by members of the definitive host population
k	Parameter of the negative binomial distribution which measures the degree of aggregation of the parasite within the definitive hosts (occasionally $(k+1)/k$ has been abbreviated to k').

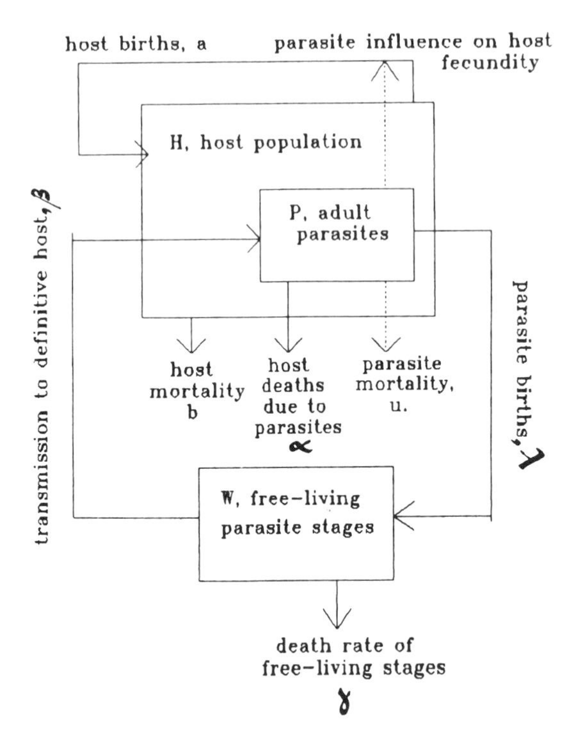

Fig. 10.1 Flow diagram for a typical direct life cycle macroparasite with a free-living infective stage. The adult parasites, P, live in the gut of infected hosts, H, where they produce free-living larvae at a rate, λ, these have a life expectancy of $1/\gamma$, and encounter potential hosts at a rate β. The life expectancy of an adult parasite is 1/u and the parasites are distributed in the host population according to the negative binomial distribution with degree of aggregation, k; the parasites lead to an increase in host mortality at a rate of ∝ per parasite. In this chapter we will ignore the potential effects of the parasite on host reproduction.

down to a single equation which describes the rate of change of parasite numbers in host populations of different constant sizes:

$$\frac{\mathrm{d}P}{\mathrm{d}t} = P\left[\frac{\lambda H}{H+H_0} - (b+u+\alpha) - \alpha \quad \frac{P}{H}\left[\frac{k+1}{k}\right]\right] \tag{4}$$

Two important epidemiological parameters may be derived from this equation: the first is R_0, the basic reproductive rate of the parasite:

$$R_0 = \frac{\beta\,\lambda\,H}{(b+u+\alpha)(\gamma+H)} = \frac{T_1}{M_1 \cdot M_2}. \tag{5}$$

Note that this expression consists of the product of the net birth and transmission rate of the parasite, T_1 ($= \beta\,\lambda\,H$), and the life expectancies of

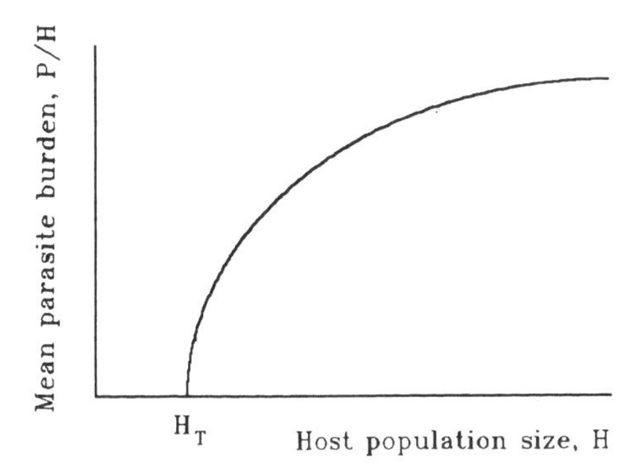

Fig. 10.2 The influence of host population size on mean parasite burden in the basic model. Parasites are only able to establish at a host population density defined by equation 6; the initial slope and asymptote of the line is given by equation 7.

the adult ($1/M_1 = 1/(b+u+\alpha)$) and free-living ($1/M_2 = 1/\gamma+\beta H$) parasite stages. When R_0 equals unity, this expression can be rearranged to give an expression for H_T the threshold number of hosts required to sustain a population of parasites:

$$H_T = \frac{\gamma \cdot M_1}{\beta(\lambda - M_1)}. \tag{6}$$

We may also rearrange equation (4) at $dP/dt = 0$, to obtain an expression for mean parasite burden in host populations of different size.

$$P/H = \left[\frac{\lambda H}{H+H_0} = (b+u+\alpha)\right] \bigg/ \alpha \left[\frac{k+1}{k}\right] = M_1 \cdot (R_0 - 1)/\alpha \cdot k' \tag{7}$$

These expressions may be used to diagrammatically represent the influence of host population density on mean parasite burden (Fig. 10.2). Here it is important to notice that similar expressions can be derived for parasites with more complex life cycles, such as the heteroxenic cestodes, digeneans and acanthocephalans (Anderson, 1978a; Dobson, 1988a), and that these may be included into the above framework in a fairly straightforward way (Table 10.3). It is also important to appreciate that in many circumstances natural selection will operate so as to increase R_0, and thus reduce H_T (Anderson and May, 1982). Any considerations of how populations of different species are assembled at the community level have first to appreciate how natural selection operates at the level of the individual. Brief examination of equation 5 and Table 10.2 suggests that R_0 is related to the various transmission and mortality terms in a very straightforward way, with selection tending to either reduce the mortality rates, M_1, or increase the transmission rates, T_1. Essentially, when considered within the graphical

Table 10.3 Expressions for basic reproductive rates, thresholds for establishment and mean parasite burdens in parasites with different life cycles

Life cycle	Ro	H_T
Monoxenic	$\frac{T_1}{M_1.M_2}$	$\frac{\delta.M_1}{\beta(\lambda - M_1)}$
Heteroxenic: one free-living stage, transmission to definitive host via a predator–prey relationship	$\frac{T_1.T_2}{M_1.M_2.M_3}$	$\frac{M_1M_2M_3}{\alpha\Gamma(T_2 - M_1M_2)}$
Heteroxenic: two free-living stages, transmission to definitive host via a cercaria or free-living larvae	$\frac{T_1.T_2}{M_1.M_2.M_3.M_4}$	$\frac{\delta M_1M_2M_3}{\Omega\beta(T_2 - M_1M_2M_3)}$

Here T_i is the transmission rate between definitive and intermediate hosts, while M_i is the total mortality rate for each stage in the life cycle. In parasites with heteroxenic life cycles Γ is the rate at which definitive hosts prey on intermediate hosts, α is the increased susceptibility to predation of infected intermediate hosts, while Ω is the rate of asexual reproduction of parasites that reproduce in their intermediate host. For a full definition of these expressions see Dobson (1988a).

framework illustrated in Fig. 10.2, both of these actions lead to an increase in both the initial slope and asymptote as well as a shift to the left of the relationship between mean parasite burden and host population size.

However, it is also important to appreciate that the stability of the interaction between a parasite and its host is dependent upon the degree of aggregation of the parasites in the host population (Crofton 1971; Anderson and May, 1978). The mechanisms causing this aggregation include behavioural and genetic differences in susceptibility of different host individuals to infection (Wakelin, 1984; Wassom *et al.*, 1984), and spatial and temporal heterogeneity in the distribution of parasite infective stages (Anderson and Gordon, 1982). Although the relative intensity with which each process contributes to observed levels of aggregation varies between different host–parasite systems, the aggregation of the parasite population into a smaller proportion of the host population always increases the stability of the relationship. Essentially, this is achieved as parasite regulatory mechanisms, such as parasite-induced host mortality or density-dependent reductions in parasite fecundity and survival, effect a higher proportion of the parasite population. Increased aggregation leads to reductions of the slope and asymptote of the line illustrated in Fig. 10.2, but has no effect on the position of H_T, the threshold number of hosts required to sustain an infection.

Dobson (1985) revisited

A second parasite species can be readily incorporated into the models for a simple one host–one parasite community (Dobson, 1985). Here I will assume that the second species requires a larger host population to sustain an endemic parasite population (Fig. 10.3); this could be due to either reduced transmission efficiency or higher mortality, or pathogenicity. Introduction of a second parasite species into a host population may lead to either increases, decreases or have no discernible effect on the population density of the parasite species already established. Both potential competitive interactions and the statistical distributions of the parasites in the two parasites are important in determining the actual outcome.

Initially let us consider the case that is equivalent to 'exploitation competition' in free-living organisms (Miller, 1967). Here we assume that the two parasite species share some common resource provided by the host, but are not antagonistic to each other while jointly exploiting this resource. If the two parasite species are relatively innocuous and tend to be independently aggregated in their distributions, then only a small proportion of hosts are likely to be concomitantly infected with both species. Under these conditions, potential competitive interactions only effect a small proportion of each parasite population and the two species are likely to persist at densities only marginally lower than those that occur in the absence of a potential competitor. More significant interactions occur when

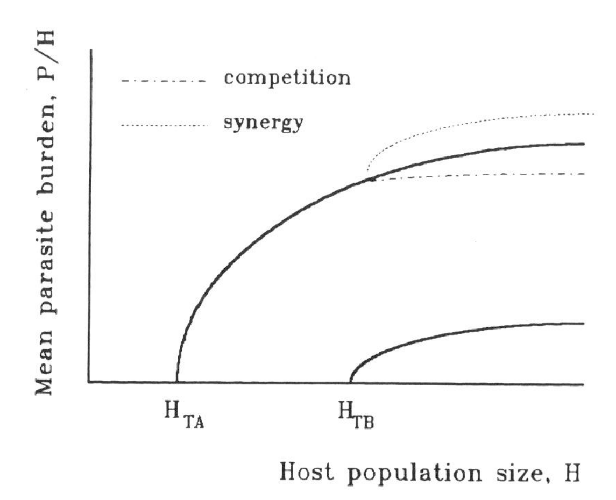

Fig. 10.3 The influence of host population size on the mean parasite burdens of two species of parasite occupying the same host at a range of population densities. For the sake of notational convenience, the parasite with the lowest threshold of establishment is labelled A, that with the higher threshold is labelled B.

either of the two parasites shows more pronounced pathogenicity against its host ($\alpha \gg 0$) and correspondingly exhibits a less aggregated distribution ($k \geq 1$) (Anderson and Gordon, 1982). Increased pathogenicity reduces the life expectancy of infected hosts; this in turn leads to an increase in the mortality rate of potential competitors in concomitantly infected hosts. However, as a lower proportion of hosts are likely to be infected with the more pathogenic parasite and a correspondingly smaller proportion of hosts simultaneously infected with both species, this may reduce the potential for competitive interactions. These effects may be expressed more quantitatively by deriving expressions for the mean parasite burden of each parasite species in terms of their basic reproductive rates, relative pathogenicities and degrees of aggregation:

$$M_A = \frac{k_A}{\alpha_A(k_A+k_B+1)} \cdot \left[(R_{0A}-1)\cdot M_{1A} - k_B\cdot(\alpha_A-\alpha_B)\right] \tag{8}$$

when both species have the same values of λ and H_0 (Dobson, 1985), and

$$M_a = \frac{k_A}{\alpha_A(k_A+k_B+1)} \cdot \left[(R_{0A}-1)\cdot M_{1A}\cdot(k_B+1) - (R_{0A}-1)\cdot M_{1B}\cdot k\right] \tag{9}$$

when λ and H_0 are different for each species. In each case the subscripts denote values for either species A or B. The dynamics of the interaction may

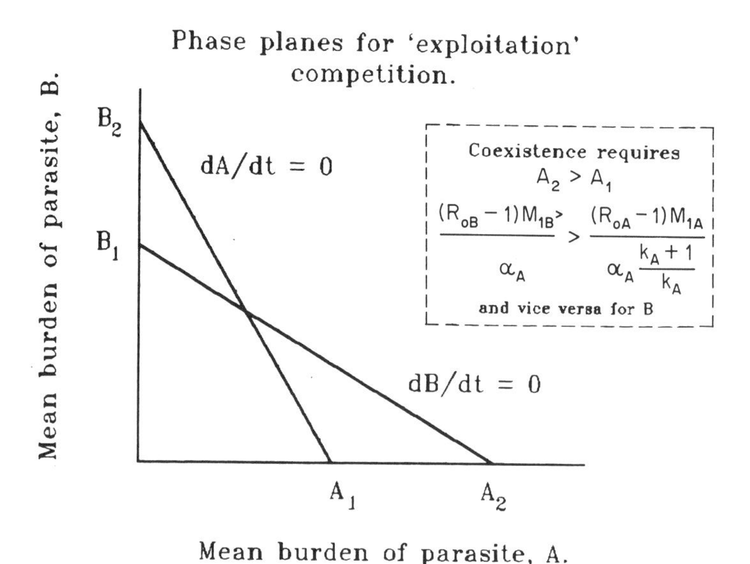

Fig. 10.4 Phase diagram for the simple 'exploitation competition' model of two parasites utilizing the same host. A stable joint equilibrium is only possible when $M_{1B}(R_{0B}-1)/\alpha_A > M_{1A}(R_{0A}-1)/\alpha_A k_A'$ and vice versa.

be most transparently appreciated graphically (Fig. 10.4); here the similarity of the phase diagram to that of the classic Lotka–Gause competiton isoclines give an important clue to the properties of the model. Essentially, persistence of the two parasite species requires them both to be aggregated in their distributions (both $M_i k_i > M$) and for the host population to be larger than the threshold density required to support either of them (both $R_{0i} > 1$). In the case where the host's dynamics are also included into the model, this expression translates into one where the intrinsic growth rate of the hosts has to exceed the summed parasite–induced host mortality rates (Dobson, 1985). An important point to emerge from this analysis is that parasite species that exhibit intermediate levels of parasite-induced host mortality are the most effective at reducing the density of another parasite species as these species produce the most pronounced reductions in host density (Anderson, 1979).

The framework can be extended to examine the more active forms of interference competition (Miller, 1967) that occur between parasite species when they are able to displace each other from preferred sites of attachment (Holmes, 1961), or manipulate the host's immune response so as to reduce

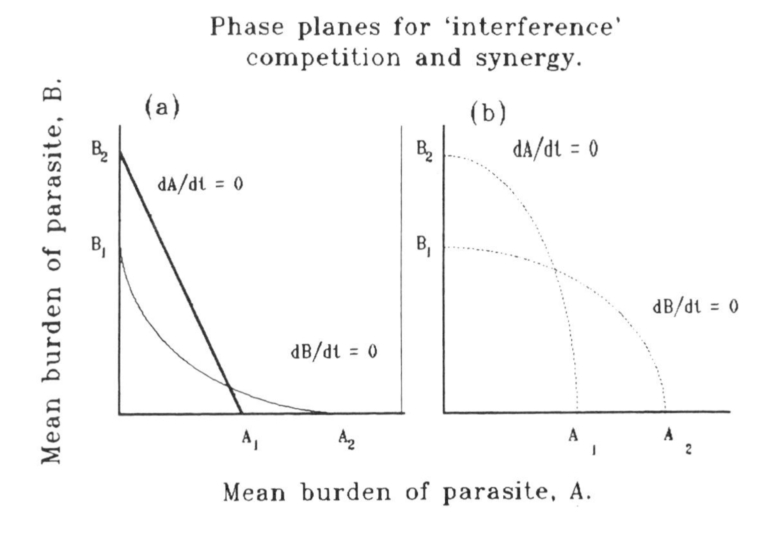

Fig. 10.5 Phase diagrams for the two parasite species competition model when parasites actively interact with each other. (a) 'interference' competition, asymmetrical competition with A actively interfering with B, but *not* vice versa (Dobson 1985); (b) the two dotted lines indicate the shape of the isoclines when the two parasite species synergistically interact with each other.

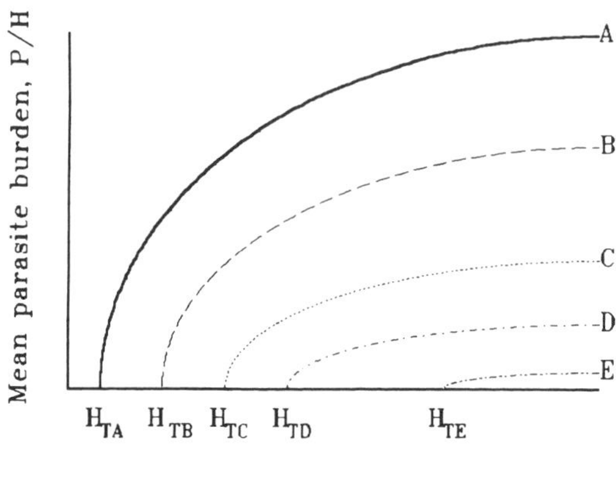

Fig. 10.6 The influence of host population size on the mean parasite burden of five species of parasite occupying the same host species at a range of population densities. Each species with a sequentially higher threshold for establishment is given a sequential letter in the alphabet. Note that here we assume that only simple exploitation competition is operating .

rates of establishment of potential competitors (Schad, 1966). Details of how this type of interaction are included into the model are given in Dobson (1985). The main dynamic consequence can best be perceived by noting that the linear isoclines in the basic phase diagram for exploitation competition (Fig. 10.4), are now likely to be curves which either bend down, if the parasites have a negative effect on each other, or possibly bend upwards, if the parasites have a synergistic effect on each other (Fig. 10.5). The latter may occur if one parasite species impairs or reduces the efficacy of the immune response of the host and thus makes it easier for a second species to become established (Jenkins and Behnke, 1977), or if heteroxenic parasites utilize a common intermediate host as a pathway to infect a common definitive host (Dobson, 1985). Here it is important to appreciate that although interactions between parasites in concomitantly infected hosts may be important in determining the survival and fecundity of the individual parasites in that host, the net effect of interactions between parasite species at the population level is ultimately still dependent upon both parasite pathogenicity and the statistical distribution of each parasite species in the host population. The individual life history tactics of each parasite species are thus as important as any direct interaction between the parasite species in determining the structure of these simple communities.

10.2 MULTI-PARASITE MODELS

The model framework developed may be extended to include further species of parasite (Fig. 10.6). Although this figure is drawn for only five species of parasites, the general trends as host population density increases and more parasite species are added should be apparent. Two important points deserve attention. First, there appears to be a 'core' of species that are present in hosts at a wide range of population densities; at higher densities the community is supplemented by species which require large host populations to support them. Second, the mean burdens of all parasite species increase asymptotically as host population size increases. Thus, those species that establish at lower densities will tend to have higher burdens in large host populations than those species that require these larger populations to establish.

These two patterns are of particular relevance to some of the recent discussions on parasite community structure. Present interest in Hanski's 'core and satellite' hypothesis of community structure in patchy environments (Hanski, 1982) stems from its application to data for parasites of ducks (Bush and Holmes, 1986a, b) and grebes (Stock and Holmes, 1987a, b). These authors define a 'core' species as one which is present in the majority of hosts examined. A more formal definition for such a species would be that its threshold for establishment is a host population of size unity. Although this may initially appear a counter-intuitive definition, it is important to remember that many of the 'core' species in Bush and Holmes's (1986a, b) study were trematodes and cestodes that are highly adapted to ephemeral definitive hosts which only spend a proportion of their life span in the habitat where transmission may occur; such species are likely to have thresholds for establishment that are effectively less than unity (see discussion in Dobson (1988a) and the reviews by Anderson (1988), Shoop (1988) and Mackiewicz (1988)).

It is also important to appreciate that the 'core and satellite' hypothesis was suggested by Hanski (1982) as a 'null model' of community structure. Hanski's analysis suggests that bimodal patterns of distribution (prevalence) and abundance (incidence) are expected in communities where the mean rates of immigration and emigration are less than about a third of the variance in these dispersal rates. The strong correlation between incidence and prevalence which produces this pattern in parasite communities is a direct consequence of the interaction between the fact that species with high transmission abilities have low thresholds for establishment and the (arti)fact that abundant species, because of their low pathogenicity, tend to be more aggregated in their distributions than either rare species or more pathogenic species with higher thresholds for establishment. In parasite communities in host populations of increasing size, cumulative mean

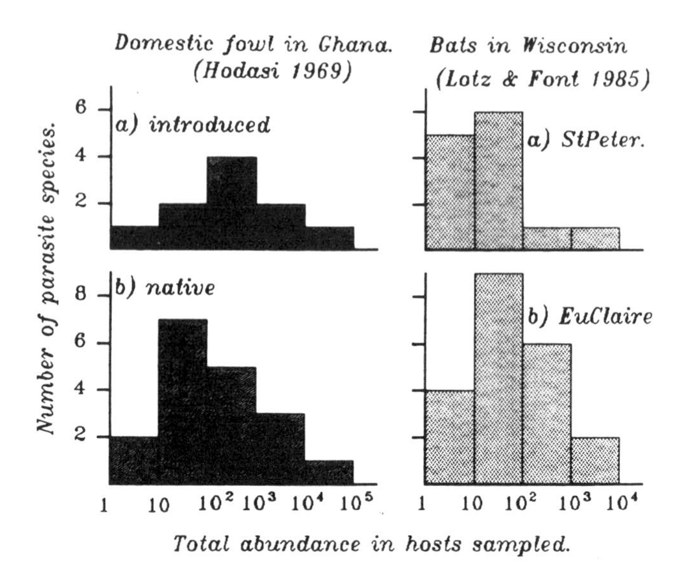

Fig. 10.7 The abundance of different parasite species in a variety of different parasite communities. The data in the upper diagram are for the parasites of domestic and introduced fowls in Ghana (from Hodasi, 1969). The data in the lower figure are for parasites from two bat (*Eptesicus fuscus*) populations (Lotz and Font, 1985). Neither distribution is significantly different from a log normal distribution.

colonization (transmission) rates increase as host population density increases; in contrast, the net variance in colonization rates, although initially high, may decline as rare or pathogenic species are added to the parasite community. This will lead to decreases in the ratio of variance to mean transmission as host population size increases and should give a relationship between prevalence and abundance that changes from unimodal to bimodal as more species are added to the parasite assemblage. Construction of the appropriate null model for the proportions of hosts infected with different parasite species in host populations of different sizes should give either a unimodel or bimodal pattern of incidence and prevalence.

In contrast to the communities studied by Hanski (1982), the species that form the 'core' of parasite communities have many of the attributes of '*r*-selected' species (MacArthur and Wilson, 1967; Pianka, 1970). However, the main thrust of Hanski's argument remains unaltered; the patterns of abundance and distribution observed in species assemblages in patchy habitats are more directly consequences of the different life history attributes of the different species in the community than they are of interactions between those species.

The models illustrated in Fig. 10.6 also provide a means of interpreting the patterns of relative abundance observed in surveys of the parasite species present in any host population. In many parasite communities, these tend to produce the approximately log-normal distribution of species abundance characteristic of many ecological communities (Fig. 10.7). As thresholds for establishment are, by definition, determined by transmission efficiency (eqns 5 and 6), those species which establish at low population densities are those whose abundance increases most rapidly as host population densities increase (Fig. 10.6). In contrast, those species with lower prevalence and incidence are likely to have lower transmission efficiencies, higher thresholds for establishment and slower increases of abundance with increasing host density. In 'species-rich' parasite communities this will ultimately give a distribution of abundances dominated by a few 'core' species and a range of less numerous rare or virulent 'satellite' species.

Kennedy revisited

Kennedy (1978; 1985) has frequently emphasized the role of chance in determining the structure of parasite communities; ecologists are achieving a greater understanding of the role that both historical and stochastic events play in determining the presence and coexistence of species in communities of free-living organisms (Chesson and Case, 1986; Ricklefs, 1987). Historical and stochastic events are likely to have two major effects on the structure of parasite communities. To a first approximation, 'historical' events are likely to be important in determining the species of parasites that are present in any habitat at any time, while the stochastic nature of the parasite transmission process is important in determining the composition of the parasite community in any host individual and the ability of these species to coexist in a population of such hosts.

In the ideal, deterministic world of our model, the addition of species into parasite communities should occur as host population density increases and successive thresholds for establishment are crossed. However, most populations of hosts are fragmented into sub-populations of finite size, each of which exhibits pronounced variation around, but usually below, some carrying capacity. The presence of any parasite species in a sub-population usually requires its introduction from a potential colonizing source and these essentially stochastic events will lead to the absence of different parasite species from each isolated sub-populations of the same host (Kennedy, 1978, 1985; Dobson and May, 1986; Dobson, 1988b). These absences are readily included into the model framework (Fig. 10.8); here it is important to notice that these 'historical' absences may change the relative abundances of the species actually present in the community.

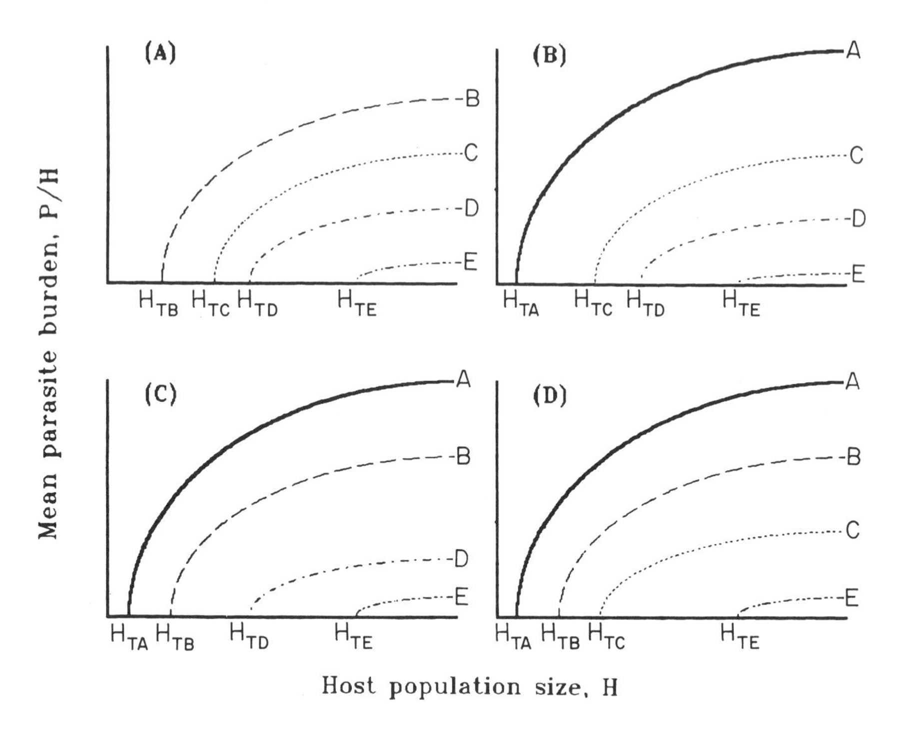

Fig. 10.8 These diagrams illustrates the same community illustrated in Fig. 10.6 with a different species missing in each case. Thus in (a) species A is missing, in (b) species B is missing, etc. Notice that removing an abundant species, such as A or B, has more pronounced an effect than removing a less common species, such as C or D.

A second level of stochasticity is encountered when we examine the number of parasite species harboured by each host. In the majority of studies for which data are available, this distribution is invariably Poisson (Fig. 10.9), the distribution we would expect if each individual host's probability of infection by any parasite species is essentially random (Macallum, personal communication). This point emphasizes an important, and perhaps under-perceived, paradox of parasite population dynamics; the stochastic nature of the infection process produces a random distribution of parasite species per host, essentially because any host has a relatively constant probability of being infected by each of the parasite species present in a finite pool of species. In contrast, the essentially random nature of the infection process for each individual parasite species is compounded by variations in the susceptibility of each host individual to infection; this produces the aggregated distributions that are characteristic of most parasite species (Anderson *et al.*, 1978; Anderson and Gordon, 1982). This has two further important consequences on the structure of parasite communities.

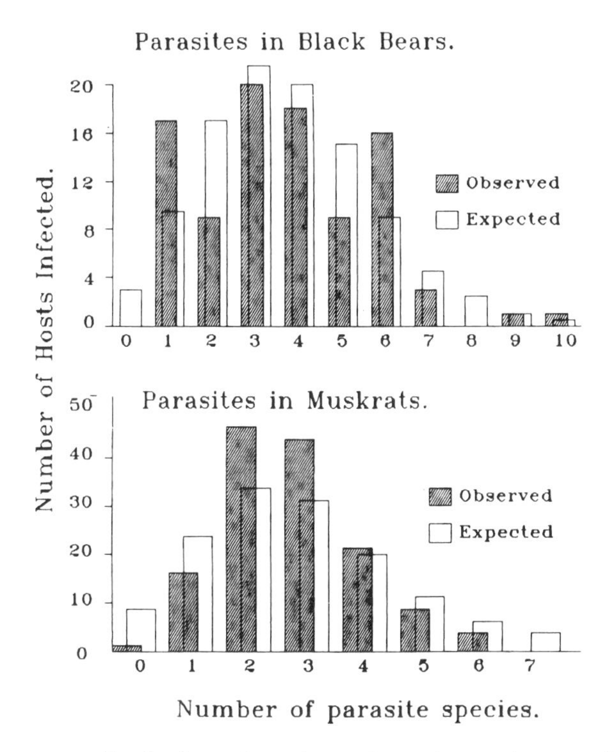

Fig. 10.9 Frequency distributions of numbers of parasite species recorded in each host for a number of different host–parasite communities. Neither observed distribution is statistically significant from the Poisson distribution. The data for black bears (*Ursus americanus*) are from Pence *et al.* (1982), from those muskrats (*Ondatra zibethica*) are from McKenzie and Welch (1979).

First, although infection by any individual parasite infective stage may be random, many parasite species share common routes of entry into their hosts (e.g. many cestodes and acanthocephalans utilize the same species of arthropods as intermediate hosts). If individual hosts selectively feed on these intermediate host species, they will increase their probability of being simultaneously infected with both parasite species. Second, many (but not all) of the properties of parasite communities correspond to those required for 'lottery models' for coexistence of species living in a patchy environment (Chesson and Warner, 1981; Chesson, 1985; Warner and Chesson, 1985). These models suggest that stochastic variation, in the form of environmentally induced variations in rates of fecundity and mortality, may be sufficient to allow the coexistence of age-structured populations of species which would otherwise be unable to coexist in the absence of variation in these demographic rates.

Failure to appreciate these points may in the past have led to premature rejection of hypotheses that seek to determine the relative importance of competition in determining parasite community structure. In particular, studies that test for the presence or absence of competition by merely comparing expected and observed numbers of hosts concomitantly infected with different combinations of parasite species, will significantly underestimate the importance of inter-parasite interactions in determining community structure. To determine if two parasites species are having an effect on each other, we not only have to examine mean parasite burden in single and mixed infections, but also the survival and fecundity of each parasite species in both classes of infection.

British lakes and Ugandan monkeys revisited

The models also suggest that we should expect to see a simple relationship between host population size and parasite diversity. This has been observed in Kennedy's (1978b) study of the parasite communities of brown trout (*Salmo trutta*) from lakes of different sizes and Freeland's (1979) study of the intestinal protozoan fauna of monkeys and baboons living in social groups of different sizes (Fig. 10.10). Both these sets of data indicate a significant relationship between host population size (or size of host habitat) and the number of parasite species found in the host population. This relationship between the size of a host population and the number of parasite species it is able to support is of particular relevance to the current debate about parasites and sexual selection (Hamilton and Zuk, 1982). These authors and Read (1987) have illustrated a significant relationship between parasite species diversity and the degree of sexual dimorphism exhibited in the host population. If the level of sexual dimorphism in the host population varies with the size of the social group in which members of the species characteristically live, then a positive relationship between sexual dimorphism and parasite diversity is an a priori expectation of the above model.

Parasite communities in Caribbean *Anolis*

The models may also be used as tools to fully dissect the structure of data collected from parasite communities. The parasites of terrestrial hosts that live on islands are ideal for this type of study, as different islands are likely to support host populations of different sizes. Similar arguments apply to the populations of aquatic hosts that live in rivers or lakes of different sizes or terrestrial hosts that live in fragmented habitats. The data presented in Fig. 10.11 are for the parasites of two 'series' of lizards from islands of the

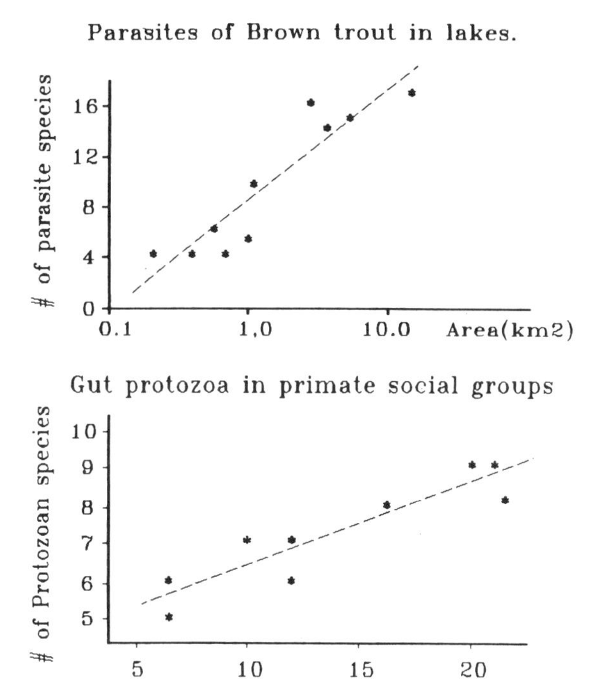

Fig. 10.10 (a) The number of parasite species recorded from brown trout sampled in British lakes (data from Kennedy, 1978). (b) The number of species of intestinal protozoa recorded from mangabey (*Cerocebus albigena* and *Cercopithecus mitis*) groups of different sizes (after Freeland, 1979).

Lesser Antilles in the Caribbean (Dobson *et al.*, in press). Patterns similar to those exhibited in Fig. 10.6 and 10.8 are apparent in these data, with one parasite species (*Thelandros cubensis*) present in lizards on all islands, while other parasites appear in the community (*Centrorhynchus* sp.) as successive transmission thresholds are crossed. However, historical events may have caused the extinction of parasite species on some islands, while other species may not yet have been able to colonize hosts on other islands (*Skrjabinodon* spp. on St Maarten).

It is important to note that the sequence of the thresholds for establishment has been determined entirely by the presence or absence of the parasite and has not been derived from any relationship between island size and host population density. Indeed, there is no simple relationship between island area and numbers of parasite species. Because the survival and fecundity of both the parasites and the hosts, as well as transmission rates, are dependent upon factors such as humidity, temperature and vegetation,

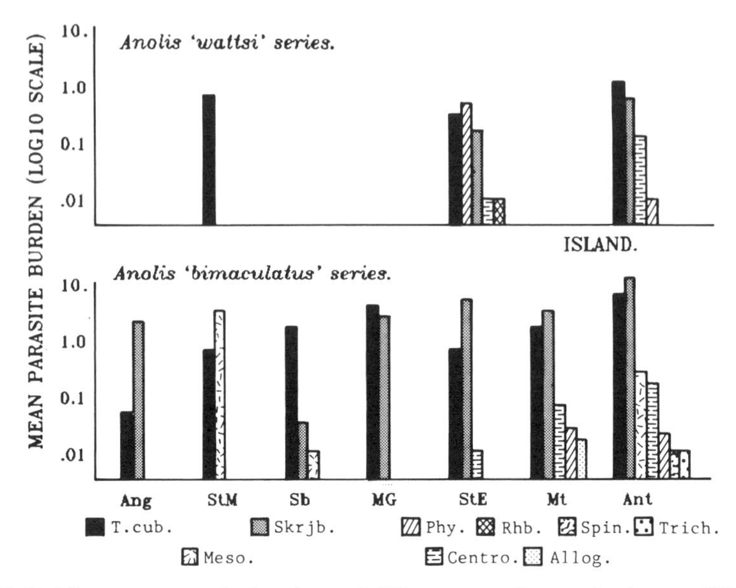

Fig. 10.11 The mean parasite burdens of different parasite species in two different species of *Anolis* lizards sampled on different islands in the Caribbean Lesser Antilles. The upper figure shows data for lizards of the *wattsi* series (only present on three islands), the lower figure shows data for the *bimaculatus* series (data from Dobson *et al.*, in press).

The seven islands sampled (and their sizes) are abbreviated as Ang – Anguilla (91 km^2); StM – St Maarten (88 km^2); Sb – Saba (13 km^2); MG – Marie Galante (155 km^2); StE – St Eustacius (31 km^2); Mt – Montserrat (85 km^2); Ant – Antigua (280 km^2). Different shadings are used for each parasite species, abbreviations used are Skrjb. – *Skrjabinodon* spp.; T.cub. – *Thelandros cubensis*; Meso. – *Mesocoelium* sp.; Phy – *Physalopteridae* gen.sp.; Rhb. – *Rhabdias* sp.; Centro – *Centrohynchus* sp.; Spin. – *Spinicaude amarili*; Allog. – *Alloglyptus crenshawi*; Trich. – *Trichospirura* sp.

variations in the areas of mesic and xeric habitat on each island are probably more important than island area *per se* in determining both the total size of host populations and parasite transmission rates (Dobson *et al.*, 1989). Only if the islands were very homogeneous in their vegetation would we expect to see any simple relationship between parasite community structure and island area.

Lewis revisited

A similar approach can be adopted to the data collated by Lewis (1968a, b) for small mammal parasites in the British Isles (Fig. 10.12). The British small mammal data show similar trends to those observed in the lizard data,

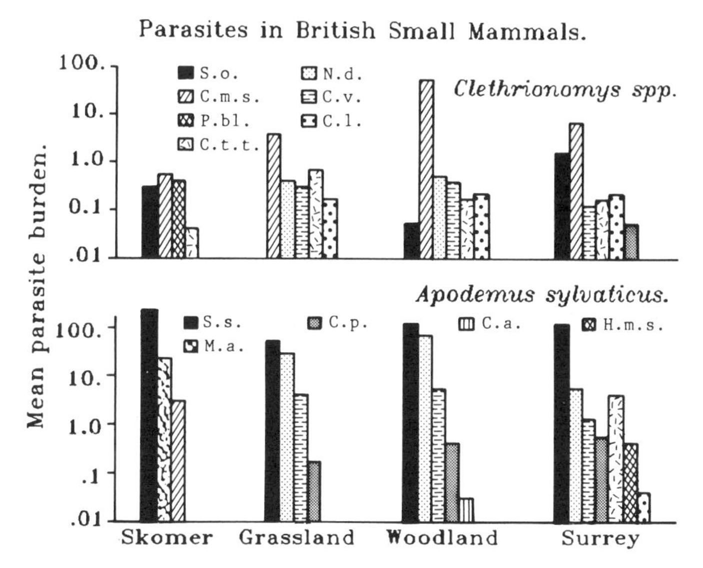

Fig. 10.12 The mean parasite burdens of two small mammal species (top – *Clethrionomys glareolus* and *C. skomerensis*; bottom – *Apodemus sylvaticus*) sampled as adults at four different sites in the British Isles: Skomer island, rough grassland and woodland in Aberystwyth and woodland in Surrey (data from Lewis, 1968a, b; Lewis and Twigg, 1972).

The abbreviations used are (Nematoda – vertical stripes) S.o. – *Syphacia obvelata*; S.s. – *S. stroma*; C.m.s. – *Capillaria muris sylvatica*; N.d. – *Nematospirodesdubius*; (Cestoda – diagonal stripes) P.bl. – *Paranoplocephala blanchardi*; C.t.t. – *Cysticercus taeniae-taeniaeformis*; C.l. – *Catenotaenia lobata*; C.p. – *C. pusilla*; H.m.s. – *Hymenolepis muris sylvatici*; (Trematoda – squares and dots) M.a. – *Maritrema apodemicum*; C.v. – *Corrigia vitta*; (Acanthocephala – solid) C.a. – *Centrorhynchus aluconis*.

with several species such as the relatively pathogenic *Nematospiroides dubius* being absent from the island population on Skomer. These data contain two important contrasts to the data from the *Anolis* lizards. First, the small mammals contain a much higher proportion of heteroxenic parasites than the Caribbean lizards. The presence and abundance of these species are dependent upon the abundance of more than one host species and we would thus not expect to see such simple relationships for the abundance of these parasites as we see for the monoxenic parasites which dominate the lizard parasite community. Second, only one of the mammal samples was from an island community where there were significant differences in host density from the populations sampled on the mainland. The higher host densities on the island lead to increased transmission rates of the parasites and this is reflected in the high burdens of *Syphacia stroma* in the Skomer sample.

Similar trends are apparent in the prevalence data collected by Pence *et al.* (1982) for parasite communities of black bears in the United States and by Holmes and Podesta (1968) for the parasites of wolves and coyotes. Perhaps the major shortcoming of the application of the model to these data, is the presence of a number of alternative host species for the parasites of these mammals. These additional species may increase the rates of transmission of some parasites, while other species may reduce the prevalence of others by acting as 'sinks' for the transmission stages. However, a coarse appreciation of the influence of alternate host species may be examined by a further extension of the basic model.

10.3 MULTI-PARASITE, TWO HOST MODELS

The models discussed so far can be further extended to examine some properties of communities that consist of a number of different parasite species and two different species of hosts (Dobson, in prep.). As expressions can be derived for R_0 and H_T for each parasite species in each host, it is possible to produce a two-dimensional phase diagram based on the thresholds for establishment of each parasite in both host species (Fig. 10.12). Here it is important to appreciate that the dynamics of the same parasite in different host species are likely to vary due to differences in the physiological and behavioural attributes of the host, due to different lengths

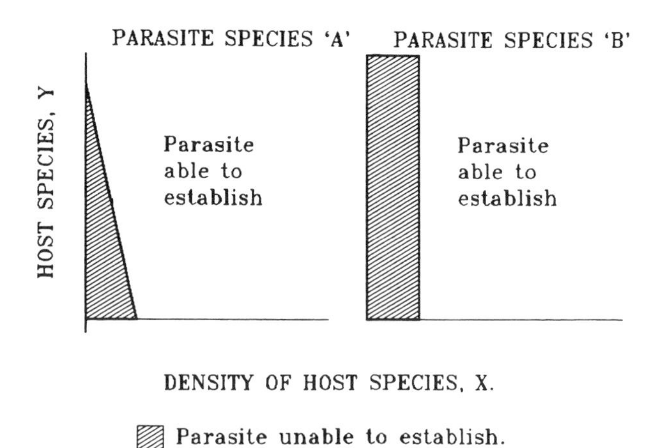

Fig. 10.13 Thresholds for establishment of parasites in two host communities: (a) 'substitutable' hosts, parasite species A is well adapted to host X, but not to host Y; (b) 'non-substitutable' hosts, parasite species A is able to establish in host species X, but only appears incidentally in species Y.

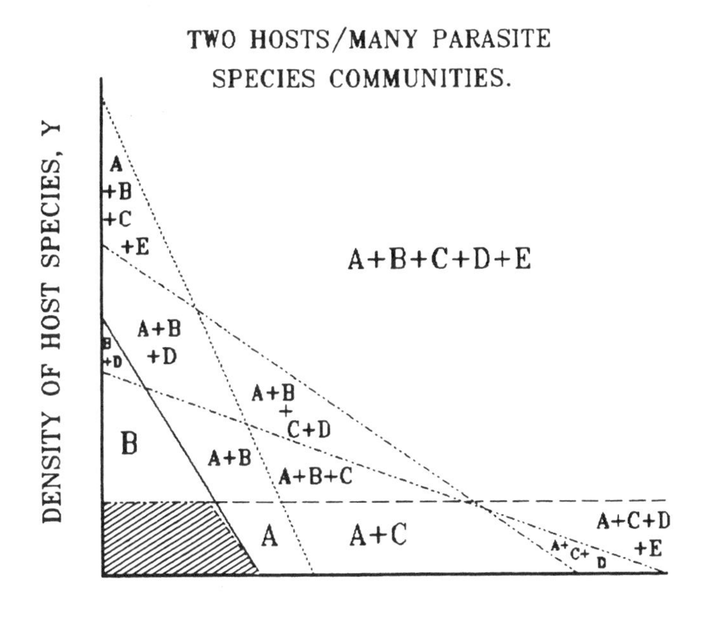

Fig. 10.14 Parasite community structure in a simple two-host community at a range of different population densities. The lines give the threholds for establishment of each of the five parasite species, A, B, C, D and E; the letters give the different parasite species able to establish in host communities of different composition.

of association in evolutionary time of the parasites with each host species, and due to the different social organizations of the host species leading to different rates of transmission between conspecifics. Thus, thresholds for establishment in these 'two host' models are likely to vary from the case where the different host species are essentially substitutable (Fig. 10.12a), through to cases where the parasite is unable to develop in a host species, but may incidentally establish in this host due to its presence in another species of host (Fig. 10.12b).

The composition of the parasite assemblage in host 'communities' of different relative abundancies may then be ascertained by sketching the sequential thresholds for establishment of each parasite species onto the phase plane of densities for each host species. Boundaries may then be delimited for the expected parasite community at each combination of host densities (Fig. 10.13). The most important point to emerge here is that parasite community structure is highly dependent upon the composition of the community of potential hosts. In particular, when densities of either host species are low, small changes in density can give rise to large changes in the structure of the parasite assemblage these hosts can potentially support.

It is also important to note that Fig. 10.13 is the 'simplest case' analysis; it neither includes interactions between the parasites nor the host species (which may be potential competitors, or linked in some predator–prey relationship); nor does it consider that the parasites may themselves effect, or even mediate, interactions between host species. All of these more complex interactions may be included into the model framework and their effects on the dynamics of the host and parasite assemblage examined (Fig. 10.14). The results of these exercises will be discussed further in Dobson (in prep.). Here we will justify the utility of this approach by illustrating how these models may be used to interpret one of the best studied long-term sets of data for a multi-parasite, two host community.

Slapton Ley revisited

Figure 10.15 coarsely illustrates the possible relative positions of the thresholds for establishment of the various parasite species that have been found in roach (*Rutilus rutilus*) and perch (*Perca fluviatilis*) in Slapton Ley by Kennedy in his studies of this community (Kennedy, 1975, 1981a, b, 1987; Kennedy and Burrough, 1981). As there have been significant changes in the abundance of roach in the course of this study it may be possible to analyse the Slapton Ley data in terms of models discussed in this chapter.

10.4 DISCUSSION AND CONCLUSIONS

The study of parasite–host communities is an important area of community ecology that deserves to receive more attention in a wider arena. Although considerable progress has been made in recent years by applying concepts and tools originally developed to examine the community structure of free-living organisms to parasite data, a full consideration of the dynamics and structure of parasite communities is likely to require a better understanding of the factors that determine the lifetime reproductive success of different parasite species and their interactions with their hosts. Thus, an understanding of the population dynamics and coevolution of simple, one host, one parasite systems is fundamental to the understanding of the structure of more complex parasite communities. This should be particularly apparent when it is appreciated that the models for these most basic systems can be used as building blocks to construct models which explore the properties of multi-species, parasite–host communities.

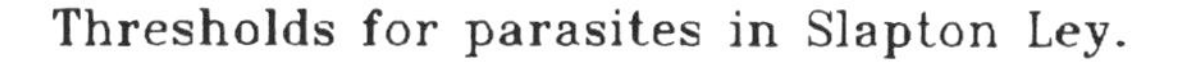

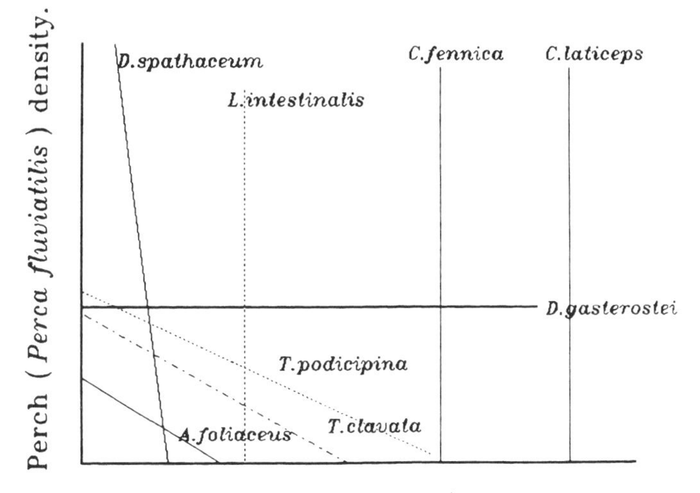

Fig. 10.15 Preliminary analysis of the parasite community in perch and roach at Slapton Ley. The position of the thresholds is mainly conjectural, but is based on the data provided by Kennedy (1975, 1981a, b, 1987; Kennedy and Burrough, 1981). The solid 'thresholds' are for parasite species which were present when the long-term study began, the broken isoclines are for species that have established during the course of the study. The population of roach in the lake increased to a peak in 1975, then declined to much lower numbers.

The above analysis suggests that the individual life history tactics of each parasite species are ultimately more important than any direct interaction between the parasite species in determining the larger scale patterns of distribution and abundance observed in surveys of parasite community structure. The significant differences in the diversity of parasite communities of birds, fish and mammals noted by Kennedy *et al.* (1986) may essentially reflect differences in the life histories of the parasite species that exploit these different classes of hosts (Esch *et al.* 1988). Further progress in our understanding of the factors which determine the constituent members of any community will require further studies of the factors which determine the life history strategies of different parasite species.

However, interactions between parasite species should not be dismissed as determinants of community structure; different parasite species may have pronounced effects on each other's densities through their potential to modify host survival and fecundity (exploitation competition) and through direct antagonism between each other when present in concomitantly infected hosts (interference competition). The stochastic nature of parasite transmission and the fact that infection processes tend to produce aggregated distributions of each parasite species in the host population, require that attempts to determine whether parasite species are 'competing' must go further than to merely test for the absence of hosts with concurrent infections. To determine if two parasites species are having an effect on each other, we not only have to examine mean parasite burden in single and mixed infections, but also must attempt to calibrate the survival and fecundity of each parasite species in both classes of infection. Similarly, we need to determine whether the survival of concomitantly infected hosts is less than that of hosts infected with comparable burdens of only a single species.

Much of the above work suggests that a fuller understanding of parasite communities requires further experimental work and a fuller understanding of the co-evolution of parasite–host relationships. In particular, as pathogenic and 'rare' species require large host populations to sustain them (eqn. 6), they may not be present when 'natural' experiments are monitored in small populations in isolated habitats; attempts should therefore be made to examine the structure of parasite assemblages in host populations at a range of population densities. Manipulative studies of terrestrial hosts on islands, or aquatic hosts in ponds, lakes and rivers, would therefore seem to be a high priority in the development of our knowledge of the forces that structure parasite communities.

ACKNOWLEDGEMENTS

None of the above work could have been undertaken without the encouragement and stimulation provided by Roy Anderson, Bryan Grenfell, Clive Kennedy, Annarie Lyles, Bob May, Steve Pacala, Jonathan Roughgarden, Marilyn Scott and Philip Whitfield. I am most grateful to them and to Robin Carper, Ed Caswell, and Annarie Lyles for comments on this manuscript. The *Anolis* work was supported by NSF grant #BSR-83-16949. The work was written while I was a guest investigator at the Woods Hole Oceanographic Institute and I am most grateful to WHOI and to Hal Caswell for making this possible.

REFERENCES

Anderson, R. C. (1988) Nematode transmission patterns. *J. Parasitol.*, **74**, 30–45.

Anderson, R. M. (1978a) The regulation of host population growth by parasitic species. *Parasitology*, **76**, 119–57.

Anderson, R. M. (1978b) Parasite pathogenicity and the depression of host population equilibria. *Nature, Lond.*, **279**, 150–52.

Anderson, R. C. (1979) The influence of parasitic infection on the dynamics of host population growth. In *Population Dynamics* (eds, R. M. Anderson, B. D. Turner and L. R. Taylor) Blackwell Scientific Publishers, Oxford, pp. 245–81.

Anderson, R. M. (1982) The population dynamics and control of hookworm and roundworm infections. In *The Population Dynamics of Infectious Diseases: Theory and Applications* (ed. R. M. Anderson) London, Chapman and Hall, pp. 67–108.

Anderson, R. M. and Gordon, D. M. (1982) Processes influencing the distribution of parasite numbers within host populations with special emphasis on parasite-induced host mortalities. *Parasitology*, **85**, 373–98.

Anderson, R. M. and May, R. M. (1978) Regulation and stability of host–parasite interactions. I. Regulatory processes. *J. Anim. Ecol.*, **47**, 219–47.

Anderson, R. M. and May, R. M. (1979) Population biology of infectious diseases: Part I. *Nature, Lond.*, **280**, 361–7.

Anderson, R. M. and May, R. M. (1982) Coevolution of parasites and hosts. *Parasitology*, **85**, 411–26.

Anderson, R. M. and May, R. M. (1985) Helminth infections of humans: mathematical models, population dynamics, and control. *Adv. Parasitol.*, **24**, 1–101.

Anderson, R. M., Whitfield, P. J. and Dobson, A. P. (1978) Experimental studies of infection dynamics: infection of the definitive host by the cercariae of *Transversotrema patialense*. *Parasitology*, **77**, 189–200.

Bush, A. O. and Holmes, J. C. (1986a) Intestinal helminths of lesser scaup ducks: patterns of association. *Can. J. Zool.*, **64**, 132–41.

Bush, A. O. and Holmes, J. C. (1986b) Intestinal helminths of lesser scaup ducks: an interactive community. *Can. J. Zool.*, **64**, 142–52.

Chesson, P. L. (1985) Coexistence of competitors in spatially and temporally varying environments: a look at the combined effects of different sorts of variability. *Theoret. Popul. Biol.*, **28**, 263–87.

Chesson, P. L. and Case, T. J. (1986) Nonequilibrium community theories: chance, variability, history and coexistence. In *Community Ecology* (eds, J. Diamond and T. J. Case), Harper and Row, New York, pp. 229–39.

Chesson, P. L. and Warner, R. R. (1981) Environmental variability promotes coexistence in lottery competition systems. *Am. Nat.*, **117**, 923–43.

Crofton, H. D. (1971) A model of host–parasite relationships. *Parasitology*, **63**, 343–64.

Crombie, J. and Anderson, R. M. (1985) Population dynamics of *Schistosoma mansoni* in mice repeatedly exposed to infection. *Nature, Lond.*, **315**, 491–3.

Dobson, A. P. (1985) The population dynamics of competition between parasites. *Parasitology*, **91**, 317–47.

Dobson, A. P. (1988a) The population biology of parasite-induced changes in host behavior. *Quart. Rev. Biol.*, **63**, 139–65.

Dobson, A. P. (1988b) Restoring island ecosystems: the potential of parasites to control introduced mammals. *Conserv. Biol.*, **2**, 31–9.

Dobson, A. P. (1989) *The Dynamics and Structure of Parasite Host Communities* (in prep.).
Dobson, A. P. and May, R. M. (1986) Patterns of invasions by pathogens and parasites. In *Ecology of Biological Invasions of the United States and Hawaii* (eds, H. A. Mooney and J. A. Drake), Springer Verlag, New York, pp. 58–76.
Dobson, A. P., Pacala, S. W., Roughgarden, J. D. and Harris, E. A. (1989) The parasites of *Anolis* lizards in the Northern Lesser Antilles. *Oecologia*, (in press).
Esch, G. W., Kennedy, C. R., Bush, A. O. and Aho, J. M. (1988) Patterns of helminth communities in freshwater fish in Great Britain: alternative strategies for colonization. *Parasitology*, **96**, 519–32.
Freeland, W. J. (1979) Primate social groups as biological islands. *Ecology*, **60**, 719–28.
Grenfell, B. T., Smith, G. and Anderson, R. M. (1987) A mathematical model of the population biology of *Ostertagia ostertagi* in calves and yearlings. *Parasitology*, **95**, 384–406.
Hodasi, J. K. M. (1969) Comparative studies on the helminth fauna of native and introduced domestic fowls in Ghana. *J. Helminthol.*, **43**, 35–52.
Holmes, J. C. (1961) Effects of concurrent infections on *Hymenolepis diminuta* (Cestoda) and *Moniliformis dubius* (Acanthocephala). 1. General effects and comparison with crowding. *J. Parasitol.*, **47**, 209–16.
Hamilton, W. D. and Zuk, M. (1982) Heritable true fitness and bright birds: a role for parasites. *Science*, **218**, 384–7.
Hanski, I. (1982) Dynamics of regional distribution: the core and satellite hypothesis. *Oikos*, **38**, 210–21.
Holmes, J. C. and Podesta, R. (1968) The helminths of wolves and coyotes from the forested regions of Alberta. *Can. J. Zool.*, **46**, 1193–204.
Jenkins, S. N. and Behnke, J. M. (1977) Impairment of primary expulsion of *Trichuris muris* in mice concurrently infected with *Nematospiroides dubius*. *Parasitology*, **75**, 71–8.
Kennedy, C. R. (1975) The natural history of Slapton Ley Nature Reserve. VIII. The parasites of fish with special reference to their use as a source of information about the aquatic community. *Field Stud.*, **4**, 177–89.
Kennedy, C. R. (1978a) The parasite fauna of resident char *Salvelinus alpinus* from Arctic Islands, with special reference to Bear island. *J. Fish Biol.*, **13**, 457–66.
Kennedy, C. R. (1978b) An analysis of the metazoan parasitocoenoses of brown trout *Salmo trutta* from British lakes. *J. Fish Biol.*, **13**, 255–63.
Kennedy, C. R. (1981a) The establishment and population biology of the eye-fluke *Tylodelphys podicipina* (Digenea: Diplostomatidae) in perch. *Parasitology*, **82**, 245–55.
Kennedy, C. R. (1981b) Long-term studies on the population biology of two species of eyefluke, *Diplostomum gasterostei* and *Tylodelphys clavata* (Digenea: Diplostomatidae), concurrently infecting the eyes of the perch, *Perca fluviatilis*. *J. Fish Biol.*, **19**, 221–36.
Kennedy, C. R. (1985) Interactions of fish and parasite populations: to perpetuate or pioneer? In *Ecology and Genetics of Host–Parasite Interactions* (eds, D. Rollinson and R. M. Anderson), Academic Press, London, pp. 1–20.

Kennedy, C.R. (1987) Long-term stability in the population levels of the eyefluke *Tylodelphys podicipina* (Digenea: Diplostomatidae) in perch. *J. Fish Biol.*, **31**, 571–81.

Kennedy, C. R. and Burrough, R. J. (1981) The establishment and subsequent history of a population of *Ligula intestinalis* in roach *Rutilus rutilus* (L.). *J. Fish. Biol.*, **19**, 105–26.

Kennedy, C. R., Bush, A. O. and Aho, J. M. (1986) Patterns in helminth communities: Why are birds and fish different? *Parasitology*, **93**, 205–15.

Kennedy, C. R., Laffoley, D. d'A., Bishop, G. *et al.* (1986) Communities of parasites of freshwater fish of Jersey, Channel Islands. *J. Fish Biol.*, **29**, 215–26.

Lewis J. W. (1968a) Studies on the helminth parasites of the long-tailed field mouse, *Apodemus sylvaticus sylvaticus* from Wales. *J. Zool., Lond.*, **154**, 287–312.

Lewis, J. W. (1968b) Studies on the helminth parasites of voles and shrews from Wales. *J. Zool., Lond.*, **154**, 313–31.

Lewis, J. W. and Twigg, G. I. (1972) A study of the intestinal parasites of small rodents from woodland areas in Surrey. *J. Zool., Lond.*, **166**, 61–77.

Lotz, J. M. and Font, W. F. (1985) Structure of the enteric helminth communities in two populations of *Eptesicus fuscus* (Chiroptera). *Can. J. Zool.*, **63**, 2969–78.

MacArthur, R. H and Wilson, E. O. (1967) *The Theory of Island Biogeography*, Princeton University Press, Princeton.

Mackiewicz, J. S. (1988) Cestode transmission patterns. *J. Parasitol.*, **74**, 60–71.

May, R. M. and Anderson, R. M. (1978) Regulation and stability of host parasite population interactions. II. Destabilizing processes. *J. Animal Ecol.*, **47**, 249–67.

May, R. M. and Anderson, R. M. (1979) Population biology of infectious diseases. II. *Nature, Lond.*, **280**, 455–61.

McKenzie, C. E. and Welch, H. E. (1979) Parasite fauna of the muskrat, *Ondatra zibethica* (Linnaeus, 1766), in Manitoba, Canada. *Can. J. Zool.*, **57**, 640–6.

Miller, R. S. (1967) Pattern and process in competition. *Adv. Ecol. Res.*, **4**, 65–72.

Pence, D..B., Crum, J. M. and Conti, J. A. (1982) Ecological analysis of helminth populations in the black bear, *Ursus americanus*, from North America. *J. Parasitol.*, **69**, 933–50.

Pianka, E. R. (1970) On *r*- and *K*-selection. *Am. Nat.*, **104**, 592–7.

Read, A. F. (1987) Comparative evidence supports the Hamilton and Zuk hypothesis on parasites and sexual selection. *Nature*, **327**, 68–70.

Ricklefs, R. (1987) Community diversity: relative roles of local and regional processes. *Science*, **235**, 167–71.

Schad, G. A. (1966) Immunity, competition, and the natural regulation of helminth populations. *Am. Nat.*, **100**, 359–64.

Shoop, W. L. (1988) Trematode transmission patterns. *J. Parasitol.*, **74**, 46–59.

Smith, G. (1984) Density-dependent mechanisms in the regulation of *Fasciola hepatica* populations of sheep. *Parasitology*, **88**, 449–61.

Stock, T. M. and Holmes, J. C (1987a) *Dioecocestus asper* (Cestoda: Diococestidae): An interference competitor in an enteric helminth community. *J. Parasitol.*, **73**, 1116–23.

Stock, T. M. and Holmes, J. C. (1987b) Host specificity and exchange of intestinal helminths among four species of grebes (Podicipedidae). *Can. J. Zool.*, **65**, 669–76.

Wakelin, D. (1984) *Immunity to Parasites. How animals Control Parasite Infections*, Edward Arnold, London, pp. 165.

Warner, R. R. and Chesson, P. L. (1985) Coexistence mediated by recruitment fluctuations: a field guide to the storage effect. *Am. Nat.*, **125**, 769–87.

Wassom, D. L., Wakelin, D., Brooks, B. O., Krco, C. J. and David, C. S. (1984) Genetic control of immunity to *Trichinella spiralis* infections in mice. Hypothesis to explain the role of H-2 genes in primary and challenge infections. *Immunology*, **51**, 625–31.

11

Free-living communities and alimentary tract helminths: hypotheses and pattern analyses

Daniel Simberloff

11.1 INTRODUCTION

A comparison of the community ecology of parasites and free-living organisms is far too large a topic for one chapter, so I will concentrate on one aspect – helminth communities of the vertebrate alimentary tract. The digestive tract and its inhabitants comprise an ecosystem (Dubos and Schaedler, 1964), one whose inhabitants might be expected to interact strongly with one another because of the limited physical extent and variety of habitats (perhaps) relative to free-living ecosystems. Differing views of these helminth communities seem to mirror debates among students of free-living organisms. Further, the nature of the evidence sought to elucidate communities and the methods used to analyse that evidence have been one focus of the free-living research and, I feel, these same issues might profitably be raised for parasite communities.

Among students of free-living systems, the entire notion of 'community' has become controversial, because the term has come to denote integration and order almost as great as that depicted in the holistic, superorganismic community of Clements (1905) and his associates, a depiction that fell out of favour in the 1940s (Simberloff, 1980). Thus, more neutral terms like 'assemblage' are now often used. This is not to say that the idea of great order and integration has no adherents, only that there are many skeptics (perhaps a majority). Many patterns classically cited as evidence for a holistic community (e.g. succession, zonation) have been reinterpreted in more 'individualistic' terms as determined largely by the independent activities of individual species (references in Simberloff, 1980). Of course, part of the debate may well be terminological and psychological; in outlining it, I have used words like 'great' and 'largely' that obviously mean different things to different observers. Similarly, 'order' and 'integration' are to some extent in the eye of the beholder. Nevertheless, the debate is not only semantic and much empirical evidence has been brought to bear.

The notion of community in parasitology is also controversial, for some of the same reasons. Kisielewska (1970a) asks whether parasite groupings qualify as 'coenoses': 'natural, integrated, actually existing units. These units have a determined organization (structure and function) which is implemented through direct and indirect correlations and is expressed by integration' (Petrusewicz, 1965). She concludes that *Clethrionomys glareolus* the gut helminths of the vole, fit this definition (Kisielewska, 1970a, b, c, d, e). At the other extreme, Kennedy (1985) finds that *Anguilla anguilla*, the intestinal acanthocephalans of the eel, form chance assemblages with little organization and few or no interactions, a depiction that Price (1980) would view as the rule rather than the exception for intestinal helminths. These different conceptions might arise from different traits and/or statistical procedures used to characterize organization and integration (as discussed below), or they might arise because the two communities really have different degrees and types of organization (possibly because they maintain different types of organisms), or both differences might be at the root of these diametrically opposite views.

First, some terminology. There are myriad jargon terms in parasite community ecology. A valiant clarification effort (Margolis *et al.*, 1982) is only partially successful. In particular, I want to distinguish between the parasites in a single host individual (which I will call 'infrapopulation' or 'infracommunity' but which has also been termed 'idiohostal grouping') and those in all individuals of a host population (which I will call 'metapopulation' or 'component community' but which has also been called 'supra-' population or community and 'synhostal grouping'). Another possible source of terminological confusion is spatial dispersion. For ecologists, 'overdispersed' means 'further apart in space, on average, than in a random arrangement'. For some parasitologists, 'overdispersion' could seem to mean exactly the opposite, namely 'contagion' (Esch, 1983). The confusion arises because parasitologists are talking about statistical contagion (mean number of parasite individuals per host less than variance) while ecologists are talking about spatial contagion. When I say 'overdispersion', I will be speaking as an ecologist. What I mean by 'random' will be discussed later. Finally, the word 'niche' will crop up repeatedly in my discussion; I intend it to encompass either Hutchinsonian (tropic) or Grinnellian (habitat) niches, or both.

11.2 FOOD WEBS AND SIZE RATIOS

Food webs are one area of historical and increasing research on free-living communities. Various perceived constraints on these webs (e.g. few levels, low connectivity, few omnivores) have been attributed to various forces (e.g. energy limitation, mathematical instability of systems missing these properties). However, the existence of the patterns themselves has been questioned (e.g. Peters, 1977; Paine, 1988), as has the cogency of the rationalizations (e.g. references in Simberloff, 1988). In particular, Paine (1988) has recently argued that many patterns said to typify trophic structure (Pimm, 1982; Cohen, 1988) are derived from caricatures of food webs, the summed interactions of several local communities rather than patterns that actually obtain in any one food web. I would make the analogous argument for helminth communities: in general, infracommunities are much more likely to be informative about the forces currently structuring communities than are component communities (cf. Bush and Holmes, 1986b). A component community can represent an assemblage never found in any individual in nature (e.g. Kennedy, 1985), so it is doubtful that the component community can tell us much about the forces acting on communities now, in ecological time (though it may suggest evolutionary scenarios).

But gut parasites are all in the same trophic level, so the potential structuring force of food web topology is inapplicable to their communities; there does not even seem to be an analogue to the trophic web. To a student of free-living communities, this absence of trophic interaction is perhaps the most striking feature of intestinal parasite communities. Predation, parasitism and disease are often thought to be key interspecific interactions structuring free-living communities, and they are largely absent from helminth associations. These worms seem rarely, if ever, to eat one another, and few, if any, species act as predators or hyperparasites on intestinal helminths. The flagellate, *Histomonas meleagridis*, parasitizes both the avian cecal nematode, *Heterakis gallinarum*, and its gallinaceous hosts (Schmidt and Roberts, 1985), though its effects on nematode populations are unknown.

Why should this 'empty niche' characterize intestinal helminth communities? One constraint, at least in some species, might be energetic. On average, there may not be enough helminths regularly occupying individual guts of a particular host to make preying on them a viable option energetically. However, in some host species, it seems that certain parasites are predictably present in substantial intensity. Another possible reason (not necessarily exclusive) would be that, to prey on intestinal helminths, a species would have to be an intestinal parasite first; otherwise there would not likely be selective pressures to evolve the ability to locate and enter a host gut. But an existing gut parasite might not be selected to specialize on helminths because the host tissues and fluids are already present in greater abundance.

Another area of intensive research on free-living communities has been the relative sizes of species within them. Since Hutchinson (1959) first suggested that pairs of species in the same trophic level would require a minimum size ratio of *c.* 1 : 3 in their trophic apparati in order to coexist ecologists have sought evidence for either minimum size ratios or (in suites of three or more species) constant size ratios between species adjacent in a size ranking. More recently this search has moved into the space age with the contention that, in order to coexist, species must be over-dispersed in a complex morphological space involving both size and shape (references in Simberloff, 1988). Price (1972) argued that ovipositor lengths in a community of wasps parasitic on a sawfly demonstrate a minimum ratio compatible with coexistence. For helminth communities of the vertebrate gut, the only such claim I have located is that of Cannon (1972), who studied three trematode species in yellow perch. He contended that one way the flukes partition their habitat and thus coexist is by having different sizes of oral suckers, and by being different widths. For free-living organisms, two general deficiencies have bedevilled contentions that certain size relationships show overdispersion because of interspecific competition: (a) absence of a null model (what size relationships would have been expected in the absence of any interaction between species?), and (b) absence of evidence that a resource is limiting and that the observed size difference lessens competition. Cannon's claim seems to typify both problems. How different would one have expected the trematodes' sizes to be if there were no interactions among them? What evidence is there that different body widths and/or oral sucker sizes cause different resources to be used? The sort of functional analysis performed by Williams (1966) on cestodes would be necessary to rationalize a hypothesis of resource partitioning by morphological difference.

11.2 COMBINATIONS OF SPECIES

Another pattern that students of free-living animals have adduced as evidence that interspecific interactions restrict membership in communities is how many combinations and which combinations of all those possible in some source region actually exist. The most flamboyant exposition was the 'assembly rules' of Diamond (1975), who aimed to show that the absence of various combinations or various numbers of combinations of birds among the islands of an archipelago is *ipso facto* evidence for interspecific competition. The lack of a null model was pointed out by Connor and Simberloff (1979) and Simberloff and Connor (1981) with repercussions that rent the ecological community (Lewin, 1983). The basic problem was that, with a relatively small number of islands and relatively large number of species, one would have expected many species combinations to be missing even if the species were arranged independently over the islands. The exact nature of the proper null hypothesis remains a bone of contention (Harvey *et al.*, 1983; Gilpin and Diamond, 1984; Connor and Simberloff, 1983, 1984), particularly the issue of whether a 'null' hypothesis may have the historical effects of some process (such as interspecific competition) covertly embedded in the supposedly null structure. But nowadays in the free-living literature few would dare talk about missing species combinations without some null hypothesis. A second problem – no demonstration that some resource is limiting – also seemed to typify Diamond's claim (Connor and Simberloff, 1984, but see Gilpin and Diamond, 1984) and others that the number of combinations present and absent can imply competition.

One might have expected frequent attempts among parasitologists to use information on numbers of combinations (or frequencies of particular combinations) to infer community-structuring forces. After all, the main drawback for application of this approach to free-living communities is too many species and too few islands, implying a high number of expected missing species combinations. For example, suppose there are 30 species. Then there are 435 species pairs, 4060 trios, etc. If there were only 30 islands, it would not be at all surprising to find many missing pairs and trios. It is easy to calculate the probability that any one combination is missing (or shares X or fewer islands). Suppose species i occupies S_i islands and species j occupies S_j islands of N islands. Let $S_i \leq S_j$. Then if all islands were equally likely to be colonized, the null probability that species i and j will share exactly X islands is (Connor and Simberloff, 1979):

$$\Pr(X) = \frac{\binom{S_i}{X}\binom{N-S_i}{S_i-X}}{\binom{N}{S_i}}$$

Then the null tail probability that two species will share X or fewer islands is:

$$\Pr(\text{tail}) = \sum_{X=K}^{S_i} \Pr(X)$$

where $K = \max(0,\ S_i + S_j - N)$ and $\Pr(X)$ is as defined above. So, for example, if there were 20 islands, species i were found at 8 of them, species j at five of them, and they did not co-occur, the null probability of this degree of exclusivity is 0.051. This might seem like a significant difference, but if there are 30 species, unless we had hypothesized in advance that this particular pair would be exclusive, we could not construe the observation as strong evidence for interaction, since there are 435 pairs. We cannot even calculate directly how many pairs should be exclusive at the 0.05 level because the pairs are not independent (the same problem obtains in the interpretation of sets of pairwise chi-squared contingency tests). However, one can construct a binary (presence–absence) species × island matrix and randomize it, subject to the constraint that each island maintains the number of species observed in nature and each species occupies the number of islands observed in nature. Then, from many randomizations, one can determine whether the observed number of pairs (or trios, etc) sharing zero, one, two, etc islands is unusual (Connor and Simberloff, 1979).

By contrast, helminth component communities of the vertebrate alimentary tract usually have few species – five or ten common ones – and, since each host individual is analogous to an island, a relatively large number of islands seems possible (cf. Holmes and Price, 1986). Also, host individuals may more closely approximate replicates of one another, as required by most statistical null models, than islands do. In any archipelago, there must be subtle habitat differences between islands such that likely colonists of one island might be much less prone to colonize a second island, no matter what other species are present. However, if for example 60 ducks are resident in a pond and the component community consists of ten helminth species, though genetic differences between individual hosts may affect prevalences and intensities (Holmes, 1983), nevertheless, it is quite likely that each duck is a potential host for each helminth species. Yet surprisingly few authors have pointed to the absence or low frequency of certain infracommunities, and even fewer have claimed such a pattern as evidence for species interaction. John (1926) studied the cestodes *Cittotaenia pectinata* and *C. denticulata* in rabbits (*Oryctolagus cuniculus*). Of 516 hosts, 358 had one or the other worm but none had both, though many infections were heavy and recent and rabbits with each of the worms were found interspersed in very small areas, suggesting that a difference in habitat or intermediate host did not cause this mutual exclusion. For example, of 20 rabbits in one 200-yard hedgerow, nine had *detriculata*, four had *pectinata*, and seven had neither. Even more remarkably, the two species occupy different parts of the intestine. John gave no potential reasons for this odd distribution, but, had he lived fifty years later, I feel certain he would have felt compelled to speculate.

Frankland (1959) studied trematodes of the mole, *Talpa europaea*. Of 593 host individuals, about 85 were infected with *Itygonimus lorum*, about 78 with *I. ocreatus*, and about 48 with *Omphalometra flexuosa*. Yet no mole had both *Itygonimus* species, and only one individual had *Omphalometra* and each of the *Itygonimus*. The apparent exclusion of the two *Itygonimus* is more mysterious given the fact that *I. lorum* is virtually restricted to the lower intestine and *I. ocreatus* to the upper intestine. *Omphalometra* has a similar distribution to that of *I. ocreatus*, which may rationalize its apparent mutual exclusion with the latter species, but renders surprising its apparent exclusion with *I. lorum*. To some extent, the apparent exclusion is an artifact of the large sampling area, which included regions where various of the flukes are absent. But parasite allopatry cannot account for the entire pattern. For example, 370 hosts came from one region where all three flukes exist; of these, 77 had *I. ocreatus*, nine had *I. lorum*, 20 had *Omphalometra*, one had *I. ocreatus* plus *Omphalometra*, one had *I. lorum* plus *Omphalometra*, and the remainder had no flukes. Because of the small numbers infected with *I. lorum*, one cannot reject the possibility that the exclusion between the two *Itygonimus* would have occurred without any interaction between them; the null probability is 0.091. Similarly, that only one mole contained *I. lorum* with *Omphalometra* could have been due to chance; the null probability of complete exclusion is 0.538. However, only one mole sharing *I. ocreatus* and *Omphalometra* would not likely have arisen if the flukes were strewn randomly and independently among moles; the null probability of only one or zero double infections is 0.034. Frankland argues that co-generic immunity is not likely to be the cause of the apparent exclusion, but gives no other explanation.

Lie *et al.* (1968) present an example of two trematode species in snails in which double infections are less frequent than expected, and argue that one species prevents infection by the other. On the other hand, Hobbs (1980) finds, among eight species of nematodes and one cestode in pika (*Ochotona* spp.) guts, that four pairs of species co-occur more frequently than expected by chance and suggests that double infections might be particularly likely because of similar life cycles of the parasites. Among 13 enteric helminth species in one population of the big brown bat (*Eptesicus fuscus*), Lotz and Font (1985) found no unusual positive or negative associations, while another host population with 21 helminths had two positively associated pairs and one negative one. The latter pair were also spatially displaced in the gut (see below), but Lotz and Font took the sum of their analyses to indicate that contemporary interactions are of little importance in structuring these infracommunities and that negative interactions are particularly uncommon. Both Hobbs and Lotz and Font used chi-squared contingency analyses of all species pairs. As noted above, one would expect a certain number of associations by chance alone, so a randomized matrix would be appropriate to see if the number of associations is surprising.

11.3 SPATIAL SEGREGATION

Overlap and overdispersion

Research on free-living communities often focuses on spatial segregation and changing patterns of spatial use as evidence for interspecific interactions, particularly competition. A classic study was that of MacArthur (1958), who found that five warblers of the genus *Dendroica* using the same trees differed, on average, in foraging height and diameter of foraging branch, as well as method of catching insects. In addition to MacArthur's study, I will discuss three others that are relevant to parasitological studies. Diamond (1978) described a number of similar situations, of which one seems particularly pertinent to studies of intestinal helminths. A thrush occupied a different elevational range on different islands (with different sets of co-existing species). Terborgh (1971) depicted a related situation: for a number of bird genera inhabiting a mountain range, each congeneric species (from two to four) occupied a limited elevational band and the bands were nearly non-overlapping. Finally, Diamond (1973, 1978) pointed to a series of fruit pigeons in which each species weighs approximately 50% more than the preceding one. This example seems out of place here but it will become clear why I include it here and not in the above discussion of size ratios.

MacArthur's study represents many in which over-dispersion or complete separation among several species is found in some n-dimensional resource or niche space. Here $n = 3$: height of foraging, diameter of foraging branch, and method of catching prey. In essence, MacArthur showed that, with this many axes, no two species occupy the same position in this space. MacArthur took this pattern as prima-facie evidence for interspecific interactions. I have already alluded to a problem that besets this observation. There was no null hypothesis. A simple null hypothesis might have been: given these three axes, some two out of five randomly placed species will coincide. One would have to decide on conventions about how the species would be strewn about the 3-dimensional space in the null model, and no doubt any set of conventions would generate an argument about whether they are the best ones, but it is clear that absence of coincidence among these five species in this 3-space is not prima-facie evidence for anything. I would argue that, unless MacArthur had specified in advance which axes would segregate these five species, even the above null hypothesis is not null enough, and the proper null hypothesis is that some set of three axes will suffice to separate the five species. In addition to foraging height, foraging mode, and diameter of foraging branch, other traits could have been measured (in fact, several were) and it might not have been surprising that some subset of three traits would yield a space in which no two species coincided.

An analogous analysis in the parasitological literature is by Schad (1963), who examined several colon-inhabiting congeneric pinworms in tortoises. He divided the colon into four linear divisions and radial distribution into two positions (ring and core). Then he summed the locations, in this 2-dimensional space, for the eight common pinworms found in six tortoises. He concluded that the overlaps seen with just the linear divisions were greatly decreased by adding the radial divisions, in the sense that the two species pairs showing the most similar patterns of linear distribution differed in radial distribution.

However, the data, reinterpreted, do not show a remarkable radial separation for pairs of species that are remarkably similar linearly. For the eight species, there are 28 pairs. I ranked (from greatest to least) each of these pairs in terms of linear similarity and, separately, radial similarity, in each case using the C_{xy} coefficient (Hurlbert, 1978). Then I performed a simulation 1000 times in which I randomly associated the linear ranks with the radial ranks and asked whether the highest ranked one, two, or three linear pairs were associated with low ranked radial pairs. The answer was no: using the sum of ranks as the test statistic, I found that 430, 799, and 882 times, respectively, the random sum of ranks was less than the observed sum of one, two, or three ranks. Schad noted that, even with both linear and radial dimensions, several broad overlaps remained among species, but proposed that different food habits, as indicated by different oral morphology, would eliminate these overlaps. An appropriate null hypothesis then, is that in a 3-space defined by these three dimensions (or perhaps just by some three dimensions), eight randomly located species would not overlap.

Stock and Holmes (1988) examined enteric helminth communities of four grebe species, focusing primarily on linear distribution in the gut. Five pairs of congeneric cestodes co-occurred in one or more host species. Although they did not calculate a null expectation of amount of range disjunction and mean location separation for each pair and for all pair–host combinations combined, Stock and Holmes noted that, in six of twelve pair–host combinations, the congeners occupied largely disjunct ranges, while in three others, ranges overlapped but mean positions were in non-overlapping portions of the ranges. In two instances, there were broadly overlapping ranges but widely separated mean locations. For only one pair were ranges and mean locations similar, and these never co-occurred in the same host individual, as they are found overwhelmingly in different host species. Stock and Holmes conclude that interactions between members of congeneric pairs are responsible for what they perceive to be a remarkable degree of dispersion in niche space, with *Schistotaenia* species being host species specialists, *Diorchis* and *Dubininolepis* species being microhabitat specialists, and *Tatria* species combining partial separation into different hosts with partial microhabitat separation. There was no null hypothesis concerning how much niche overlap in these two dimensions would have been expected given twelve parasite pair–host combinations.

Parasitological analogues to Terborgh's observation of differing elevational ranges of congeneric birds are numerous. Crompton (1973) says that the localization of alimentary trace parasite species was recognized at least by 1892, tabulates many such localizations, and rationalizes some broad patterns. For example, acanthocephalans are absorbers and tend to be in sites where host absorption of nutrients occurs, especially the small intestine; cestodes, also absorbers, are mostly in the small intestine. Nematodes, because they have an alimentary tract, feed in many ways and are found in correspondingly many sites; they feed on tissue, suck blood, chew on the mucosa, swallow lumen contents, etc. In other words, adaptations to habitat differences along the gut (exemplified by the cestodes of skates discussed below) might explain why spatial ranges of helminths are distinct or nearly so. At least in the range shift studies I will discuss, an experiment (laboratory or 'natural') has been performed; the problem is in control and interpretation. For the static observation of amount of difference in a community, the null hypothesis might have been that each species would be localized to some extent (because of different habitat requirements) and the ranges would have little overlap (because the ranges tend to be small). In other words, if one took a long line segment (representing the gut) and randomly and independently distributed along it a bunch of smaller segments (each representing the range of one species), how far apart should the segments have been and how much overlap would one have expected?

Without experiment, even improbably low overlap cannot strongly implicate the present action of competition. Abramsky and Sellah (1982) studied two gerbils in the coastal dunes of Israel. These dunes are separated into northern and southern strips by a mountain. In both strips, sand dunes are interspersed among other soil types. *Meriones tristrami* has invaded from the north and is now associated with the dunes in both northern and southern strips. *Gerbillus allenbyi* is also associated with dunes, but has invaded from the south and is not found in the northern strip. In the northern strip, *M. tristrami*, in allopatry, is found on other soil types as well as on sand. In the southern strip, in sympatry with *G. allenbyi*, it occupies other soil types but not sand dunes, which are occupied by *G. allenbyi*. This would appear to be a case of competitive exclusion, so Abramsky and Sellah (1982) attempted to induce a niche shift (see below) by removing *G. allenbyi* from plots including dunes in the southern strip. *M. tristrami* did not increase its numbers; in this region, its distaste for dunes is apparently genetically fixed, and its absence is not the result of present competition.

One may look at overlap by comparing metapopulations of pairs of species (Henricson, 1977), but this is not likely to be as informative as looking at infrapopulations. To take an extreme example, one could imagine two parasites, A and B, and many hosts. In half the hosts, all A are in the anterior of the small intestine and all B are in the posterior, while in the other hosts all A are posterior and all B anterior. A comparison of metapopulations would show complete overlap, while a comparison of infrapopulations would show no overlap whatsoever. Of course, even looking at infrapopulations, one must bear in mind that statistically significant large or small overlap of a particular species pair can be simply an artifact of the large number of pairs examined. This is exactly the problem mentioned above with respect to exclusive occupancy between two species. For example, if there were 30 species, there would be 435 pairs, so one or a few results with $p < 0.05$ would not necessarily be surprising. Unfortunately, since the pairwise overlaps are not independent, one cannot easily tell how many of them would be expected at some probability level in a null model (Connor and Simberloff, 1979).

Bush and Holmes (1986b) avoided the above pitfalls by using infracommunity data and by computing a summary overlap statistic for the entire community, rather than focusing on particular species pairs. They first computed the pairwise overlaps for gut helminths of 45 ducks by a standard ecological measure (C_{xy} in Hurlbert (1978)). Then they used the variance test (Schluter, 1984) and found an overall tendency towards less overlap than expected for all 52 worm species and for the eight 'core' species alone. However, sample-to-sample variation independent of species interactions would likely confound the results of Schluter's test (McCulloch, 1985), so this analysis of Bush and Holmes (1986b) is questionable.

Lotz and Font (1985) took a different approach with the helminth community of the big brown bat. For each bat, they divided the gut into ten segments. Then for each pair of worm species, they maintained the infrapopulation of one worm as they found it in nature, but randomized the infrapopulation of the second worm (that is, simulated a shuffling of the ten gut segments) and calculated the overlap C_{xy}. Thus they derived a null distribution of overlap for comparison to the observed value. They found four pairs of species (of 13 species) at one site and eight pairs (of 21 species) at another site with greater overlap than expected, and one pair at each site with less (the excess of positive associations is very similar to those analogous studies done with birds of the Bismarck Archipelago (Gilpin and Diamond, 1984) and the Hawaiian Islands (Mountainspring and Scott, 1985)). Because of the large number of pairs (78 for one site, 210 at the other) it is impossible to say which, if any, of the significantly large or small overlaps are real. Because the pairs are not independent (if A and B have high overlap, and B and C have high overlap, we can predict that B and C will have high overlap), we cannot even be sure there are more positive associations than the null expectation. A simple change in their procedure would solve this problem. Instead of simulated shuffling of one species' locations with respect to another's, one could randomize all species locations together. In other words, with *N* species and ten gut segments, the observed data for each infracommunity comprise an $N \times 10$ matrix m, with each entry m_{ij} the number of individuals of the ith species in the jth segment. The row entries for all rows save the first would then be randomly permuted and all pairwise overlaps computed; in addition to comparing the simulated overlaps to the corresponding observed ones, one could tabulate summary statistics such as the number of overlaps greater than observed and number less than observed.

Bush and Holmes (1986b) presented two other patterns they felt implicated interspecific competition in the spacing of helminths in their ducks. First, they argued (as had Rohde (1979)) that, as infrapopulation sizes increase, if the distribution of each worm species is independent of those of other species, overlaps between adjacent species should increase. In fact, they found no such pattern, and reasoned that the distributions must not be independent – the worms must have been actively excluding one another.

The final major spacing pattern they sought was evenness of spacing along the gut. This is where the fruit pigeon example is relevant. Diamond (1973, 1978) saw each species approximately 50% bigger than the next smallest one. In other words, on a log-scaled line, he saw the mean sizes approximately equally spaced. Bush and Holmes (1986b) saw equal spacing on an arithmetic line. Several tests are available for whether points on a line segment can be reasonably construed to have been randomly placed, as opposed to the alternative hypothesis that they are more equally spaced than random. Three deserve particular attention because they are widely cited in ecology and some misinformation has been published about them. First is a test of the variance (or standard deviation) of the segments, employed by Bush and Holmes (1983, 1986b). If *N* points are placed on a line and the end-points are included, they form $N+1$ segments, and, if they are equally spaced, the variance among these segments is zero. There does not appear to be an analytic determination of the distribution of the variance for randomly strewn points. Cohen (1966) did not derive this distribution or its expectation but showed that previous estimations of the latter by ecologists were incorrect. Bush and Holmes (1983) calculated the expected size of each size-ranked segment (these distributions *are* known) and used the variance of these expectations as a slightly biased estimator of the expected variance. Poole and Rathcke (1979) gave an approximation to the distribution of the variance, but did not state the basis for this approximation. However, it is a simple matter to simulate this distribution, and that is what I did in the subsequent examples, repeatedly randomly breaking a stick and calculating the variance of the sizes of resulting segments.

Barton and David (1956) suggested a family of statistics, any one (or group) of which could serve as a test for equal spacing of points on a line. For N points, they size-ranked the $N+1$ segments from smallest to largest. Then, for every $i < j$, they formed the ratio (G_{ij}) of the size of segment i to the size of segment j, and gave the distribution of this ratio. For example, one could use the ratio of the smallest to largest segment as test statistic. Simberloff and Boecklen (1981) used Barton–David (B–D) statistics to test for equality of size ratios.

Hopf and Brown (1986) introduced yet another test, the 'bull's eye', for equality of segments generated by points on a unit line. For N points ($N+1$ segments), they conceived of a target in $N+1$ space. The centre of the target is the point whose coordinates are all $1/(N+1)$, that is, all segments equal. The periphery of the target is defined by points whose coordinates are $(1,0,0, \ldots ,0)$, $(0,1,0, \ldots ,0)$, $\ldots$,$(0,0,0, \ldots ,1)$, that is points in which one segment is as large as possible and the N others are vanishingly small. The target is then divided into concentric bands of equal probability. A point near the centre has extraordinarily equal spacing, while a point near the periphery has extraordinarily unequal spacing.

Hopf and Brown (1986) for the bull's eye and Bush and Holmes (1983) for the variance test argued against B–D statistics on two grounds: (a) the B–D test 'uses' only a small amount of the data, and (b) the B–D test is less powerful than their preferred tests. They misunderstand the significance of the first point and have not demonstrated the second. It is true that any one B–D test is calculated from only two segments, thus at most four points (though one can use up to $(N^2+N)/2$ B–D statistics for N points). However, it is not necessarily a virtue if all the data are submerged in a single statistic; for example, often a single value (say, the largest) can falsify a hypothesis while a summary statistic such as a mean cannot. Neither Hopf and Brown (1986) nor Bush and Holmes (1983) have done a comprehensive test of power of their preferred test against a B–D test, and it is clear from simple examples that a B–D test can, in some circumstances, detect a non-random arrangement where neither the bull's eye nor the variance test can. Furthermore, by using several B–D statistics concurrently, one can detect what part of a data set is inconsistent with the null hypothesis.

Suppose there were four species in the gut, so the four mean positions plus the ends of the gut divided the gut into five segments. Suppose the five segments had lengths 0.4, 0.001, 0.3, 0.2, and 0.099. Then $G_{15} = 0.0025$ and the null probability of a value that large is 0.974. In other words, one would reject the hypothesis of random placement because only 0.026 of the time would one have got a G_{15} that small. Similarly, G_{14} rejects the null hypothesis, whereas G_{25} does not, so we know that it is the smallest distance that is extraordinary. Now, the sample variance for these five segments is

0.025, and a simulation run 500 times shows that the null probability of a variance that high is 0.556, so one would not have rejected the null hypothesis of random placement. The bull's eye test shows a probability of 0.45 of greater evenness among the segments, so again one would not have rejected the hypothesis of random placement. It is also not automatic that the B–D test would fail to detect a remarkably even distribution of segment sizes. For example, suppose there were 19 species dividing the gut into 20 segments, of which 18 had length 0.05, one had length 0.005, and one had length 0.095. The variance of these segments is 0.0002, and all 500 runs of the simulation gave larger variances – in other words, the segments are improbably similar in length. Similarly, the bull's eye test shows this set of segment lengths to be in the extreme 0.05 tail of the evenness distribution. But $G_{1,20} = 0.053$, and the null probability of a larger value is only 0.037, so again the set of segments would be detected as having unusually even lengths: only 3.7% of the time would the smallest and largest segments be as similar as they are.

After this digression on how to test for remarkably equally spaced locations, what do the data show? Using the variance test on the mean locations in the component community, Bush and Holmes (1986b) found the eight core species to be remarkably evenly spaced, while the 44 other species were not, being concentrated in two groups, one anterior and the other posterior. Given the rather arbitrary categorization into core, secondary, and satellite species (Bush and Holmes, 1986a), it would be interesting to see if the same result would have obtained if the core group had been expanded to encompass 9–12 species. For infracommunities, Bush and Holmes lumped the eight core and eight secondary species and, for each bird, calculated variance and determined whether it was greater or less than would have been expected for 16 randomly arranged points. For 39 of 45 birds, the variance was less than the expected (greater evenness); Bush and Holmes (1986b) take this disparity as evidence that competition is structuring the infracommunities. However, the distribution of the variance of the broken stick is highly skewed; the expected value is much greater than the median. For one thousand simulation runs for 16 points, I found the expected value to be in the 61st percentile. That is, according to the null hypothesis one would have expected 61% (27.45 of 45) of individual birds to have less than the expected variance. The observed 39 is still highly significant by the binomial test ($p < 0.0005$), but not so striking as it would have been if the distribution were symmetric; it would not have been surprising to find 34 of 45 birds with variance less than expected. It would be informative to know just how remarkably even the individual infracommunities were. In sum, Bush and Holmes (1986b) view this system as one with remarkable displacement of the core species, in which overlap is minimized and evenness maintained by asymmetric linear range extensions and/or minor range shifts at high densities.

Niche shift

Now to the example (Diamond, 1978) of the thrush that occupies a different elevational range on different mountains, which Diamond interprets as a response to the different sets of species present. Numerous parasitological studies divide the alimentary tract into several sections, then note that species A is distributed differently in single infections than in joint infections with species B (Lang, 1967; Chappell, 1969; MacKenzie and Gibson, 1970; references in Crompton, 1973; references in Holmes, 1973; references in Halvorson, 1976; Evans, 1977; Grey and Hayunga, 1980; Hobbs, 1980; Silver *et al.*, 1980; Pojmanska, 1982; Holland, 1984; Stock and Holmes, 1987).

With respect to his shifting thrush, Diamond (1978) noted contemporary criticism of such evidence for competition on the grounds that the different sites might differ in some key way other than presence or absence of the putative competitor. However, he felt 'this objection strains one's credulity'. Perhaps the degree of strain is directly proportional to the degree of credulity, which probably varies among scientists. I have never understood why, categorically, such an objection is invalid, though I concede that some patterns of apparent 'niche' shift seem more cogent evidence for competitive release or compression than others, at least partly depending on how much empirical evidence has been gathered on alternative hypotheses.

For the parasitological analogs to the shifting thrush, potential differences among hosts would confound matters particularly in those studies (Chappell, 1969; MacKenzie and Gibson, 1970; Evans, 1977; Grey and Hayunga, 1980; Hobbs, 1980; Pojmanska, 1982; Stock and Holmes, 1987) in which comparisons are made in wild-caught, naturally infected hosts, rather than those experimentally infected in the laboratory. If one catches in the field some hosts containing A alone and others containing both A and B, it seems possible that some biological difference (rather than chance) predisposed the former not to have B. If that were so, it would not strain credulity (mine, anyway) to imagine that the same biological difference or a related one might cause A to locate in a different part of the gut. An analogous problem has been pointed out by Strauss (1988) for studies on the effects of animals on plants in which a comparison is made between plants eaten in nature by animals and 'controls' that are not attacked. The controls might not have been eaten because they were different in ways that might affect the traits examined as potentially animal-induced. In parasitology, problems of interpretation can be lessened by using laboratory-infected animals, where this technique is possible.

Another problem with interpreting locational shifts along the gut is that the spatial distribution of some parasites changes with changes in their own density. For example, the mean position of the nematode *Trichinella spiralis* in rats shifts posteriorly as its density increases (Silver *et al.*, 1980). An analogous problem has cropped up in free-living systems, where it is also difficult to disentangle the effects of interspecific and intraspecific competition (Connell, 1983; Underwood, 1986). The key is to demonstrate that an observed change in species A (say, a niche shift) is a response to a change in species B's density, and would not have arisen simply from a similar change in density of species A itself. Underwood (1986) points out that many recent studies have argued that competition between two species is asymmetrical – species A appears to affect species B more than vice versa. Numerous studies of helminth communities in the gut claim such an effect on position (Lang, 1967; Evans, 1977; Grey and Hayunga, 1980; Bush and Holmes, 1986b), numbers (Lang, 1967; Dash, 1981; Gordon and Whitfield, 1984; Holland 1984), or cross-immunity (Alghali and Grencis, 1986). Underwood (1986) distinguishes three types of potential asymmetry between two competing species. First, and the one generally considered in studies on both free-living species and parasites, is the degree of similarity of effect of each species on the other. Second is the relative effect of members of species A on other A compared to the effect of members of species A on members of species B. Third is the comparative effect of members of species A on other A relative to the effect of species B on species A. The second and third types have rarely been examined. Underwood (1986) provides an ANOVA design that distinguishes between inter- and intraspecific competition and detects all three types of asymmetry, but notes that interpretation is difficult if natural or relevant densities of two species are different. An apparent asymmetry of the first type can be an artifact of using different densities of the two species in a mixed treatment, a common feature in parasitological research.

Fundamental faunas

Another version of the idea that communities are over-dispersed in some niche space (which may include physical space) is the contention that there are 'fundamental faunas', groups of species that are pre-adapted to pre-existing niches and so can coexist, while sets of species cannot persist if they do not, as a group, use the available resources efficiently and without too much interspecific overlap. For example, McNab (1971) argued that, on small Caribbean islands, there is a fundamental bat fauna consisting of one fruit bat, two nectar bats (of different sizes), two insectivorous bats, and one fishing bat. Nowadays the idea that niches exist and then species fill them has fallen from favour in ecology with the realization that one species can occupy many different trophic niches at different sites, but versions of the concept of a fundamental fauna still abound. In general, these arguments do not test a null hypothesis; one would wish to know whether sets of random draws from some species pool would have produced a set of faunas with niche space as similarly structured as those observed.

In the parasitological literature, Kisielewska (1970a) proposed that a fundamental fauna is manifest in helminth infracommunities in the alimentary tract of the vole, *Clethrionomys glareolus*. Usually there is one nematode species of the genus *Heligmosomum* in the small intestine (if there are any congeners, they are in low density) and a cestode is the second most important species in the small intestine (and remaining cestodes quite rare). In the stomach, one of two nematodes is numerous (the other rare), and in the caecum the nematode *Syphacia obvelata* is found. The sort of null hypothesis just proposed seems appropriate here. Holmes (1973) cites an example for the monogenean, *Gyrocotyle*, in chimaerid fishes. Eight species form four species pairs (each pair specialized on one fish species). Each pair consists of one large species with particular set of morphological features (complex rosette, numerous lateral ruffles, etc) and one small species with the opposite morphological features. Each pair comprises the component community of one host species, but each infracommunity contains just one of the two species. Holmes views this fundamental fauna as competitively determined, but this pattern seems as if it might have resulted from allopatric speciation in each of two lineages of *Gyrocotyle*, in a model similar to that proposed by Brooks (1980) (see below).

Carvajal and Dailey (1975) suggest that cestodes of the genus *Echeneibothrium* may comprise fundamental faunas in skates (*Raja* spp.). In two different skate species, there were three cestode species of strikingly different size and with other characteristic morphological traits, each occupying a distinct part of the gut. Williams (1960, 1966), upon close morphological examination of both the worms and the host intestine, has convincingly argued that scolex structure and other features in this genus closely adapt each species not only to a particular host species but to a particular region of the gut. Thus the hypothesis of competitive exclusion is plausible, though there is no direct evidence that more than one worm species cannot occupy the same region of the gut in the same host species.

Causes of unusual segregation

For comparisons of parasitic and free-living communities, it is useful to distinguish between proximate and ultimate causes. By 'proximate' I mean the present mechanistic causes for an observed phenomenon, analogous to Aristotelian 'efficient' cause (Kuhn, 1977). By 'ultimate' cause, I mean something analogous to Aristotelian 'formal' cause: the evolutionary pressures that led to the existence of the proximate cause. Among the parasitological studies cited so far, it seems clear that, despite many instances of poorly framed and/or tested hypotheses, there are some examples of niche shifts and unusually low overlaps in infracommunities, and possibly of over-dispersion in component communities. As I noted above, niche shifts have been cited in the free-living literature, and some studies already mentioned (e.g., Terborgh, 1971; Diamond, 1978) note low overlap. The constancy of size ratios provides a statistical analogue for over-dispersion along the gut, but I know of no real spatial analogue to this phenomenon in the free-living literature.

For unusually low overlaps in free-living communities, there are at least two potential proximate causes: niche specificity and present-day competition. For niche specificity – where niches do not overlap because of genetic predispositions and there is no shift or broadening in the absence of other species, as in the gerbil example described above – one can question why this should be so. In other words, granted that helminths actively select specific sites (Ulmer, 1971; Crompton, 1973), one can ask for the ultimate cause of this selection. One obvious explanation is competitive character displacement. 'Character displacement' originally referred to morphological divergence between two species in zones of sympatry relative to zones of allopatry (Brown and Wilson, 1956); one selective force that could generate such displacement would be interspecific competition: species evolve to differ because the most different phenotypes would compete less with the other species. Although subsequent examination of the original examples and many others suggests that such morphological displacement is a rare phenomenon (Grant, 1972, 1975; Connell, 1980; Arthur, 1982), the concept is rational and could be extended to traits other than morphological ones – preferred habitats, for example. Thus one could argue that species observed to have genetically determined non-overlapping niches have been selected to behave this way by competition in the past. Connell (1980) points out that hypotheses invoking the 'ghost of competition past' are frequently unfalsifiable, plausible though they may be.

Another possible explanation for genetically determined non-overlapping niches is reproductive character displacement (Bossert, 1963): the evolution of increased difference between two species was engendered by selection for avoidance of hybridization. Again, the evolutionary scenario is difficult to falsify.

Yet a third possible ultimate explanation for niche specificity seems not to have been considered. All species manifest some degree of niche specificity; for most, niches are quite narrow. It could be that the genetic integrity of a species would make it difficult to be sufficiently flexible to occupy a large range of niches. Then niche specificity *per se* would evolve simply by selection against individuals so genetically different that they are suited to a niche outside the usual range of niches used by the species. Under what conditions a multiple niche polymorphism can evolve is an unanswered question. Certainly exposure to a wide range of niches engenders selection for genetic variability within a population. For example, ascaridoid nematodes with life cycles encompassing two host species have greater genetic variability than those using one host (Bullini *et al.*, 1986). However, all species, both free-living and parasitic, have niches that are restricted to some extent, and underlying genetic differences might explain why some are very restricted and others are not. Of course, an observation that different species' niches have little overlap might demand a further explanation. Given N species with niches represented by solids in M-dimensional niche space, what is the null probability that a group of N niches of the same sizes but randomly arranged in niche space would have as little overlap as observed?

The parasitological literature on causes of overdispersion, niche shifts, and low overlaps parallels that for free-living organisms. A continuing criticism of competitive explanations for over-dispersion among free-living animals is that limitation of resources is rarely demonstrated (Simberloff, 1982). Price (1980) argued similarly that food is not usually limiting for parasites in general, citing a variety of studies suggesting resources are usually available, because of low population densities combined with niche specificity. Price (1980) felt that densities of parasites were usually too low for significant competitive interaction, so that competitive character displacement is unlikely. He suggested that random locations of each species' niche among intestinal helminths, plus niche specificity, would usually yield a system in which overlap is low. However, this scenario to some extent begs the question; what keeps niches specific (and population sizes low)? (Price (1980) feels that low colonization rates usually limit population sizes.) If resources were available, would not selection lead to niche expansion?

Holmes (1973) argued that most proximate interactions among intestinal helminths are minimized by site-specificity that evolved through competitive character displacement, though Bush and Holmes (1986b) saw a role for present-day competition as well, particularly in minimizing overlap in infracommunities. Rohde (1979), focusing on gill parasites, felt site specificity evolved to facilitate reproduction (but not to avoid hybridization, so his explanation was not a version of character displacement). However, other interpretations of site-specificity are versions of reproductive character displacement (prevention of hybridization) as discussed above for free-living animals (references in Holmes, 1983).

Finally, Brooks (1980) inveighed against all these scenarios on the grounds that parasites speciate allopatrically and one must know the history of speciation and reinvasion in order to understand such matters as why parasites occupy overlapping or non-overlapping regions within one host population. Certainly in the free-living literature there is the same de-emphasis of historical biogeography that Brooks bemoans in parasitology, though occasionally such an explanation is raised for a pattern that might otherwise be attributed to present-day competition. For example, Connor and Simberloff (1979) suggested that at least part of an apparently surprisingly large number of missing species combinations among birds of the West Indies could be attributed to the existence on different islands of sister-species that had recently speciated and had not yet re-invaded one another's ranges.

Price (1980) may well be correct in his contention that the majority of parasite communities are non-interactive, whether or not there is substantial niche overlap. For one thing, studies detecting interaction (particularly as evidenced by niche shift) are probably more likely to be published than those that fail to find evidence of it, while people probably choose to look for interactions in systems that on average have a high a priori probability of interactions. Connor and Simberloff (1986) made analogous arguments with respect to the controversy between Connell (1983) and Schoener (1983, 1985) about the proportion of experiments on free-living systems that have manifested competition and what that proportion says about the likely importance of competition in nature.

Nevertheless, an impressive number of intestinal helminth studies indicate short-term competitive niche shift and, occasionally, other changes. So, whatever the ultimate causes for frequent non-interaction, it is worth considering the proximate causes for interaction where it occurs. For free-living organisms, competition for space and for food have received the most attention, with the former demonstrated particularly well in elegant experiments in the rocky intertidal and in terrestrial plants (references in Simberloff, 1982). Ecologists of free-living systems generally divide interspecific competition into 'resource' (or 'exploitation') competition and 'interference' competition. In the former, a population of one species inimically affects another simply by removing a resource in short supply. In the latter, one species actively hinders access to the resource (say, by aggressive behaviour or allelochemicals). Among rocky intertidal species and terrestrial plants, space competition is both resource competition (a plant or intertidal organism, simply by occupying space, prevents individuals of other species from settling there) and interference competition (a plant deposits allelochemicals, or a marine animal prize another one off a surface, or overgrows and smothers it). Resource competition for food, on the other hand, has been notoriously difficult to document. Though good laboratory demonstrations exist (Tilman *et al.*, 1981), and some field examples seem cogent (e.g. several references in Connell (1983) and Schoener (1983)), most claimed demonstrations in the field have been inferential rather than direct and many have been controversial.

The situation for parasites seems similar. Short-term locational niche shifts certainly demonstrate competition for space and, although many studies cannot demonstrate conclusively a single mechanism, results consistent with both physical crowding (resource competition) (Grey and Hayunga, 1980) and inflammation of the gut by one species forcing another to move or decreasing its numbers (interference competition) (Loach, 1962; Thomas, 1964; Kennedy, 1980; Silver *et al.*, 1980) have been reported. Another sort of interference competition, quite analogous to allelopathy, has been suggested but not conclusively demonstrated (Keeling, 1961; Lang, 1967; Joysey, 1986) – toxic secretion by one species lowering the numbers of another. As with free-living organisms, interspecific resource competition for nutrients is not as clear. Several studies cannot rule it out (e.g. Keeling, 1961; Gordon and Whitfield, 1984) but widely cited claimed demonstrations (Read and Phifer, 1959; Holmes, 1959) do not conclusively distinguish between the effects of inter- and intraspecific competition. The experimental design was generally X individuals of species A as one treatment, Y individuals of species B as a second treatment, and X A's plus Y B's as a third treatment. Consequently, an observed difference between, say, A in the first treatment (monospecific infection) and A in the third treatment (concurrent infection) could have been due to the increased density, and not the addition of species B (cf. Underwood, 1986). A treatment with $X + Y$ individuals of A is required.

Another form of interference competition is unique to parasites: one species can induce cross-immunity by the host to a second species' concurrent or subsequent infection (Damian, 1964; Bruna and Xenia, 1976; Moqbel and Wakelin, 1979; Gordon and Whitfield, 1984; Holland, 1984; Alghali and Grencis, 1986; Joysey, 1986; reviews by Kazacos, 1975; Holmes, 1983). Interactions mediated by immune responses have great variety. They can be symmetric or asymmetric. They can arise through convergent evolution of antigens by two unrelated species or can be a synapomorphic trait of two closely related species. They can even be facilitatory – one parasite species can depress the host immune system and thereby enhance survival and/or reproduction of another parasite (Jenkins and Behnke, 1977; references in Kennedy, 1980; Bristol *et al.*, 1983).

Although immune-mediated interactions do not occur between free-living species, there are analogues. Holmes (1973), for example, suggests that a host immune response, if it weighs disproportionately on a parasite that would otherwise have been a competitive dominant, can allow more parasite species to coexist, analogously to the situation in free-living communities in which a predator or grazer on an otherwise dominant species can reduce the effect of the latter and thereby allow more competitor species (Paine, 1966; Harper, 1969). A study of *Dactylogyrus* trematodes on carp gills (Paperna, 1964) exemplifies this scenario, though I know of no similar example for intestinal helminths. Herbivores sharing host plants can affect one another either favourably or inimically through chemical and/or structural modification of the plant. For example, early season damage to *Quercus emoryi* by leaf-chewing insects increases defensive tannins and lowers protein content in the leaves, leading to lower densities and greater mortality of subsequent leaf-mining moths on damaged than on undamaged leaves (Faeth, 1986). By contrast, the aphid, *Eulachnus agilis*, grows faster and is more likely to survive on pine shoots and needles previously fed on by another aphid, *Schizolachnus pineti*, apparently because of increased nutritive quality induced by the latter (Kidd *et al.*, 1985). Finally, S. Strauss (personal communication) finds that two beetles on sumac act on one another through altered distribution of shoot types on the tree. The damage of each characteristically changes subsequent tree growth in a way that affects the other beetle.

Interference competition through disease (Price *et al.*, 1986) may also be a free-living analogue to cross-immunity, though the phenomenon has not been documented often. Holmes (1982) points out that the meningeal worm *Parelaphostrongylus tenuis*, of white-tailed deer is usually lethal in moose, caribou, and other ungulates. In Minnesota and Ontario, moose and infected deer overlap over large areas, but moose density is inversely related to prevalence of the worm in deer. In many areas, moose are apparently restricted to uplands because they die of the worm if they descend; the deer stay in the lowlands because of upland snow.

11.4 CONCLUSION

At the outset, I stated that the most fundamental debate about free-living communities is not about what forces structure them but about whether there is much structure at all, and what properties manifest that structure. Frequently community ecologists espouse an ecumenical solution to the problem of community structure: different communities have different degrees of structure and are governed by different forces, hence the apparent differences. Whittaker (1969) first proposed such a solution, arguing that bird communities may differ characteristically from plant and insect communities, in that the latter have virtually unlimited possibilities for niche division and diversification, whereas bird communities are easily saturated and competition sets a severe limit on diversity. This idea accorded well with the general pattern that the most prominent proponents of saturation, order, and competitive structure studied birds (e.g. MacArthur, 1958), while botanists and entomologists often failed to perceive these features. A more up-to-date version of the same idea is given by Begon *et al.* (1986), who argue that unsaturated niche space and a relative dearth of interspecific competition typify phytophagous insects (and perhaps herbivores as a whole) relative to other groups. Schoener (1986) concludes that terrestrial arthropod communities are less regulated than terrestrial vertebrate communities. Other versions of this solution talk about equilibrium vs non-equilibrium communities. Schoener (1986) perceptively observes that the debate on whether communities are structured or appear regulated is largely isomorphic to the debate about whether interspecific competition is a dominant ecological process. Adherents of the view that communities are not very highly structured tend to feel that physical factors such as disturbances or a variable environment usually keep population sizes low enough that competition does not often occur (e.g. Wiens, 1977; Strong, 1984). Thus the argument comes strikingly to resemble the arguments of the 1950s about whether populations were generally regulated by density-dependent or density-independent forces.

As noted above, the same debate about order and integration or lack thereof characterizes parasite community ecology. Studies depicting order and regularity come primarily from birds and mammals (e.g. Kisielewska, 1970a, b, c, d, e; Hobbs, 1980; Bush and Holmes, 1986a, b), while those depicting haphazard, disorganized systems are from fish (e.g. Thomas, 1964; Kennedy, 1985). There are exceptions (e.g. Lotz and Font, 1985), however. J. Moore and I (Moore and Simberloff, 1988) deliberately chose an avian host (bobwhite quail) for comparison to the general results of Holmes and his associates in their extensive study of the helminth community of lesser scaup. Four nematodes and two cestodes had high prevalences, and the cestodes and two of the nematodes frequently had high intensities (though not as high as those in scaup). Our results differed greatly from Bush and Holmes's (1986a, b). We found a largely non-interactive community. There was one asymmetric spatial shift relationship (between two cecal nematodes) and a strongly negative numerical correlation between the two tapeworms. Kennedy *et al.* (1986) argue that alimentary canal helminth communities of birds and fishes are fundamentally different. Birds (and mammals) have higher species richness and diversity and more individual worms than fishes, possibly because of endothermy and a more differentiated intestine. Goater *et al.* (1987) add salamander communities to those of fish and bat communities to those of birds in this dichotomy, agreeing that endothermy and a more differentiated intestine are likely reasons for the greater parasite species richness and intensities in the latter category. It is tempting to think that the greater species richness and intensities would lead to competitive interactions and thus to coevolution. In the same vein, two studies on the same three tapeworms of the genus *Diorchis* in coots are suggestive. Oszewska (1975) found no spatial shifts in mixed infections and no evidence for competition. Pojmanska (1982), working with a greater range of intensities, found that each species extends its locational range in heavy monospecific infections, and in heavy concurrent infections there is increased site segregation (though the data and statistical assessment she presents do not make the latter assertion clear). Chappell (1969) proposes a general sequence, as infection intensity increases, from no interaction through spatial separation to elimination of one species from the infracommunity.

Such an explanation for the apparent diversity of structures seems premature to me, just as the ecumenical approach to resolving the debate about free-living communities is not yet convincing. It is a truism that every parasite community is different from every other one, but that does not mean there are easily discerned patterns to these differences. I have pointed out a number of ways in which pattern assessment for parasite communities has not been sufficiently rigorous, in much the same way that it has not been sufficiently rigorous for free-living communities. Until there is a larger body of experiments in both sorts of communities, and the application of more appropriate statistical pattern analyses, it probably casts more shadow than light to make broad generalizations about community structure.

ACKNOWLEDGEMENTS

Drs Barbara Downes and Albert Bush made many insightful suggestions on drafts of this manuscript. Dr Janice Moore introduced me to many problems in parasite community ecology, continually raised issues on the ecology of both parasites and communities, and commented on the manuscript. Sharon Strauss let me see an important preprint.

REFERENCES

Abramsky, Z. and Sellah, C. (1982) Competition and the role of habitat selection in *Gerbillus allenbyi* and *Meriones tristrami*: a removal experiment. *Ecology*, 63, 1242–7.

Alghali, S. T. O. and Grencis, R. K. (1986) Immunity to tapeworms: intraspecific cross-protective interactions between *Hymenolepis citelli, H. diminuta* and *H. microstoma* in mice. *Parasitology.*, **92**, 665–74.

Arthur, W. (1982) The evolutionary consequences of interspecific competition. *Adv. Ecol. Res.*, **12**, 127–87.

Barton, D. E. and David, F. N. (1956) Some notes on ordered random intervals. *J. Royal Stat. Soc., B*, **18**, 79–94.

Begon, M., Harper, J. L. and Townsend, C. R. (1986) *Ecology: Individuals, Populations and Communities*, Sinauer, Sunderland, Mass.

Bossert, W. H. (1963) Simulation of Character Displacement in Animals. PhD Thesis, Harvard University, Cambridge, Mass.

Bristol, J. R., Pinon, A. J. and Mayberry, L. F. (1983) Interspecific interactions between *Nippostrongylus brasiliensis* and *Eimeria nieschulzi* in the rat. *J. Parasitol.*, **69**, 372–4.

Brooks, D. R. (1980) Allopatric speciation and non-interactive parasite community structure. *Syst. Zool.*, **29**, 192–203.

Brown, W. L. and Wilson, E. O. (1956) Character displacement. *Syst. Zool.*, **5**, 49–64.

Bruna, C. D. and Xenia, B. (1976) *Nippostrongylus brasiliensis* in mice: reduction of worm burden and prolonged infection induced by the presence of *Nematospiroides dubius*. *J. Parasitol.*, **62**, 490–1.

Bullini, L., Nascetti, G., Paggi, L. *et al.* (1986) Genetic variation of ascaridoid worms with different life cycles. *Evolution*, **40**, 437–40.

Bush, A. O. and Holmes, J. C. (1983) Niche separation and the broken-stick model: Use with multiple assemblages. *Am. Nat.*, **122**, 849–55.

Bush, A. O. and Holmes, J. C. (1986a) Intestinal helminths of lesser scaup ducks: patterns of association. *Can. J. Zool.*, **64**, 132–41.

Bush, A. O. and Holmes, J. C. (1986b) Intestinal helminths of lesser scaup ducks: an interactive community. *Can. J. Zool.*, **64**, 142–52.

Cannon, L. R. G. (1972) Studies on the ecology of the papillose allocreadid trematodes of the yellow perch in Algonquin Park, Ontario. *Can. J. Zool.*, **50**, 1231–9.

Carvajal, J. and Dailey, M. D. (1975) Three new species of *Echeneibothrium* (Cestoda: Tetraphyllidea) from the skate, *Raja chilensis* Guichenot, 1848, with comments on mode of attachment and host specificity. *J. Parasitol.*, **61**, 89–94.

Chappell, L. H. (1969) Competitive exclusion between two intestinal parasites of the three-spined stickleback, *Gasterosteus aculeatus* L. *J. Parasitol.*, **55**, 775–8.

Clements, F. E. (1905) *Research Methods in Ecology*, University Publishing Company, Lincoln, Nebr.

Cohen, J. E. (1966) *A Model of Simple Competition*, Harvard University Press, Cambridge, Mass.

Cohen, J. E. (1988) Food webs and community structure. In *Perspectives in Theoretical Ecology*, (eds, S. Levin, R. M. May and J. Roughgarden), (in press).

Connell, J. H. (1980) Diversity and the coevolution of competitors, or the ghost of competition past. *Oikos*, **35**, 131–8.

Connell, J. H. (1983) On the prevalence and relative importance of interspecific competition: Evidence from field experiments. *Am. Nat.*, **122**, 661–96.

Connor, E. F. and Simberloff, D. (1979) The assembly of species communities: Chance or competition? *Ecology*, **60**, 1132–40.

Connor, E. F. and Simberloff, D. (1983) Interspecific competition and species co-occurrence patterns on islands: Null models and the evaluation of evidence. *Oikos*, **41**, 455–65.

Connor, E. F. and Simberloff, D. (1984) Neutral models of species' co-occurrence patterns. In *Ecological Communities: Conceptual Issues and the Evidence* (eds, D. R. Strong, D. Simberloff, L. G. Abele and A. B. Thistle), Princeton University Press, Princeton, New Jersey, pp. 316–31.

Connor, E. F. and Simberloff, D. (1986) Competition, scientific method, and null models in ecology. *Amer. Sci.*, **74**, 155–62.

Crompton, D. W. T. (1973) The sites occupied by some parasitic helminths in the alimentary tract of vertebrates. *Biol. Rev.*, **48**, 27–83.

Damian, R. T. (1964) Molecular mimicry: Antigen sharing by parasite and host and its consequences. *Am. Nat.*, **98**, 129–49.

Dash, K. M. (1981) Interaction between *Oesophagostomum columbianum* and *Oesophagostomum renulosum* in sheep. *Intern. J. Parasitol.*, **11**, 210–17.

Diamond, J. M. (1973) Distributional ecology of New Guinea birds. *Science*, **179**, 759–69.

Diamond, J. M. (1975) Assembly of species communities. In *Ecology and Evolution of Communities* (eds, M. L. Cody and J. M. Diamond), Harvard University Press, Cambridge, Mass., pp. 342–444.

Diamond, J. M. (1978) Niche shifts and the rediscovery of interspecific competition. *Amer. Sci.*, **66**, 322–31.

Dubos, R. and Schaedler, R. W. (1964) The digestive tract as an ecosystem. *Amer. J. Med. Sci.*, **248**, 267–71.

Esch, G. W. (1983) The population and community ecology of cestodes. In *Biology of the Eucestoda* (eds, C. Arme and P. W. Pappas), Academic Press, London, pp. 81–133.

Evans, N. A. (1977) The site preferences of two digeneans, *Asymphylodora kubanicum* and *Sphaerostoma bramae*, in the intestine of the roach. *J. Helminthol.*, **51**, 197–204.

Faeth, S. H. (1986) Indirect interactions between temporally-separated herbivores mediated by the host plant. *Ecology*, **67**, 479–94.

Frankland, H. M. T. (1959) The incidence and distribution in Britain of the trematodes of *Talpa europaea. Parasitology.*, **49**, 132–42.

Gilpin, M. E. and Diamond, J. M. (1984) Are species co-occurrences on islands non-random, and are null hypotheses useful in community ecology? In *Ecological Communities: Conceptual Issues and the Evidence* (eds, D. R. Strong, D. Simberloff, L. G. Abele and A. B. Thistle), Princeton University Press, Princeton, New Jersey, pp. 297–315.

Goater, T. M., Esch, G. W. and Bush, A. O. (1987) Helminth parasites of sympatric salamanders: Ecological concepts at infracommunity, component and compound community levels. *Am. Midl. Nat.*, **118**, 289–300.

Gordon, D. M. and Whitfield, P. J. (1984) Interactions of the cysticercoids of *Hymenolepis diminuta* and *Raillietina cesticillus* in their intermediate host, *Tribolium confusum. Parasitology*, **90**, 421–31.

Grant, P. R. (1972) Convergent and divergent character displacement. *Biol. J. Linnean Soc.*, **4**, 39–68.

Grant, P. R. (1975) The classical case of character displacement. *Evol. Biol.*, **8**, 237–337.

Grey, A. J. and Hayunga, E. G. (1980) Evidence for alternative site selection by *Glaridacris laruei* (Cestoidea: Caryophyllidea) as a result of interspecific competition. *J. Parasitol.*, **66**, 371–2.

Halvorsen, O. (1976) Negative interaction amongst parasites. In *Ecological Aspects of Parasitology* (ed. C. R. Kennedy), North-Holland, Amsterdam, pp. 99–114.

Harper, J. (1969) The role of predation in vegetational diversity. In *Diversity and Stability in Ecological Systems* (eds, G. M. Woodwell and H. H. Smith), Brookhaven National Laboratory, Upton, New York, pp. 48–61.

Harvey, P. H., Colwell, R. K., Silvertown, J. W. and May, R. M. (1983) Null models in ecology. *Annu. Rev. Ecol. Syst.*, **14**, 189–211.

Henricson, J. (1977) The abundance and distribution of *Diphyllobothrium dendriticum* (Nitzsch) and *D. ditremum* (Creplin) in the char *Salvelinus alpinus* (L.) in Sweden. *J. Fish. Biol.*, **11**, 231–48.

Hobbs, R. P. (1980) Interspecific interactions among gastrointestinal helminths in pikas of North America. *Am. Midl. Nat.*, **103**, 15–25.

Holland, C. (1984) Interactions between *Moniliformis* (Acanthocephala) and *Nippostrongylus* (Nematoda) in the small intestine of laboratory rats. *Parasitology*, **88**, 303–16.

Holmes, J. C. (1959) Competition for carbohydrates between the rat tapeworm, *Hymenolepis diminuta*, and acanthocephalan *Moniliformis dubius*. *J. Parasitol.*, **45**, (Suppl.), 31.

Holmes, J. C. (1973) Site selection by parasitic helminths: Interspecific interactions, site segregation, and their importance to the development of helminth communities. *Can. J. Zool.*, **51**, 333–47.

Holmes, J. C. (1982) Impact of infectious disease agents on the population growth and geographical distribution of animals. In *Population Biology of Infectious Diseases* (eds, R. M. Anderson and R. M. May), Springer-Verlag, New York, pp. 37–51.

Holmes, J. C. (1983) Evolutionary relationships between parasitic helminths and their hosts. In *Coevolution* (eds, D. J. Futuyma and M. Slatkin), Sinauer, Sunderland, Mass., pp. 161–85.

Holmes, J. C. and Price, P. W. (1986) Communities of parasites. In *Community Ecology: Pattern and Process* (eds, J. Kikkawa and D. J. Anderson), Blackwell Scientific Publications, Melbourne, Australia, pp. 187–213.

Hopf, F. A. and Brown, J. H. (1986) The bull's-eye method for testing randomness in ecological communities. *Ecology*, **67**, 1139–55.

Hurlbert, S. H. (1978) The measurement of niche overlap and some relatives. *Ecology*, **59**, 67–77.

Hutchinson, G. E. (1959) Homage to Santa Rosalia, or why are there so many kinds of animals? *Am. Nat.*, **93**, 145–59.

Jenkins, S. N. and Behnke, J. M. (1977) Impairment of primary expulsion of *Trichuris muris* in mice concurrently infected with *Nematospiroides dubius*. *Parasitology*, **75**, 71–8.

John, D. D. (1926) On *Cittotaenia denticulata* (Rudolphi 1804), with notes as to the occurrence of other helminthic parasites of rabbits found in the Aberystwyth area. *Parasitology*, **26**, 436–54.

Joysey, H. S. (1986) Suppression of *Taenia crassiceps* during concurrent infections with *Mesocestoides corti* in mice. *Parasitology*, **92**, 199–207.

Kazacos, K. R. (1975) Increased resistance in the rat to *Nippostrongylus brasiliensis* following immunization against *Trichinella spiralis*. *Vet. Parasitol.*, **1**, 165–74.

Keeling, J. E. D. (1961) Experimental Trichuriasis. I. Antagonism between *Trichuris muris* and *Aspicularis tetraptera* in the albino mouse. *J. Parasitol.*, **47**, 641–6.

Kennedy, C. R. (1985) Site segregation by species of Acanthocephala in fish, with special reference to eels, *Anguilla anguilla*. *Parasitology.*, **90**, 375–90.

Kennedy, C. R., Bush, A. O. and Aho, J. M. (1986) Patterns in helminth communities: Why are birds and fish different? *Parasitology*, **93**, 205–15.

Kennedy, M. W. (1980) Immunologically mediated non-specific interactions between intestinal phases of *Trichinella spiralis* and *Nippostrongylus brasiliensis* in the mouse. *Parasitology.*, **80**, 60–72.

Kidd, N. A. C., Lewis, G. B. and Howell, C. A. (1985) An association between two species of pine aphid, *Schizolachnus pineti* and *Eulachnus agilis*. *Ecol. Entomol.*, **10**, 427–32.

Kisielewska, K. (1970a) Ecological organization of intestinal helminth groupings in *Clethrionomys glareolus* (Schreb.)(*Rodentia*). I. Structure and seasonal dynamics of helminth groupings in a host population in the Bialowieza National Park. *Acta Parasitol. Pol.*, **18**, 121–47.

Kisielewska, K. (1970b) Ecological organization of intestinal helminth groupings in *Clethrionomys glareolus* (Schreb.)(*Rodentia*). II. An attempt at introduction of helminths of *Clethrionomys glareolus* from the Bialowieza National Park into an island of the Beldany Lake (Mazurian Lakeland). *Acta Parasitol. Pol.*, **18**, 149–62.

Kisielewska, K. (1970c) Ecological organization of intestinal helminth groupings in *Clethrionomys glareolus* (Schreb.)(*Rodentia*). III. Structure of helminth groupings in *Clethrionomys glareolus* populations of various forest biocoenoses in Poland. *Acta Parasitol. Pol.*, **18**, 163–76.

Kisielewska, K. (1970d) Ecological organization of intestinal helminth groupings in *Clethrionomys glareolus* (Schreb.)(*Rodentia*). IV. Spatial structure of helminth groupings within the host population. *Acta Parasitol. Pol.*, **18**, 177–96.

Kisielewska, K. (1970e) Ecological organization of intestinal helminth groupings in *Clethrionomys glareolus* (Schreb.)(*Rodentia*). V. Some questions concerning helminth groupings in host individuals. *Acta Parasitol. Pol.*, **18**, 197–208.

Kuhn, T. S. (1977) *The Essential Tension: Selected Studies in Scientific Tradition and Change*, University of Chicago Press, Chicago.

Lang, B. Z. (1967) *Fasciola hepatica* and *Hymenolepis microstoma* in the laboratory mouse. *J. Parasitol.*, **53**, 213–14.

Lewin, R. (1983) Santa Rosalia was a goat. *Science*, **221**, 636–9.

Lie, K. J., Basch, P. F., Heyneman, D. *et al.* (1968) Implications for trematode control of interspecific larval antagonism within snail hosts. *Trans. Roy. Soc. Trop. Med. and Hyg.*, **62**, 299–319.

Loach, C. D. (1962) Increased resistance to *Trichinella spiralis* in the laboratory rat following infections with *Nippostrongylus muris*. *J. Parasitol.*, **48**, 24–6.

Lotz, J. M. and Font, W. F. (1985) Structure of enteric helminth communities in two populations of *Eptesicus fuscus* (Chiroptera). *Can. J. Zool.*, **63**, 2969–78.

MacArthur, R. H. (1958) Population ecology of some warblers of northeastern coniferous forests. *Ecology*, **39**, 599–619.

MacKenzie, K. and Gibson, D. (1970) Ecological studies on some parasites of plaice, *Pleuronectes platessa* (L.) and flounder, *Platichthys flesus* (L.). *Symposia of the Brit. Soc. for Parasitol.*, **61**, 55–63.

Margolis, L., Esch, G. W., Holmes, J. C. *et al.* (1982) The use of ecological terms in parasitology (report of an *ad hoc* committee of the American Society of Parasitologists). *J. Parasitol.*, **68**, 131–3.

McCulloch, C. E. (1985) Variance tests for species associations. *Ecology*, **66**, 1676–81.

McNab, B. K. (1971) The structure of tropical bat faunas. *Ecology*, **52**, 352–8.

Moore, J. K. and Simberloff, D. (1988) Intestinal helminth communities of bobwhite quail. *Ecology*, subm.

Moqbel, R. and Wakelin, D. (1979) *Trichinella spiralis* and *Strongyloides ratti*. Immune interaction in adult rats. *Exp. Parasitol.*, **47**, 65–72.

Mountainspring, S. and Scott, J. M. (1985) Interspecific competition among Hawaiian forest birds. *Ecol. Monogr.*, **55**, 219–39.

Oszewska, G. M. (1975) Topospecificity of three cestode species of the genus *Diorchis* (Clerc, 1903) parasitizing *Fulica atra* (L.). *Acta Parasitol. Pol.*, **23**, 329–38.

Paine, R. T. (1966) Food web complexity and species diversity. *Am. Nat.*, **100**, 65–75.

Paine, R. T. (1988) On food webs: Road maps of interactions or the grist for theoretical development? *Ecology*, **69**, (in press).

Paperna, I. (1964) Competitive exclusion of *Dactylogyrus extensus* by *Dactylogyrus vastator* (Trematoda: Monogenea) on the gills of reared carp. *J. Parasitol.*, **50**, 94–8.

Peters, R. H. (1977) The unpredictable problems of tropho-dynamics. *Env. Biol. Fishes*, **2**, 97–101.

Petrusewicz, K. (1965) Dynamika liczebności, organizacja i struktura ekologiczna populacji. *Etol. pol. B.*, **11**, 299–316.

Pimm, S. L. (1982) *Food Webs*, Chapman and Hall, London.

Pojmanska, T. (1982) The co-occurrence of three species of *Diorchis* Clerc, 1903 (Cestoda: Hymenolepididae) in the European coot, *Fulica atra* L. *Parasitology*, **84**, 419–29.

Poole, R. W. and Rathcke, B. J. (1979) Regularity, randomness, and aggregation in flowering phenologies. *Science*, **203**, 470–71.

Price, P. W. (1972) Parasitoids utilizing the same host: Adaptive nature of differences in size and form. *Ecology*, **53**, 190–5.

Price, P. W. (1980) *Evolutionary Biology of Parasites*, Princeton University Press, Princeton, New Jersey.

Price, P. W., Westoby, M., Rice, B. *et al.* (1986) Parasite mediation in ecological interactions. *Ann. Rev. Ecol. Syst.*, **17**, 487–505.

Read, C. P. and Phifer, K. (1959) The role of carbohydrates in the biology of cestodes. VII. Interactions between individual tapeworms of the same and different species. *Exp. Parasitol.*, **8**, 46–50.

Rohde, K. (1979) A critical evaluation of intrinsic and extrinsic factors responsible for niche restriction in parasites. *Am. Nat.*, **114**, 648–71.

Schad, G. A. (1963) Niche diversification in a parasite species flock. *Nature*, **198**, 404–6.

Schluter, D. (1984) A variance test for detecting species associations, with some applications. *Ecology*, **65**, 998–1005.

Schmidt, G. D. and Roberts, L. S. (1985) *Foundations of Parasitology*, 3rd edn, Times Mirror/Mosby College Publishing, St. Louis.

Schoener, T. W. (1983) Field experiments on interspecific competition. *Am. Nat.*, **122**, 240–85.

Schoener, T. W. (1985) Some comments on Connell's and my reviews of field experiments on interspecific competition. *Am. Nat.*, **125**, 730–40.

Schoener, T. W. (1986) Patterns in terrestrial vertebrate versus arthropod communities: Do systematic differences in regularity exist? In *Community Ecology* (eds, J. Diamond and T. J. Case), Harper and Row, New York, pp. 556–86.

Silver, B. B., Dick, T. A. and Welch, H. E. (1980) Concurrent infections of *Hymenolepis diminuta* and *Trichinella spiralis* in the rat intestine. *J. Parasitol.*, **66**, 786–91.

Simberloff, D. (1980) A succession of paradigms in ecology: Essentialism to materialism and probabilism. *Synthese*, **43**, 3–39.

Simberloff, D. (1982) The status of competition theory in ecology. *Ann. Zool. Fennici*, **19**, 241–53.

Simberloff, D. (1988) Pattern analysis and the detection of interactions in natural communities. In *Proc. of III Congresso Societa Italiana* (ed. A. Renzoni), Societa Italiana di Ecologica, Pisa.

Simberloff, D. and Boecklen, W. J. (1981) Santa Rosalia reconsidered: Size ratios and competition. *Evolution*, **35**, 1206–28.

Simberloff, D. and Connor, E. F. (1981) Missing species combinations. *Am. Nat.*, **118**, 215–39.

Stock, T. M. and Holmes, J. C. (1987) *Dioecocestus asper* (Cestoda: Dioecocestidae): An interference competitor in an enteric helminth community. *J. Parasitol.*, **73**, 1116–23.

Stock, T. M. and Holmes, J. C. (1988) Functional relationships and microhabitat distributions of enteric helminths of grebes (Podicipedidae): The evidence for interactive communities. *J. Parasitol.*, **74**, 214–27.

Strauss, S. Y. (1988) Determining the effects of herbivory using naturally-damaged plants. *Ecology*, **69**, (in press).

Strong, D. R., Jr (1984) Density-vague ecology and liberal population regulation in insects. In *A New Ecology: Novel Approaches to Interactive Systems*, (eds, P. W. Price, C. N. Slobodchikoff and W. S. Gaud), Wiley, New York, pp. 313–27.

Terborgh, J. (1971) Distribution on environmental gradients: Theory and a preliminary interpretation of distributional patterns in the avifauna of the Cordillera Vilcabamba, Peru. *Ecology*, **52**, 23–40.

Thomas, J. D. (1964) Studies on populations of helminth parasites in brown trout (*Salmo trutta*, L.). *J. Anim. Ecol.*, **33**, 83–95.

Tilman, D., Mattson, M. and Langer, S. (1981) Competition and nutrient kinetics along a temperature gradient: an experimental test of a mechanistic approach to niche theory. *Limnol. and Oceanogr.*, **26**, 1020–33.

Ulmer, M. J. (1971) Site-finding behavior in helminths in intermediate and definitive hosts. In *Ecology and Physiology of Parasites* (ed. A. M. Fallis), University of Toronto Press, Toronto, pp. 123–59.

Underwood, A. J. (1986) The analysis of competition by field experiments. In *Community Ecology: Pattern and Process* (eds, J. Kikkawa and D. J. Anderson), Blackwell Scientific Publications, Melbourne, Australia, pp. 240–68.

Whittaker, R. H. (1969) Evolution of diversity in plant communities. In *Diversity and Stability in Ecological Systems* (eds, G. M. Woodwell and H. H. Smith), Brookhaven National Laboratory, Upton, New York, pp. 178–96.

Wiens, J. (1977) On competition and variable environments. *Am. Sci.*, **65**, 590–7.

Williams, H. H. (1960) The intestine in members of the genus *Raja* and host-specificity in the Tetraphyllidea. *Nature*, **188**, 514–16.

Williams, H. H. (1966) The ecology, functional morphology and taxonomy of *Echeneibothrium* Beneden, 1850 (Cestoda: Tetraphyllidea), a revision of the genus and comments on *Discobothrium* Beneden 1871, *Pseudanthrobothrium* Baer, 1956, and *Phormobothrium* Alexander, 1963. *Parasitology*, **56**, 227.

12

Concluding remarks

Albert O. Bush and John M. Aho

Most of the theories and paradigms in ecology result from observations of, and/or experimentation with, free-living organisms. A quick perusal through any general ecology text will provide little information on parasites or parasitism. If parasitism is mentioned, it is usually anecdotal or in the context of the impact that parasites can have on the fitness of free-living organisms. But what about the ecology and evolution of the parasites themselves? Parasites are fascinating. They represent what is probably the most prevalent lifestyle, they exhibit an extraordinary diversity of form and function and they are ubiquitous in distribution. However, the potential role of parasitic systems as models for understanding patterns and processes in community ecology has not been fully explored. (Although we take heart in the number of chapters on parasitic organisms which have recently appeared in edited volumes on community and evolutionary ecology.)

Ecological studies on free-living systems will not necessarily produce an understanding of the ecology of parasitic systems and the reverse is equally true. However, for some ecological questions, studies on parasitic systems may yield answers more readily than those on their free-living counterparts. There are several reasons for such an assertion. First, helminths are confined to a single host individual during specific stages of their life history. This provides unambiguous boundaries within which any potential interactions will be completed. Second, because individual hosts are discrete entities, communities are easily replicated, allowing variability in organizational patterns to be examined at several hierarchical levels (i.e. infra-, component and compound community). Third, the effects of predation cannot readily be invoked to explain community structure in most host systems. Only for some digeneans in their molluscan intermediate hosts does predation occur. Fourth, helminths within their hosts are buffered through the host's homeostatic mechanisms. As a consequence, helminth communities, in appropriate hosts, are usually not affected by direct, external disturbances or environmental vagaries. Finally, with the exception of some nematodes and alariid trematodes in a few species of mammals, hosts are born helminth-free. Thus, they are the Krakatau's upon which we can investigate colonization and community development. Uninfected, susceptible hosts are much more numerous than newly-created or denuded islands. Therefore, colonization can be examined more rigorously using comparative and experimental approaches. Collectively, these five features make helminth communities attractive complements to free-living organisms for investigating some of the major questions in community ecology. On the other hand, parasites also provide some additional complications. Perhaps the most distinct, and one for which there is no direct free-living parallel, is the host's immune system. Though, as we've noted above, the parasite may gain an advantage by being buffered from the external environment, the host may concurrently be mounting an immune response which will kill the parasite.

As Esch *et al.* suggest in the Introduction to this volume, the roots of ecological parasitology can be traced to the books by Dogiel and his colleagues in the early 1960s (*Parasitology of Fishes* and *General Parasitology*). Although these books, and the primary literature which followed, focused on populations, the seeds of community analyses were present. Within the past ten years, there has been a marked increase in the study of parasite communities, stimulated largely we believe, by Price's provocative book (*Evolutionary Biology of Parasites*) in the early 1980s. The intent of the present book, was to assess current perspectives regarding our knowledge of helminth communities. The mandate was to identify patterns and processes which might explain structure in helminth communities of both intermediate and definitive hosts. The questions posed by the contributors were deceptively simple. To what extent do similar patterns characterize communities in similar hosts; in different kinds of hosts? If patterns are similar, do they result from common processes? How well do studies on helminth communities complement studies on free-living communities? Answers to these questions proved to be much more complex, and controversial, than anticipated.

For the different host groups, we know the least about organizational processes in intermediate hosts. This is unfortunate since this level is central to many parasite transmission strategies and such hosts are clearly an alternative resource base upon which to examine community structure. What little detailed information is available on communities in intermediate hosts is primarily restricted to digenean infections on molluscs. The chapters by Kuris and Sousa, on infracommunities in the same snail host, but at different localities, provide similar conclusions as to how such infracommunities are organized. Both agree that infracommunities can be structured by hierarchical, antagonistic interactions among the larval digeneans, with large, actively feeding, redial forms dominating. They also agree that indirect forms of antagonism, such as interference competition, may occur (but is more difficult to document). At the component community level, they agree that resource monopolization does not occur and that community structure (as measured by species diversity) may be independent of the antagonistic interactions they find at the infracommunity level. These observations suggest that very different organizing processes operate at different spatial scales.

There is considerably more information available on helminth communities in a variety of vertebrate definitive hosts. Despite this (or perhaps because of it), there remains considerable divergence of opinion with respect to common patterns and processes structuring helminth communities in the various vertebrate groups. Based on a diversity of approaches by the various authors, and admittedly restricted to only a few host examples for each major group, infracommunities in freshwater fishes, amphibians, reptiles, and mammals appear to be depauperate (suggestive of isolationist communities) while those in marine fishes and birds are comparatively species-rich (suggestive of interactive communities). At the level of the component community, freshwater fishes, amphibians, reptiles, and mammals have helminth communities with low faunal similarity, while those in marine fishes and birds have much higher values. All authors imply that the animals in their respective host groups are independent samplers of the environment; all agree that there appears to be a predictable pattern (though the 'pattern' is not the same for all host groups). Where they differ most is in identifying the processes leading to those patterns. One major reason for these dichotomies may be that, for most cases, we still lack sufficient data to fully understand and appreciate the inherent variability in recruitment potential for different combinations of helminths and their hosts. Dobson's chapter attempts to bridge some of these gaps and demonstrates the need for autecological studies in the development of community models. Similar suggestions for better integration of population and community dynamics are raised throughout the book.

We envy the authors of final chapters in other books which usually conclude with definitive statements. In comparison, the present effort may appear a rather weak commentary with it's 'maybe it is – maybe it's not, it could be – it's probably not' conclusions, particularly to a book which attempts to synthesize information on a central theme. Perhaps the initial objective was too ambitious and Simberloff is correct when he concludes his chapter, 'Until there is a larger body of experiments in both sorts of communities [*sic* free-living and parasitic], and the application of more appropriate statistical pattern analyses, it probably casts more shadow than light to make broad generalizations about community structure'. However, the problem might be even more fundamental than he envisages. As we suggested earlier, given the diversity of hosts and their parasites, we are possibly asking far too much to expect any holistic explanations for different patterns, simply because the organisms under discussion are collectively called 'parasites'. Many ecologists studying free-living organisms seem to have been encumbered with finding a 'fundamental' law or theory which would be widely applicable within the discipline (e.g. competition theory). In our opinion, their recent series of polarized debates has been constructive, albeit acrimonious at times, and has led to the recognition of the need for pluralism. Ecologists studying parasites should avoid the same pitfall of searching for some fundamental truth and appreciate, at the outset, that many of the answers will invoke pluralism. We certainly do not deny that some theories may be applicable across different host species or different parasitic taxa, we just caution against making, or searching for, unfounded generalizations. Nevertheless, we do see common ground. The host is a resource for parasites (Chapter 2) and, the manner in which the host exploits it's habitat, coupled with the manner in which parasites exploit their hosts, will ultimately determine the most fundamental level for collective discussions on patterns and processes in helminth communities.

Where do we go from here? We could write an entire book on that subject, but will limit ourselves to those areas where we feel more immediate attention should be focused. On a global scale, we need considerably more data on the dynamics of invertebrate helminth communities *per se*, and of how these community relationships may ultimately affect the dynamic processes of parasitism at higher trophic levels. Such data may, in large part, determine the kinds of questions we should be asking. The parasitological literature is dominated by studies on hosts from north temperate biomes. There is a desperate need for studies from other biomes, particularly the tropics. There is also a general need for long-term studies on helminth communities in order to address questions on persistence and stability. There is definitely a need for more studies at the infracommunity level. Only then will we be able to evaluate the relative importance of abiotic versus biotic events in structuring communities. Strange as it may seem, one thing we really need is 'negative data' (i.e. data on host populations or species which are uninfected). Unfortunately, editors seem loathe to publish such 'analyses'.

But, truly moving forward and advancing our understanding of the ecology of parasites will require more than simply increasing the number of surveys. It will require many more experimental studies and this will necessitate an interdisciplinary approach so that, for example, when we argue for competition as an organizing mechanism, we can also identify limiting resources. This will require joint efforts among several disciplines (e.g. physiological, molecular, and evolutionary biologists). Such collaborative efforts are rare to non-existent in the literature. To advance will clearly require experimentation and much more rigorous statistical analyses, coupled with the development of *appropriate* null models. The same can be said for component communities (and, ultimately, compound communities), as this is where we will find the requisite data to expand on such concepts as host specificity, the significance of host phylogeny and the importance of colonization. Finally, and perhaps most important, we must not lose sight of the potentially overriding importance of the biology of the host(s) and of the environments which they inhabit.

We draw no central conclusion from these chapters. Nevertheless, we believe that the unifying question which emerges – what happens to empty space? – is of central importance to understanding community structure (be it free-living or parasitic) even though the answer(s) may not be applicable to 'all' parasites, or to 'all' host types, 'all' of the time.

Index